HSC Year 12
CHEMISTRY

COL HARRISON

+ summary notes
+ revision questions
+ detailed sample answers
+ study and exam preparation advice

STUDY NOTES

A+ HSC Chemistry Study Notes
1st Edition
Col Harrison
ISBN 9780170465281

Publisher: Sarah Craig and Cathy Beswick-Davison
Series editor: Catherine Greenwood
Copyeditor: Sally Woollett
Series text design: Nikita Bansal
Series cover design: Nikita Bansal
Series designer: Cengage Creative Studio
Artwork: MPS Limited
Production controller: Karen Young
Typeset by: Nikki M Group Pty Ltd

ISBN 978 0 17 046528 1

Cengage Learning Australia
Level 7, 80 Dorcas Street
South Melbourne, Victoria Australia 3205

Cengage Learning New Zealand
Unit 4B Rosedale Office Park
331 Rosedale Road, Albany, North Shore 0632, NZ

For learning solutions, visit **cengage.com.au**

Printed in China by 1010 Printing International Limited.
1 2 3 4 5 6 7 26 25 24 23 22

CONTENTS

CHAPTER 3

MODULE 7: ORGANIC CHEMISTRY

CHAPTER 4

MODULE 8: APPLYING CHEMICAL IDEAS

CONTENTS

HOW TO USE THIS BOOK

The A+ HSC Chemistry resources are designed to be used year-round to prepare you for your HSC Chemistry exam. *A+ HSC Chemistry Study Notes* includes topic summaries of all key knowledge in the NSW HSC Chemistry syllabus that you will be assessed on during your exam. Each chapter of this book addresses one module. This section gives you a brief overview of each chapter and the features included in this resource.

Module summaries

The module summaries at the beginning of each chapter give you a high-level overview of the essential knowledge and key science skills you will need to demonstrate during your exam.

Concept maps

The concept maps at the beginning of each chapter provide a visual summary of each module outcome.

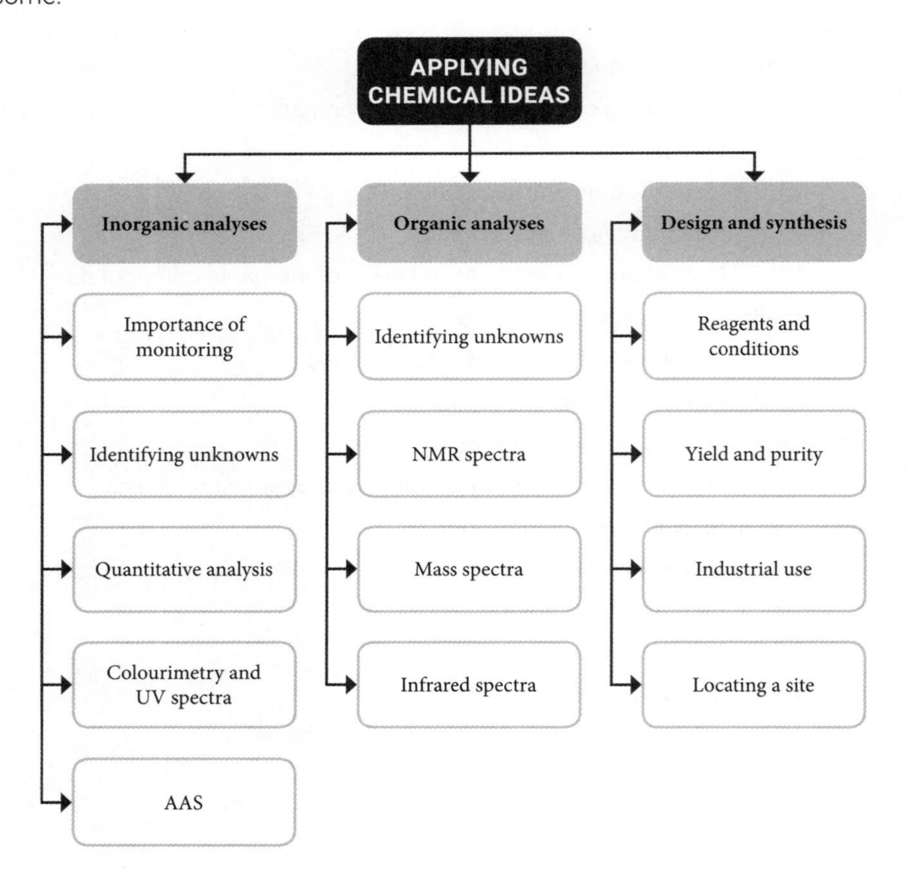

Inquiry question summaries

All of the dot points under each inquiry question are summarised sequentially throughout inquiry section summaries.

Exam practice

Exam practice questions appear at the end of each chapter to test you on what you have just reviewed in the chapter. These are written in the same style as the questions you will find in the actual HSC Chemistry exam. The questions include some official past exam questions.

Multiple-choice questions

Each chapter has 10 multiple-choice questions.

Short-answer questions

There are six short-answer questions in each chapter, often broken into parts. These questions require you to apply your knowledge across multiple concepts. Mark allocations have been provided for each question.

Solutions

Solutions to practice questions are supplied in Chapter 5. They have been written to reflect a high-scoring response and include explanations of what makes an effective answer.

Explanations

The solutions section includes explanations of each multiple-choice option, both correct and incorrect. Explanations of written response items include what a high-scoring response looks like and indications of potential mistakes.

SOLUTIONS

CHAPTER 1 MODULE 5

Equilibrium and acid reactions

Multiple-choice questions

1 B

When magnesium reacts with acid, it generates hydrogen. In a beaker, the hydrogen escapes from the system; hence, this is an open system.

A saturated solution of sodium chloride with a visible precipitate in a test tube is open, but the equilibrium system exists between the solid and the ions in solution; hence, there is no exchange of matter between the system and the surroundings, and **A** is not correct. A combustion reaction taking place inside a bomb calorimeter is ideally an isolated system because neither matter nor energy can escape, so **C** is not correct. An equilibrium established between iron(III) ions, thiocyanate ions and ferric thiocyanate ions in a sealed flask is a closed system, so **D** is not correct.

Short-answer questions

11 The carbon dioxide equilibrium in a soft drink is affected whenever the bottle is opened. Some of the gas escapes and the system becomes an open system. As carbon dioxide is lost from the system, carbon dioxide in solution (aq) is converted to gas (g), according to Le Chatelier's principle. The equilibrium shifts to the right as the system adjusts to restore the lost carbon dioxide. As long as the top of the bottle is off, the system remains open and equilibrium can never be re-established. However, when the bottle top is replaced, the system becomes a closed system again and it adjusts to produce a new equilibrium.

Mark breakdown

- 1 mark for correctly identifying the change to the system (i.e. drop in pressure, loss of carbon dioxide)
- 1 mark for quoting and applying Le Chatelier's principle to counter the change (shift towards the gaseous side; right)
- 1 mark for re-establishing a closed system when the bottle is sealed again

HOW TO USE THIS BOOK

Icons

The following icons occur in the summaries and exam practice sections of each chapter.

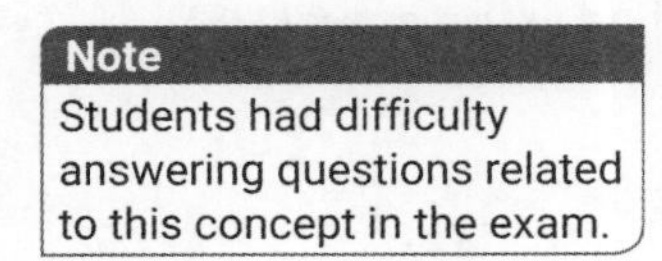

Note boxes appear throughout the summaries to provide additional tips and support.

©NESA 2018 SI Q4

This icon appears with official past NESA questions.

These icons indicate whether the question is easy, medium or hard.

A+ HSC Chemistry Practice Exams

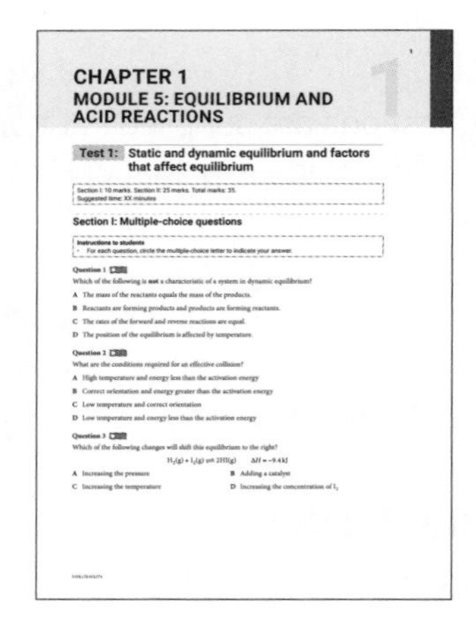

A+ HSC Chemistry Study Notes can be used independently or alongside the accompanying resource *A+ HSC Chemistry Practice Exams*. *A+ HSC Chemistry Practice Exams* features topic tests comprising original HSC-style questions and official HSC questions. Each topic test includes multiple-choice and short-answer questions and focuses on one inquiry question of the New South Wales HSC Chemistry syllabus. There are two complete practice exams following the tests. Like the *A+ Chemistry Study Notes*, detailed solutions are included at the end of the book, demonstrating and explaining how to craft high-scoring exam responses.

A+ DIGITAL

Just scan the QR code or type the URL into your browser to access:

- A+ Flashcards: revise key terms and concepts online
- Revision summaries of all concepts from each inquiry question.

Note: You will need to create a free NelsonNet account.

https://get.ga/aplus-hsc-chemistry-u34

PREPARING FOR THE END OF YEAR EXAM

Exam preparation is a year-long process. It is important to keep on top of the theory and consolidate often, rather than leaving work to the last minute. You should aim to have the theory learned and your notes complete so that by the time you reach STUVAC, the revision you do is structured, efficient and meaningful.

Effective preparation involves the following steps.

Study tips

To stay motivated to study, try to make the experience as comfortable as possible. Have a dedicated study space that is well lit and quiet. Create and stick to a study timetable, take regular breaks, reward yourself with social outings or treats, and use your strengths to your advantage. For example, if you are great at art, turn your Chemistry notes into cartoons, diagrams or flow charts. If you are better with words or lists, create flashcards or film yourself explaining tricky concepts and watch the videos back.

Another strong recommendation is to engage with the Performance Band Descriptors published by NESA. Clear information is provided as to what is expected of a student performing at a band 6, band 5 and so on, mapped against the knowledge and understanding and Working Scientifically course outcome. Have an honest conversation with yourself as to what level you are currently performing at. This will in turn provide you with guidance on what you need to do to improve. For example, a band 6 student must:

- demonstrate an extensive knowledge and understanding of complex and abstract ideas
- apply knowledge and information to unfamiliar situations to propose comprehensive solutions and explanations
- communicate scientific understanding succinctly, logically, and consistently using correct and precise scientific terms.

Revision techniques

Here are some useful revision methods to help information '**STIC**'.

Spaced repetition	This technique helps to move information from your short-term memory into your long-term memory by spacing out the time between your revision and recall flash card sessions. As the time between retrieving information is slowly extended, the brain processes and stores the information for longer periods.
Testing	Testing is necessary for learning and is a proven method for exam success. If you test yourself continually before you learn all the content, your brain becomes primed to retain the correct answer when you learn it. As part of this process, engage with the marking criteria provided to help decide on areas where improvement is needed.
Interleaving	This is a revision technique that sounds counterintuitive but is very effective for retaining information. Most students tend to revise a single topic in a session, and then move onto another topic in the next session. With interleaving, you choose three topics (1, 2, 3) and spend 20–30 minutes on each topic. You may choose to study 1-2-3 or 2-1-3 or 3-1-2, 'interleaving' the topics and repeating the study pattern over a long period of time. This strategy is most helpful if the topics are from the same subject and are closely related.
Chunking	An important strategy is breaking down large topics into smaller, more manageable 'chunks' or categories. Essentially, you can think of this as a branching diagram or mind map where the key theory or idea has many branches coming off it that get smaller and smaller. By breaking down the topics into these chunks, you will be able to revise the topic systematically.

These strategies take cognitive effort, but that is what makes them much more effective than re-reading notes or trying to cram information into your short-term memory the night before the exam!

9780170465281

Time management

It is important to manage your time carefully throughout the year. Make sure you are getting enough sleep, that you are getting the right nutrition, and that you are exercising and socialising to maintain a healthy balance so that you don't burn out.

To help you stay on target, plan out a study timetable. One way to do this is to:

1 Assess your current study time and social time. How much are you dedicating to each?

2 List all your commitments and deadlines, including sport, work, assignments, etc.

3 Prioritise the list and re-assess your time to ensure you can meet all your commitments.

4 Decide on a format, whether weekly or monthly, and schedule in a study routine.

5 Keep your timetable somewhere you can see it.

6 Be consistent.

Studies suggest that 1-hour blocks with a 10-minute break are most effective for studying, and remember you that can interleave three topics during this time! You will also have free periods during the school day you can use for study, note-taking, assignments, meeting with your teachers and group study sessions. Studying does not have to take hours if it is done effectively. Use your timetable to schedule short study sessions often.

The exam

The examination is held at the end of the year and contributes 50% to your HSC mark. You will have 180 minutes plus 5 minutes of reading time. You are required to attempt multiple-choice questions (Section I) and short-answer questions (Section II), covering all areas of study in modules 5–8. The following strategies will help you prepare for the exam conditions.

Practise using past papers

To help prepare, download the past papers from the NESA website and attempt as many as you can in the lead-up to the exam. These will show you the types of questions to expect and give you practice at writing answers. It is a good idea to make the trial exams as much like the real exam as possible (conditions, time constraints, materials etc.). You can also use *A+ HSC Chemistry Practice Exams*.

Use trial papers, school-assessed coursework, and comments from your teacher to pinpoint weaknesses, and work to improve these areas. Do not just tick or cross your answers; look at the suggested answers and try to work out why your answer was different. What misunderstandings do your answers show? Are there gaps in your knowledge? Read the examiners' reports to find out the common mistakes students make.

Make sure you understand the material, rather than trying to rote learn information. Most questions are aimed at your understanding of concepts and your ability to apply your knowledge to new situations.

The day of the exam

The night before your exam, try to get a good rest and avoid cramming, as this will only increase stress levels. On the day of the exam, arrive at the venue early and bring everything you will need with you. If you must rush to the exam, your stress levels will increase, thereby lowering your ability to do well. Further, if you are late, you will have less time to complete the exam, which means that you may not be able to answer all the questions or may rush to finish and make careless mistakes. If you are more than 30 minutes late, you may not be allowed to enter the exam. Do not worry too much about exam jitters. A certain amount of stress is required to help you concentrate and achieve an optimum level of performance. However, if you are feeling very nervous, breathe deeply and slowly. Breathe in for 6 seconds and out for 6 seconds until you begin to feel calm.

Important information from the syllabus

Sixty per cent of your school-based assessment will have addressed the skills required for Working Scientifically. This will have included a mandatory depth study. You are strongly encouraged to engage with the Working Scientifically syllabus outcomes as part of your study, as they will also constitute a significant component of the HSC exam.

Outcome	Description
CH11/12-1	**Questioning and predicting** develops and evaluates questions and hypotheses for scientific investigation
CH11/12-2	**Planning investigations** designs and evaluates investigations in order to obtain primary and secondary data and information
CH11/12-3	**Conducting investigations** conducts investigations to collect valid and reliable primary and secondary data and information
CH11/12-4	**Processing data and information** selects and processes appropriate qualitative and quantitative data and information using a range of appropriate media
CH11/12-5	**Analysing data and information** analyses and evaluates primary and secondary data and information
CH11/12-6	**Problem solving** solves scientific problems using primary and secondary data, critical thinking skills and scientific processes
CH11/12-7	**Communicating** communicates scientific understanding using suitable language and terminology for a specific audience or purpose

Section I of the exam

Section I consists of a question book and an answer sheet. The answers for multiple-choice questions must be recorded on the answer sheet provided. A correct answer scores 1, and an incorrect answer scores 0. There is no deduction for an incorrect answer, so attempt every question. Read each question carefully and underline key words. If you are given a graph or diagram, make sure you understand the graphic before you read the answer options. You may make notes on the diagrams or graphs.

Section II of the exam

Section II consists of a question book with space to write your answers. The space provided is an indication of the detail required in the answer. Most questions will be broken down into several parts, and each part will be testing new information; so, read the entire question carefully to ensure you do not repeat yourself. Use correct chemical terminology and make an effort to spell it correctly. Look at the mark allocation. Generally, if there are two or three marks allocated to the question, you will be expected to make two or three relevant points. If you make a mistake, cross out any errors but do not write outside the space provided; instead, ask for another booklet and re-write your answer. Mark clearly on your paper which questions you have answered where.

Make sure your handwriting is clear and legible and attempt all questions. Marks are not deducted for incorrect answers, and you might get some marks if you make an educated guess. You will definitely not get any marks if you leave a question blank.

Do not be put off if you do not recognise an example or context; questions will always be about the concepts that you have covered. In fact, top performing students are expected to apply learned knowledge to an unfamiliar context.

Reading time

Use your time wisely, *Do not* use the reading time to try and figure out the answers to any of the questions until you have read the whole paper. The exam will not ask you a question testing the same knowledge twice, so look for hints in the stem of the question and avoid repeating yourself. Plan your approach so that when you begin writing you know which section, and ideally which question, you are going to start with. You do not have to start with Section I.

Strategies for answering Section I

Read the question carefully and underline any important information to help you break the question down and avoid misreading it. Read all the possible solutions and eliminate any clearly wrong answers. You can annotate or write on any diagrams or infographics and make notes in the margins. Fill in the multiple-choice answer sheet carefully and clearly. Every few questions, check that your question number matches the answer number you are filling in, check your answer and move on. Do not leave any answers blank.

Strategies for answering Section II

The examiners' reports always highlight the importance of planning responses before writing. Remember you have 3 hours to complete 100 marks. This means you have an average of 1.8 minutes per mark. For a 5-mark question this equals 9 minutes. You should spend a good proportion of this time planning your response.

To do this, **CUBE** the question.

Circle the verb (identify, describe, explain, evaluate).

Underline the key chemical concepts to be covered in your response (e.g. equilibrium, Le Chatelier's principle, limiting agents, buffers, nomenclature of organic compounds).

Box important information.

Elaborate on depth required to answer question.

Many questions require you to apply your knowledge to unfamiliar situations, so it is okay if you have never heard of the context before. You should, however, know which part of the course you are being tested on and what the question is asking you to do. Plan your response in a logical sequence based on the level of detail required by the verb of the question.

Another useful acronym to remember is based on the **ALARM** scaffold (**A** **L**earning **A**nd **R**esponding **M**atrix) developed by Max Woods. We typically accumulate knowledge in a hierarchical nature. For example, before you can evaluate the impact of an industrial chemistry process on society and the environment, you must be able to identify and describe the specific chemical synthesis involved, as well as the desired product or products and any by-products, and then explain how and why they may affect either society or the environment.

Plan your responses by following the same logic, and scaffold your responses using **IDEA/E**. If the question requires an assessment or evaluation, first **Identify**, then **Describe**, then **Explain** and finish with the **Assessment/Evaluation**. If the question requires an explanation, stop at IDE.

Rote-learned answers are unlikely to receive full marks, so you must relate the concepts of the syllabus back to the question and ensure that you answer the question that is being asked and *not* the question you think they are asking. Planning your responses to include the relevant information and the key terminology will help you avoid writing too much, contradicting yourself, or 'waffling on' and wasting time. If you have time at the end of the paper, go back and re-read your answers.

ABOUT THE AUTHOR

Col Harrison

Col Harrison is the Head of Science and the Head of Professional Learning at St Philip's Christian College in Port Stephens. He has spent 35 years in education, both in New South Wales public schools in Western Sydney and in independent schools in Sydney, Newcastle and Port Stephens. He also spent 2 years working with the Abu Dhabi Education Council. Col has marked HSC Biology and Chemistry papers and has held previous roles as a pilot marker and senior marker.

Col completed his Master of Science in 1995 and Doctor of Philosophy in 2017. He has presented at science teacher and technology integration conferences in Sydney and Dubai, and was a previous convener of the Association of Independent Schools (AIS) Science Professional Learning Advisory Council. He is currently involved in the AIS Experienced Teacher Inquiry Project, and is a member of the Royal Australian Chemical Institute NSW Chemical Education Group.

He is married to Jess, has three adult children and two grandchildren, and devotes some of his spare time to a burgeoning YouTube channel providing videos for flipped and blended learning.

CHAPTER 1 MODULE 5: EQUILIBRIUM AND ACID REACTIONS

1

Chapter 1
Module 5: Equilibrium and acid reactions

Module summary

Chemical reactions are vital to our modern way of life. They provide the energy we need for generating electricity and operating transport. They produce cement, plastics, popular beverages and even some flavourings and perfumes. There are reactions to produce ammonia, a vital ingredient in fertilisers and explosives; and sulfuric acid, a key catalyst for many chemical reactions. Many of these chemical reactions stop when one or more reactants are used up.

Some reactions do not go to completion. Instead, they reach an **equilibrium** – a balance between the formation of products and the re-formation of reactants. We will examine the conditions that affect equilibrium reactions, and how the reactions respond to change. We will calculate when a particular reaction reaches equilibrium, and consider equilibrium when a precipitate comes out of a solution.

In Module 5, we explore systems at equilibrium to:

- determine why some chemical systems reach equilibrium and some do not
- describe the factors that indicate a system has reached equilibrium, including a consideration of enthalpy and entropy changes
- analyse systems at equilibrium in the context of collision theory and reaction rates

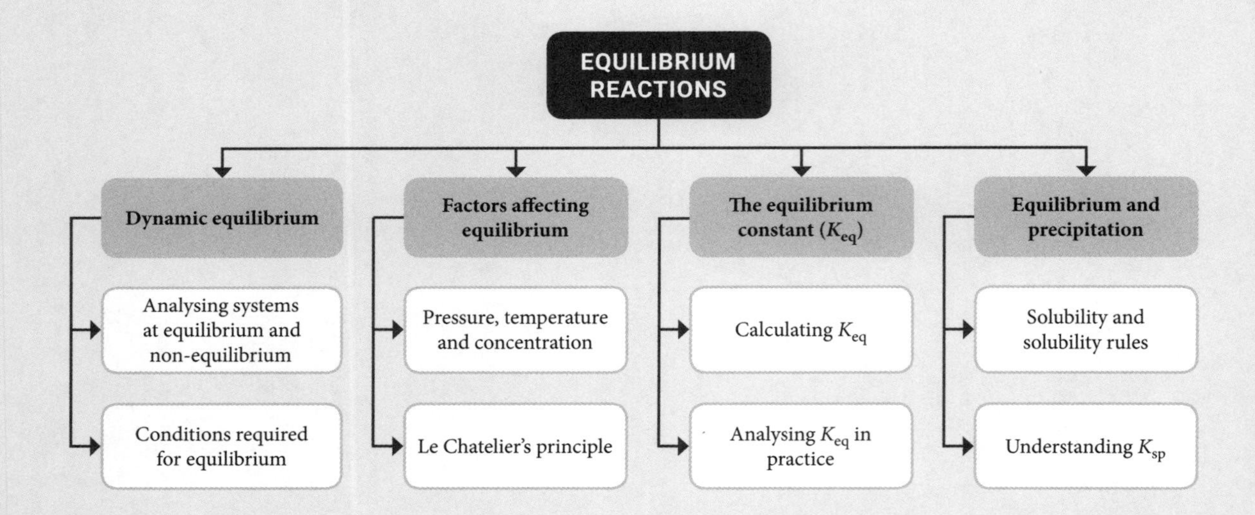

Outcomes

On completing this module, you should be able to:

- explain the characteristics of equilibrium systems, and the factors that affect these systems

Working Scientifically skills

In this module, you are required to demonstrate the following Working Scientifically skills:

- select and process appropriate qualitative and quantitative data and information using a range of appropriate media
- analyse and evaluate primary and secondary data and information
- solve scientific problems using primary and secondary data, critical thinking skills and scientific processes
- communicate scientific understanding using suitable language and terminology for a specific audience or purpose

1.1 Static and dynamic equilibrium

The chemical reactions covered during the Year 11 course included precipitation reactions, decomposition and synthesis reactions, redox (reduction and oxidation) reactions, combustion reactions, and reactions involving acids, bases and metals. Most of these reactions have several characteristics in common. They:

- involve energy
- produce at least one new substance
- may involve a colour change
- may produce a gas (observed as bubbles)
- may produce an insoluble solid
- continue until at least one reactant has been completely used.

In these types of reaction, when one or more reactants have been completely used, the reaction stops – it has gone to completion. This is **static equilibrium**.

If the product particles have enough energy to spontaneously re-form the reactants, the chemical reaction can proceed in both directions. At a certain point, the rate of the forward reaction (reactants → products) is the same as the rate of the reverse reaction (products → reactants). At this point, the reaction has reached **dynamic equilibrium**.

1.1.1 Reversibility of chemical reactions

The reactions we considered in the Year 11 course proceed until one or more of the reactants is completely used up.

1. Two or more reactants interact.
2. Reactants become products as reaction proceeds.
3. Concentrations of reactants decrease and rate of reaction decreases.
4. Concentrations of products increase.
5. Reaction proceeds to completion if one or more reactants is used up.

We used mole ratios to determine the limiting reagent (the one that is used up) and the reactant(s) in excess (any that remain after all the limiting reagent has been used).

As these reactions proceed, they include some reactant(s) and some product(s). In none of these reactions do the products re-form to be reactants.

Collision theory indicates that, as a reaction proceeds and the number of product particles increases, at least some of those particles are more likely to have enough energy to collide and combine to re-form the reactants.

Reactions in which products can re-form to be reactants are **reversible reactions**. These reactions do not proceed to completion. Instead, they reach a dynamic equilibrium somewhere between all reactants and all products. This is usually referred to as just 'equilibrium'.

A reversible reaction can start with the reactants, with the products or with a mixture of reactants and products. It is easiest to think about reversible reactions in terms of the reaction steps for an irreversible reaction, except that in step 5 the reactants are never used up, because the products can re-form them.

As a reversible reaction proceeds, the rate at which the reactants form products (the forward reaction) decreases, and the rate at which the products re-form the reactants (the reverse reaction) increases. The forward and reverse reactions continue, and eventually their rates are the same – they reach equilibrium. This can be drawn as a graph, as shown in Figure 1.1.

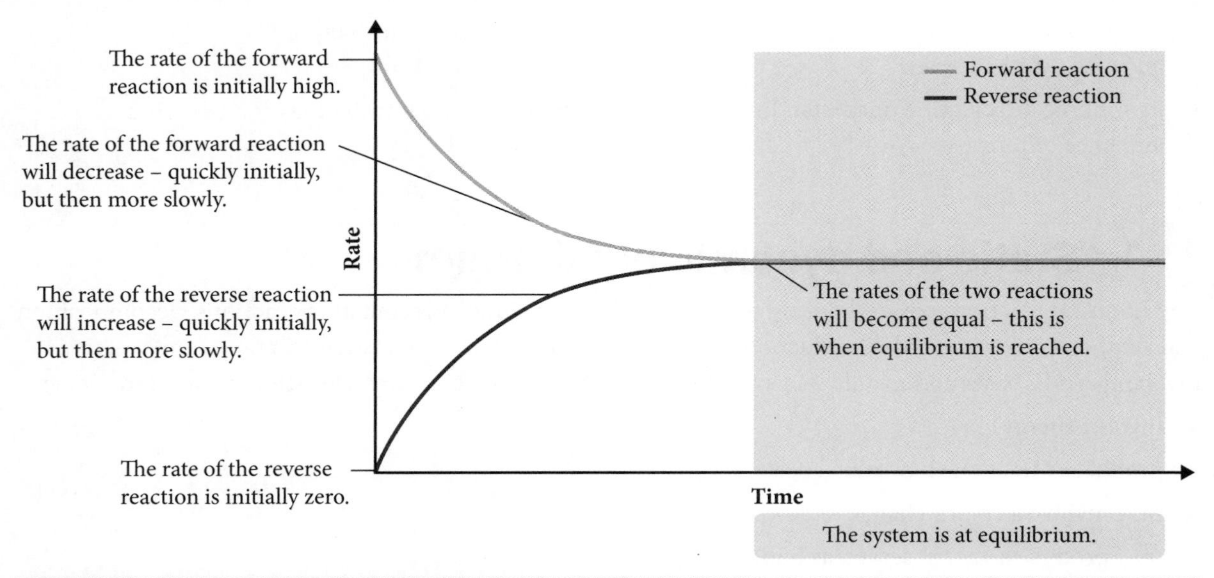

FIGURE 1.1 Changes in reaction rates over time for an equilibrium system

In the reaction shown in Figure 1.2, equilibrium is reached when the rate of the forward reaction ($CO + NO_2 \rightarrow CO_2 + NO$) is the same as the rate of the reverse reaction ($CO_2 + NO \rightarrow CO + NO_2$).

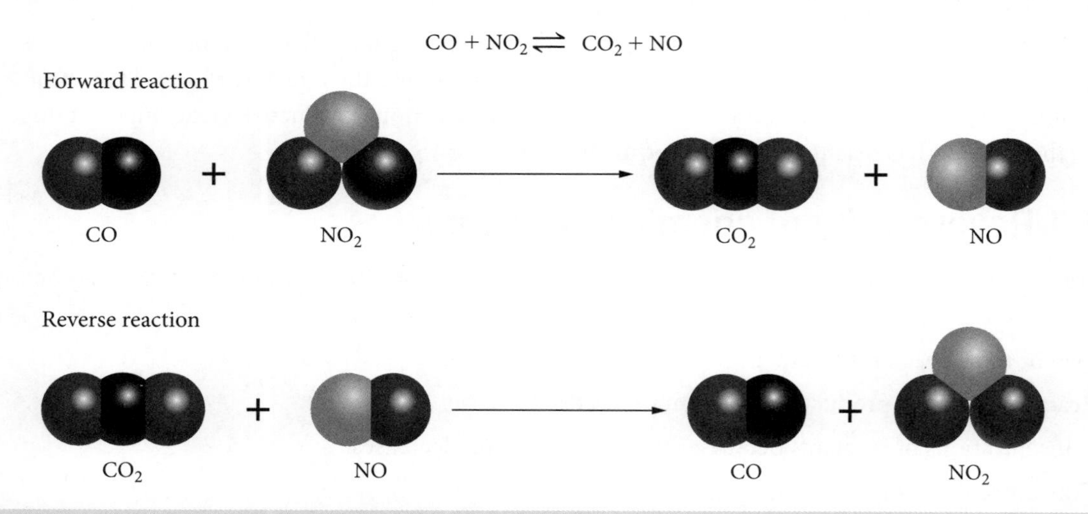

FIGURE 1.2 Equilibrium is reached when the rates of the forward and reverse reactions are the same.

Equilibrium does *not* usually indicate equal amounts of products and reactants. Equilibrium is the equivalence of the rates of the forward and reverse reactions.

When writing equations for reactions that proceed to completion, we use a single arrow: $\rightarrow$. A reversible reaction that will reach an equilibrium is shown by a double arrow: $\rightleftharpoons$. The following examples of reversible and irreversible reactions show how the two different arrow types are used.

Hydrated and anhydrous cobalt(II) chloride

Solid cobalt(II) chloride has several **hydrated** forms (with water molecules within the crystal structure) and an **anhydrous** form (without water). Heating a hydrated form drives water molecules out of the solid. Adding water to the anhydrous form produces the hydrated form.

Water molecules in the hydrated forms of cobalt(II) chloride change the way light interacts with it; hence, its colour. This property change can be used to determine which form of cobalt(II) chloride is present.

When solid anhydrous cobalt chloride (which is blue) dissolves in water, the colour of the solution is pink. The colour change indicates a new substance has formed – cobalt(II) chloride hexahydrate. When the hydrated cobalt(II) chloride solution is heated, the water evaporates and the blue anhydrous form returns.

These observations suggest that the reaction is reversible, and so we use the equilibrium arrow when writing the chemical equation:

$$\underset{\text{blue}}{[CoCl_4]^{2-}(aq)} + 6H_2O(l) \rightleftharpoons \underset{\text{pink}}{[Co(H_2O)_6]^{2+}(aq)} + 4Cl^-(aq)$$

Iron(III) nitrate and potassium thiocyanate

Iron(III) nitrate (where iron(III) is also known as ferric) is soluble in water, forming a pale orange solution. Potassium thiocyanate is a colourless solution. These two substances react in solution to produce iron(III) thiocyanate, which is a blood-red colour.

This system can be adjusted to change the intensity of the red colour, which means it is a reversible reaction. The equation for the reaction can be simplified by cancelling out the spectator ions (NO_3^- in iron(III) nitrate and K^+ in potassium thiocyanate). These ions do not contribute to the colour of the solution.

The simplified equation is:

$$\underset{\text{orange}}{Fe^{3+}(aq)} + \underset{\text{colourless}}{SCN^-(aq)} \rightleftharpoons \underset{\text{red}}{FeSCN^{2+}(aq)}$$

Burning magnesium

The burning of magnesium is a very exothermic reaction ($\Delta H = -601\,kJ\,mol^{-1}$; see section 1.1.3), even though some energy is needed to initiate it. A strip of magnesium ribbon in a Bunsen burner flame combines with oxygen in the air to produce a cloud of white magnesium oxide ash. The ash is a product of this synthesis reaction. The amount of energy released is much greater than the amount added. However, the magnesium oxide cannot be changed back to magnesium.

You might predict an increase in the order of this system because the solid particles and gas particles combine to form a stable, solid structure. However, this reaction leads to a very slight increase in disorder and hence has a positive entropy value ($\Delta S = 26.8\,J\,mol^{-1}\,K^{-1}$). So, this is a spontaneous reaction in the forward direction and it is irreversible, depicted in the equation by a single arrow:

$$2Mg(s) + O_2(g) \rightarrow 2MgO(s)$$

Burning steel wool

Like the burning of magnesium, the burning of steel wool requires the addition of heat and oxygen. It produces more energy than it absorbs (exothermic, negative enthalpy, $\Delta H = -826\,kJ\,mol^{-1}$) and slightly less order in the formation of iron(III) oxide (positive entropy, $\Delta S = 90\,J\,mol^{-1}\,K^{-1}$). The burning of steel wool is also an irreversible reaction, depicted in the equation by a single arrow:

$$4Fe(s) + 3O_2(g) \rightarrow 2Fe_2O_3(s)$$

1.1.2 Open and closed systems

A chemical system may be classified as:

- an **open system**: both matter and energy can enter or leave the system
- a **closed system**: matter is conserved, but energy (such as heat and light) can enter or leave the system
- an **isolated system**: both energy and matter are conserved within the system.

A certain minimum quantity of products must be produced in a reversible reaction before any reactants can re-form, and this cannot happen in an open system. For example, gas produced in an open system – even if the system is reversible – will leave the system.

So, a reversible reaction can only happen in the reverse direction if the system is a closed system (or isolated system), as shown in Figure 1.3.

Charles D. Winters / Science Photo Library

FIGURE 1.3 A closed system conserves matter but not energy.

1.1.3 Non-equilibrium systems

In the Year 11 course, we examined the concepts of **enthalpy** and **entropy** and how they relate to Gibbs free energy in a closed system:

$$\Delta G = \Delta H - T\Delta S$$

where:

ΔG is change in Gibbs free energy (kJ mol^{-1})

ΔH is change in enthalpy (kJ mol^{-1})

T is temperature of the system (K)

ΔS is change in entropy of the system (J mol^{-1} K^{-1}, which should be converted to kJ mol^{-1} K^{-1}).

Table 1.1 summarises the relationships between changes in enthalpy and entropy in such systems.

TABLE 1.1 Relating enthalpy and entropy changes in closed chemical systems

Entropy change, ΔS	Enthalpy change, ΔH	
	Positive	**Negative**
Positive	Spontaneous in forward direction at high temperatures ∴ reversible	Always spontaneous in forward direction ∴ irreversible
Negative	Always spontaneous in reverse direction ∴ irreversible	Spontaneous in forward direction at low temperatures ∴ reversible

Considering ΔG in closed systems:

- a negative ΔG indicates that a forward reaction is spontaneous
- a positive ΔG indicates that a forward reaction is not spontaneous
- a ΔG of 0 indicates that a reaction is at equilibrium; if temperature change favours the forward or reverse reaction, the reaction could be reversible.

We can further investigate Gibbs free energy in chemical systems by examining two very important biochemical processes: photosynthesis and respiration.

Photosynthesis

The overall equation for photosynthesis is:

$$6CO_2(g) + 6H_2O(l) \rightarrow C_6H_{12}O_6(aq) + 6O_2(g) \quad \Delta H \text{ is positive}$$

A lot of energy (in the form of light) is required to initiate photosynthesis, which means that this reaction has a positive enthalpy change. Overall, there is a decrease in entropy (ΔS is negative) because a large molecule (glucose) is produced. We see from Table 1.1 that positive enthalpy change and negative entropy change favour the reverse reaction; hence, the forward reaction is not spontaneous.

Photosynthesis is the basis of all food chains on Earth; life would not exist without it. How can we explain this system that is not at equilibrium?

Photosynthesis is not a single reaction, but a series of reactions. These reactions are an *open* system, not one of the closed systems summarised in Table 1.1. Exchange of matter – gases (such as carbon dioxide and oxygen), water and glucose – is occurring.

During the light stage of photosynthesis, a significant input of energy from sunlight, as well as a catalyst (chlorophyll), enables water molecules to be split, producing oxygen molecules and hydrogen ions. The other product at this stage is ATP (adenosine triphosphate), which is the chemical energy source that drives the subsequent stages of photosynthesis, including synthesis of glucose.

The overall equation for photosynthesis summarises a complex series of reactions involving the chloroplasts inside plant cells. Using Hess's law, we can determine the overall enthalpy and entropy change for photosynthesis, which can then be used to determine that the Gibbs free energy value is positive.

Combustion reactions

Combustion reactions happen between a fuel source and oxygen. Most combustion reactions we have examined have been reactions involving a **hydrocarbon**, such as methane, propane or octane. Combustion reactions release energy (they are exothermic). This is sometimes used for heating, such as heat from a log fire, or to drive other processes, such as in a coal-fired power station. Heat energy released by the combustion of coal can change water to steam, which turns a turbine to generate electricity.

A solid or liquid fuel combined with sufficient oxygen releases energy and produces water and carbon dioxide. If oxygen levels are low, carbon monoxide and carbon (as soot) may be produced.

In addition to being exothermic, combustion reactions in closed systems generally increase entropy because they release many gaseous products. A negative enthalpy change and positive entropy change indicate a spontaneous, irreversible reaction. A combustion reaction that occurs in all of our cells is respiration.

Respiration

The overall equation for respiration is:

$$C_6H_{12}O_6(aq) + 6O_2(g) \rightarrow 6CO_2(g) + 6H_2O(l) \quad \Delta H \text{ is negative}$$

This reaction has a negative enthalpy change because substantial energy is produced by this reaction.

Entropy is increasing (ΔS is positive) because of oxidation of the large glucose molecule. Table 1.1 shows that negative enthalpy change and positive entropy change indicate a spontaneous forward reaction, but this is an over-simplification.

How can we explain this system that is not at equilibrium?

The exchange of gases, water and glucose happens in and beyond animal cells, which means that respiration is an open system.

The overall equation for respiration summarises a complex series of reactions in the mitochondria of animal cells. Chemical energy stored in molecules such as ATP fuels the different stages. As for photosynthesis, Hess's law can be used to determine the overall enthalpy and entropy change for respiration. The Gibbs free energy value for respiration is of the same magnitude and opposite sign as the value for photosynthesis.

9780170465281

1.1.4 Reviewing collision theory

Collision theory can help us to predict whether chemical reactions will happen and whether they are reversible.

Two key requirements for a reaction are:

- *activation energy*, E_a: particles must have enough energy when they collide to break the bonds within them
- *orientation*: particles must collide in such a way that new bonds can form between them.

Particles with enough energy under certain conditions will chemically react to form particles with new arrangements of atoms. This means collision theory predicts that the greater the number of successful collisions, the greater the reaction rate.

The stages of a reversible reaction can also be described in terms of collision theory.

Start of reaction

1 There are many reactant particles and no product particles.

2 The rate of the forward reaction is high because many reactant particles with sufficient activation energy collide to form product particles.

Middle of reaction

3 The concentration of *reactant* particles with sufficient energy to react decreases, so successful collisions happen less often – the rate of the forward reaction decreases.

4 The concentration of *product* particles with sufficient energy to react increases, so successful collisions happen more often – the rate of the reverse reaction increases.

At equilibrium

5 The rates of the forward reaction (reactants → products) and the reverse reaction (products → reactants) are the same.

Energy profile diagrams

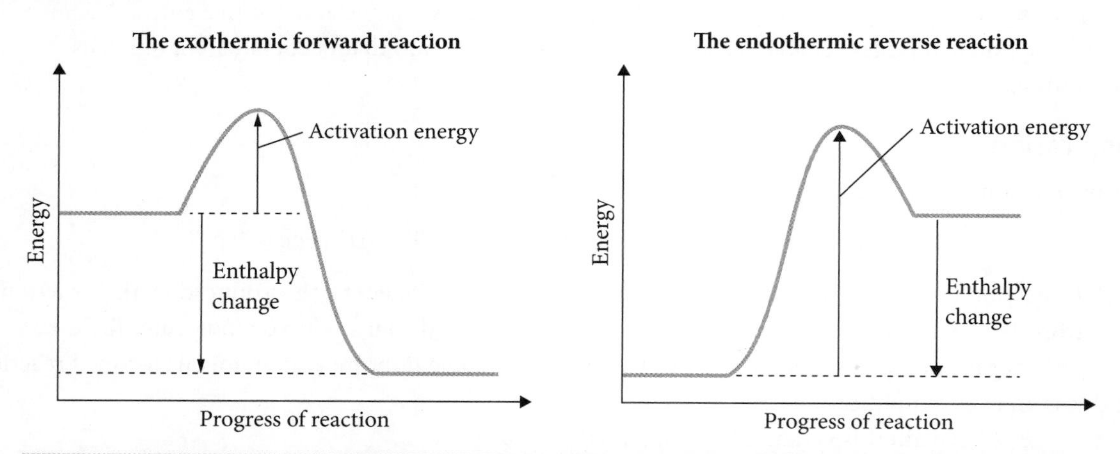

FIGURE 1.4 Energy profile diagrams indicate enthalpy change and activation energy for exothermic and endothermic reactions.

We can use energy profile diagrams such as those in Figure 1.4 to compare the forward and reverse reactions in a system at equilibrium.

Any reversible reaction will have mirror-image energy profiles for the forward and reverse reactions. If the forward reaction is exothermic, it will have a negative enthalpy change and an observable or obvious activation energy. In the reverse direction, the enthalpy change will have the same magnitude but a positive sign, and the activation energy will be equal to the original activation energy, E_a, plus the magnitude of the enthalpy change, ΔH. Figure 1.4 clearly shows the additional energy necessary to initiate the endothermic reaction.

Temperature and volume changes

Sometimes, temperature determines whether a forward or a reverse reaction is favoured and when the system will reach equilibrium.

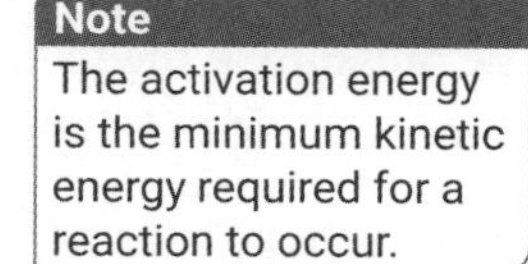

Note

The activation energy is the minimum kinetic energy required for a reaction to occur.

If the temperature of a system is increased, the particles move faster (they have not changed mass) as heat energy transforms to kinetic energy: $E_k = \frac{1}{2}mv^2$. It becomes more likely for the particles to have enough energy to react (E_a) (Figure 1.5).

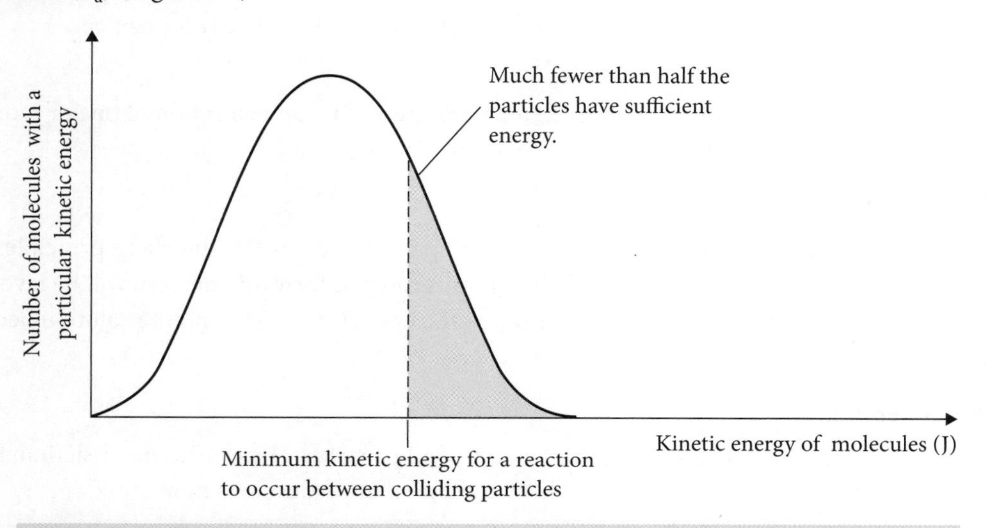

FIGURE 1.5 The Maxwell–Boltzmann distribution shows that only some molecules in chemical systems have enough energy to react.

As the proportion of particles with higher energy increases, so does the reaction rate. This happens to the product particles and reactant particles. Both forward and reverse reaction rates will increase until a new equilibrium is reached.

If the particles are gases, a change in volume also affects reaction rate. For example, a decrease in volume increases the probability of successful collisions because the gas particles are closer together.

1.2 Factors that affect equilibrium

Some factors that affect systems in equilibrium are:

- change in temperature
- change in pressure
- change in volume
- change in concentration of one or more substances
- use of a catalyst.

We need to be able to identify these factors and predict how a system might respond to them.

When examining these factors, keep in mind the importance of collision theory in predicting the rate of a chemical reaction.

1.2.1 Predictions using Le Chatelier's principle

Le Chatelier's principle is used to predict the effect of change on a system at equilibrium. Le Chatelier's principle has two parts:

- a change to a system at equilibrium
- a response by the system to the change.

When a system at equilibrium is changed, the system responds by adjusting to a new equilibrium that minimises the change. For example, if the pressure of a system increases, the system responds to reduce the pressure; if the temperature increases, the system adjusts by absorbing heat. The following examples show some of the changes to and responses by systems at equilibrium.

Cobalt(II) chloride hydrate/dehydrate equilibrium

The equation for this reaction is:

$$\underset{\text{pink}}{[Co(H_2O)_6]^{2+}(aq)} + 4Cl^-(aq) \rightleftharpoons \underset{\text{blue}}{[CoCl_4]^{2-}(aq)} + 6H_2O(l) \quad \Delta H \text{ is positive}$$

For this system, we can investigate changes in temperature and in concentration of one or more of the substances. (There are no gas particles, so pressure is not a factor.)

Increase temperature

Increasing temperature increases the kinetic energy of all particles. Using Le Chatelier's principle, we can correctly predict that the endothermic reaction, in this case the forward reaction, will be favoured. The forward reaction absorbs heat, driving water out of the crystal structure, and the solution becomes more blue.

Decrease temperature

A decrease in temperature decreases the kinetic energy of all particles. The exothermic reaction, in this case the reverse reaction, is favoured, releasing heat and the solution becomes more pink.

Increase concentration of chloride ions (add NaCl)

Adding sodium chloride (NaCl) will increase the number of sodium and chloride ions in the solution. The sodium ions do not affect equilibrium because there were no sodium ions originally. However, the concentration of chloride ions increases, which means there are more successful collisions between reactant particles. The forward reaction rate increases until a new equilibrium is reached, and the solution becomes more blue.

Nitrogen dioxide/dinitrogen tetroxide equilibrium

The equation for this reaction is:

$$\underset{\text{brown}}{2NO_2(g)} \rightleftharpoons \underset{\text{colourless}}{N_2O_4(g)} \quad \Delta H \text{ is negative}$$

This is a gaseous system, so it can be used to show the effect of changes in pressure.

Increase pressure

The simplest way to increase pressure is to decrease volume. This brings the gaseous molecules closer together and hence there are more collisions. The only way to decrease the number of collisions between the particles is to decrease the total number of particles. An increase in the forward reaction will decrease the number of molecules in the system: for every two nitrogen dioxide molecules that react, only one dinitrogen tetroxide molecule is formed. As the forward reaction increases, more of the colourless product is formed and the brown colour will fade.

Increase temperature

Heating increases the average kinetic energy of all particles. Le Chatelier's principle suggests that the system will shift to counter this change. The reaction absorbs some of this extra energy by favouring the endothermic reaction. In this case, the reverse reaction is endothermic. A shift to favour the endothermic reaction will increase the concentration of the nitrogen dioxide. Hence, the gas mixture will become darker brown.

Iron(III) thiocyanate equilibrium

The equation for this reaction is:

$$\underset{\text{brown}}{Fe^{3+}(aq)} + \underset{\text{colourless}}{SCN^{-}(aq)} \rightleftharpoons \underset{\text{blood red}}{FeSCN^{2+}(aq)} \quad \Delta H \text{ is negative}$$

This is an example of ions in solution, so we will investigate changes in temperature and in the concentration of one or more of the ions.

Decrease temperature

Removing heat decreases the average kinetic energy of all particles. Le Chatelier's principle suggests that the system will release heat by shifting towards the exothermic reaction. The exothermic reaction is the forward reaction, so this shift will increase the concentration of iron(III) thiocyanate ions and deepen the red colour.

Add iron(III) nitrate

Adding iron(III) nitrate will add iron(III) ions (Fe^{3+}) to the solution. This will increase the concentration of Fe^{3+}. Le Chatelier's principle suggests that the system will counter the change by decreasing the concentration of Fe^{3+}. This happens when the forward reaction is favoured, using up more of the reactants. This will deepen the red colour.

1.2.2 Collision theory and Le Chatelier's principle

Collision theory can be used to explain the behaviour of some equilibrium systems, and Le Chatelier's principle can then be applied to help predict how the system will respond to a change.

Table 1.2 summarises changes to reactions at equilibrium, the effects in terms of collision theory and the system response as predicted by Le Chatelier's principle.

TABLE 1.2 Change and response in equilibrium systems: Le Chatelier's principle and collision theory

Change	Predicted system response (Le Chatelier's principle)	Explanation (collision theory)
Increase pressure (gaseous systems only)	System shifts to reduce pressure	• Pressure relates to the number of collisions between gaseous particles and their container. • The reaction rate increases in the direction that reduces the number of gaseous particles.
Increase temperature	System shifts towards endothermic reaction	• Particles gain energy when heated. • Heat energy transforms to kinetic energy, which means particles move more quickly. They are more likely to collide with other particles and/or the walls of their container. • The proportion of particles with enough kinetic energy to overcome activation energy for a reaction will be higher for an exothermic reaction (lower E_a) than for an endothermic reaction. All particles have extra energy, but a higher proportion of reactant particles for the endothermic reaction will have enough energy to react. • Initially, the endothermic reaction has a greater rate than the exothermic reaction, and the reactants are favoured.
Decrease temperature	System shifts towards exothermic reaction	• Particles decrease in kinetic energy when cooled. • Less kinetic energy means particles move more slowly. They are less likely to collide with other particles and/or the walls of their container. • The proportion of particles with enough kinetic energy to overcome activation energy is higher for an exothermic reaction (which has a lower E_a) than for an endothermic reaction. All particles are cooling, but a higher proportion of reactant particles for the exothermic reaction still have enough energy to react.
Increase in concentration of a reactant or product ion	System shifts away from the side containing the added ions	• Adding more of an ion already involved in a reaction makes collisions with other ions more likely. • There is a consequent decrease in the rate of the reaction that will produce the added ion.
Decrease in concentration of a reactant or product ion	System shifts towards the side containing the removed ion	• Removing an ion (e.g. by precipitation) involved in a reaction makes collisions with other ions less likely. • There is a consequent increase in the rate of the reaction that will produce the removed ion.

1.2.3 Position of equilibrium

We have used the phrases 'shift towards the right' to indicate an equilibrium response that favours the products (forward reaction) and 'shift towards to the left' to indicate a response favouring the reactants (reverse reaction).

A very important characteristic for all systems at equilibrium is:

rate of forward reaction = rate of reverse reaction

This does not usually mean that the *concentration* of reactants and products is the same. Equilibrium reactions exist in a range of systems with a range of concentrations of reactants and products.

We can distinguish the position of equilibrium by identifying whether the concentration of reactants or products is higher.

- If the concentration of *reactants* at equilibrium is higher than the concentration of products, the equilibrium lies to the left, i.e. it favours the reactants.
- If the concentration of *products* at equilibrium is higher than the concentration of reactants, the equilibrium lies to the right, i.e. it favours the products.

If a reaction proceeds almost to completion but reaches equilibrium with a small concentration of reactants and a very large concentration of products, the equilibrium is far to the right. If the reaction does not proceed very far, there may be a very high concentration of reactants and little of the products. This equilibrium is far to the left.

Activation energy and position of equilibrium

Our study of energy profile diagrams gave us an understanding of the way to calculate enthalpy change and activation energy. The diagrams also made clear that the enthalpy change has the same magnitude (size) for both the forward and reverse directions in a chemical reaction. The only difference is the sign, which is positive for an endothermic reaction and negative for an exothermic reaction. The diagrams show that the exothermic reaction has a smaller activation energy than the endothermic reaction.

Changes to the temperature of the system increase the kinetic energy of all particles and the average kinetic energy of the particles, and hence increase the rate of both forward and reverse reactions because successful collisions are more common. The Maxwell–Boltzmann distribution curves in Figure 1.6 illustrate the relationship between temperature, activation energy and position of equilibrium.

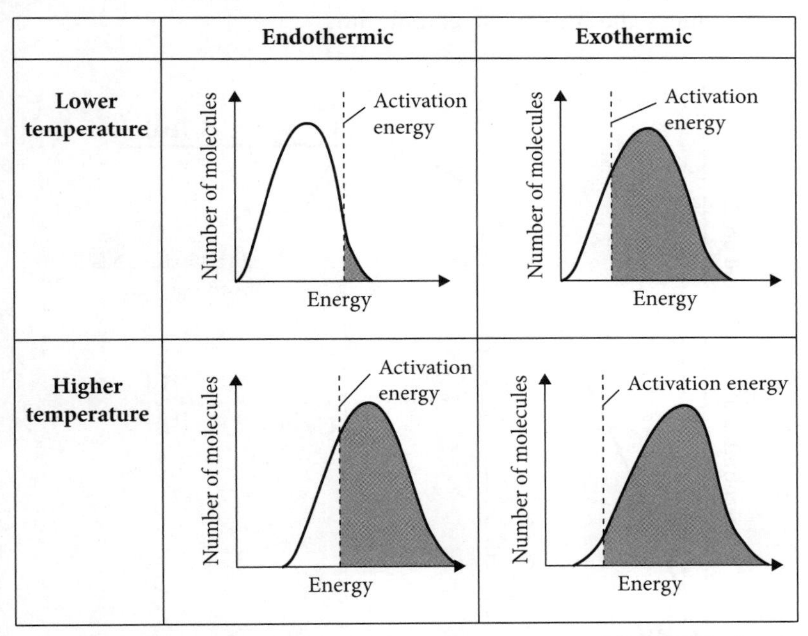

FIGURE 1.6 Adding heat to an equilibrium reaction increases the proportion of molecules with enough kinetic energy to react.

The curves in Figure 1.6 show that at higher temperatures a greater proportion of particles have enough kinetic energy to overcome the activation energy needed to react. Also, at both high and low temperatures, the proportion of particles in an exothermic reaction that have enough energy to react is higher than for an endothermic reaction.

Concentration change in equilibrium systems

It seems from the Maxwell–Boltzmann distribution curves that most reactions favour the exothermic reaction over the endothermic reaction, and it is logical that the position of equilibrium would reflect this. However, equilibrium systems are affected by pressure, temperature and concentration, and a catalyst can provide a pathway of lower activation energy so that more particles can react (although the catalyst will not change the position of the equilibrium).

With so many variables, how can we analyse an equilibrium system?

In Figure 1.1, we saw that an equilibrium system can be represented by plotting reaction rate against time. Figure 1.7 shows the change in concentrations of reactants and products over time for the following equilibrium system:

$$H_2(g) + I_2(g) \rightleftharpoons 2HI(g)$$

1. The forward reaction begins with reactants (H_2 and I_2) but not product (HI), so H_2 and I_2 concentration initially is at a maximum and HI concentration is zero.
2. As the forward reaction proceeds, the concentration of HI increases quickly and the concentrations of H_2 and I_2 decrease quickly.
3. As H_2 and I_2 are used, the rate of the forward reaction does not increase as quickly. There are enough HI particles for them to collide and to re-form H_2 and I_2, which is the reverse reaction. The concentrations of H_2 and I_2 are no longer decreasing as quickly.
4. The forward and reverse reactions continue to adjust in this way until the concentrations of reactants and products are not changing over time (although they are not equal). The forward and reverse reaction rates are the same – the system is at equilibrium.

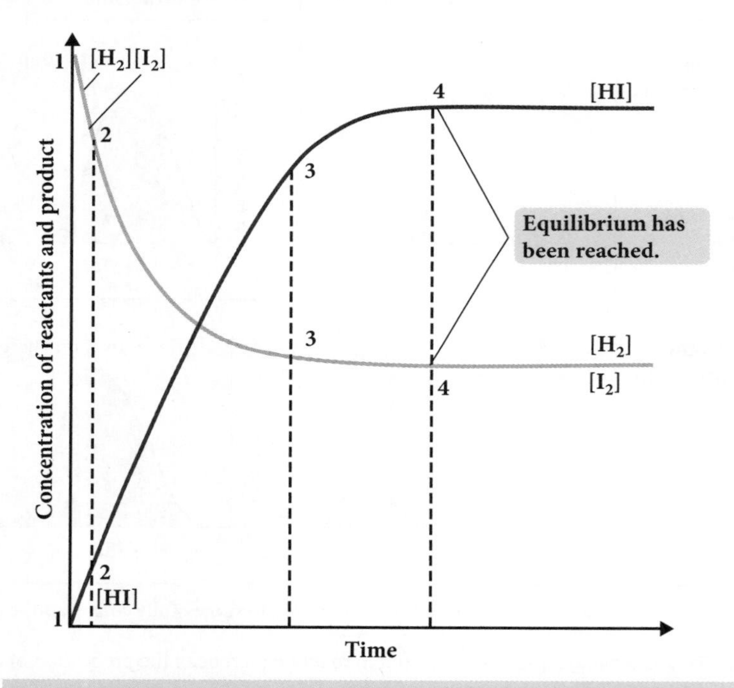

FIGURE 1.7 Changes in concentration over time for $H_2(g) + I_2(g) \rightleftharpoons 2HI(g)$

The section of Figure 1.7 where the solid lines are parallel indicates equilibrium. Parallel horizontal lines indicate no further change in concentration. The reaction is still proceeding, but the rates of the forward reaction and reverse reaction are equal. The equilibrium for this reaction is likely to be to the right given the high concentration of HI (the product in this reaction because its starting concentration was zero).

Graphs of concentration against time can help show not just when equilibrium is reached, but proportions of reactants and products. They can also be used to analyse any changes to an equilibrium system and how the system responds.

What happens if there is a sudden decrease in the concentration of HI after equilibrium is reached (e.g. because some of this product has been liquefied and removed from the system)?

We can apply Le Chatelier's principle to work out how the system would respond to the change. The concentration of HI has decreased, so the system will adjust to increase the concentration of HI. The reaction rates will be quicker at first, then quickly level off as equilibrium is re-established. Figure 1.8 shows these changes as dotted lines.

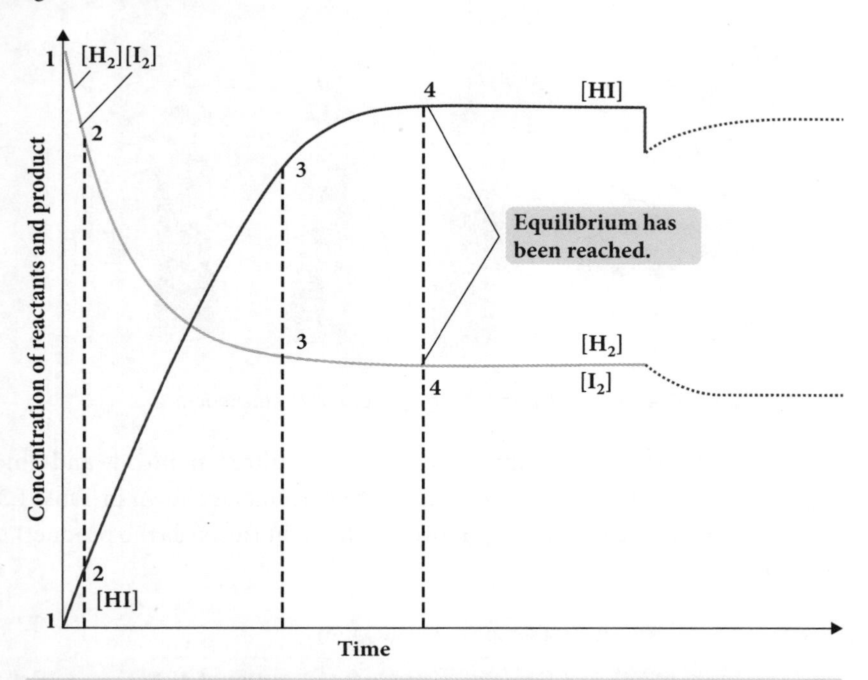

FIGURE 1.8 Response of the equilibrium system $H_2(g) + I_2(g) \rightleftharpoons 2HI(g)$ to a sudden change in the concentration of HI

These graphs are a good way to represent general changes to temperature, pressure or concentration in an equilibrium system. As long as the general pattern is correct, the graph is sufficient. But what if we wanted to add a numerical scale to the *y*-axis? It is possible to accurately calculate the values expected from such a change.

1.3 Calculating the equilibrium constant (K_{eq})

Le Chatelier's principle can be applied to equilibrium systems to determine how systems will respond to change. These changes include colour changes, evolution and formation of precipitates, all of which are **qualitative** measures.

Mathematical equations are a **quantitative** way to describe equilibrium systems to predict the effect of changes and the consequences of the application of Le Chatelier's principle. The equilibrium constant expression is calculated using the relative concentrations of reactants and products in a reaction.

1.3.1 The equilibrium expression

Consider the equation for the decomposition reaction of nitrogen dioxide (Figure 1.9):

$$2NO_2(g) \rightleftharpoons N_2O_4(g)$$

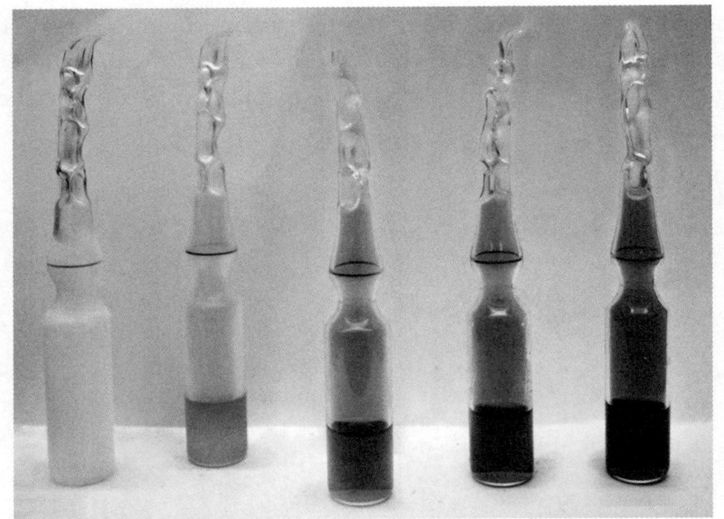

FIGURE 1.9 The nitrogen dioxide–dinitrogen tetroxide equilibrium

Consider a 1 L closed container that initially contains 2 mol L^{-1} nitrogen dioxide and 0 mol L^{-1} dinitrogen tetroxide. We can model some ratios of product to reactant, as shown in Table 1.3.

As the reaction continues and the equilibrium moves to the right (towards the product), the ratio of product to reactant increases.

TABLE 1.3 Modelling ratios of product to reactant for $2NO_2(g) \rightleftharpoons N_2O_4(g)$

Initial amount of reactant, NO_2 (mol)	Amount of NO_2 that reacts (mol)	Amount of product, N_2O_4 (mol) (mole ratio 2 : 1)	Amount of NO_2 remaining (mol)	Ratio of product to reactant
2.0 mol L^{-1} in 1 L = 2 mol	0	0	2.0 – 0 = 2.0	–
	0.5	0.25	2.0 – 0.5 = 1.5	1 : 6 (0.25 : 1.5)
	1.0	0.5	2.0 – 1.0 = 1.0	1 : 2 (0.5 : 1.0)
	1.5	0.75	2.0 – 1.5 = 0.5	3 : 2 (0.75 : 0.5)

A mathematical expression based on the ratio of product concentrations to reactant concentrations is a useful quantitative measure of an equilibrium system.

The equilibrium expression is used to calculate the equilibrium constant, K_{eq}:

$$K_{eq} = \frac{[\text{products}]}{[\text{reactants}]}$$

This can be remembered as PoRK: Products over Reactants = K.

Equilibrium constants are usually reported as numbers without units, because the units of the numerator and denominator often cancel out. The important thing to remember is that all concentrations in the equilibrium expression must be expressed in mol L^{-1}.

Consistent with the modelling in Table 1.3, the expression calculates that:

- a system with equilibrium to the *left* will have a *low concentration of product* (small numerator in the equation) and a *high concentration of reactant* (large denominator in the equation). A small numerator and large denominator mean a *small equilibrium constant*
- a system with equilibrium to the *right* will have a *high concentration of product* (large numerator in the equation) and a low concentration of reactant (small denominator in the equation). A large numerator and small denominator mean a *large equilibrium constant.*

We can use these general rules when considering values for the equilibrium constant:

- $K_{eq} < 0.1$: equilibrium is to the left (the smaller the number, the further to the left)
- $0.1 < K_{eq} < 10$: equilibrium is neither to the left nor right
- $K_{eq} > 10$: equilibrium is to the right (the larger the number, the further to the right).

We need to take the general equilibrium expression a step further. Using our previous example, we see that the calculation of K_{eq} for $2NO_2(g) \rightleftharpoons N_2O_4(g)$ needs to account for the 2 : 1 mole ratio:

$$K_{eq} = \frac{[\text{products}]}{[\text{reactants}]}$$

$$= \frac{[N_2O_4]}{[NO_2] \times [NO_2]}$$

or

$$K_{eq} = \frac{[N_2O_4]}{[NO_2]^2}$$

Note
If one of the substances in a reaction is a solid, its concentration will not change. For this reason, we do *not* include solids when calculating an equilibrium constant.

Note
Whether we include water in an equilibrium expression depends on its state. The concentration of water in the gaseous state can change, so we must include it. We do not consider the concentration of water as a liquid (e.g. an aqueous solution containing ions or polar substances) because it will not change.

This expression can be written in general form. For the equilibrium $aA + bB \rightleftharpoons cC + dD$:

$$K_{eq} = \frac{[\text{products}]}{[\text{reactants}]} = \frac{[C]^c \times [D]^d}{[A]^a \times [B]^b}$$

The numbers of moles of the products and reactants in the balanced equation are shown as superscript lowercase letters in the equilibrium expression.

Calculating an equilibrium constant

Every reaction that reaches an equilibrium has a corresponding equilibrium constant that is specific to a particular temperature. This is because changing the temperature favours the endothermic or exothermic reaction and hence changes the relative concentrations of reactant and product.

When we have the concentrations for the reactants and products in a reaction, we can use the equilibrium expression to calculate the equilibrium constant.

For example, a 4 L container contains 0.2 mol N_2O_4 and 12 mol NO_2. The reaction is:

$$2NO_2(g) \rightleftharpoons N_2O_4(g)$$

To calculate K_{eq}, we first calculate the equilibrium concentration of each reactant and product:

For NO_2:

$$c = \frac{n}{V}$$

$$= \frac{12}{4}$$

$$= 3 \text{ mol L}^{-1}$$

For N_2O_4:

$$c = \frac{n}{V}$$

$$= \frac{0.2}{4}$$

$$= 0.05 \text{ mol L}^{-1}$$

Second, we substitute values into the equilibrium expression:

$$K_{eq} = \frac{[\text{products}]}{[\text{reactants}]}$$

$$= \frac{[N_2O_4]}{[NO_2]^2}$$

$$= \frac{0.05}{3^2}$$

$$= 0.005\,555\,55$$

$$= 0.006 \quad \text{(1 significant figure)}$$

9780170465281

1.3.2 Equilibrium when pressure or concentration changes

An equilibrium constant for a reaction also indicates whether a system has reached equilibrium and, if not, which way it will move.

Determining whether a system has reached equilibrium

To determine whether a system has reached equilibrium, we first need to calculate the **reaction quotient**, Q.

We calculate the reaction quotient using the equilibrium expression but with the values of the system:

$$Q = \frac{[\text{products}]}{[\text{reactants}]}$$

We can then compare the reaction quotient with the equilibrium constant at the same temperature.

- If $Q = K_{eq}$, the system has reached equilibrium.
- If $Q < K_{eq}$, the system needs more product and less reactant (a shift to the right) to reach equilibrium.
- If $Q > K_{eq}$, the system needs more reactant and less product (a shift to the left) to reach equilibrium.

Consider again the reaction $2NO_2(g) \rightleftharpoons N_2O_4(g)$ for which K_{eq} is 0.0056 at 298 K. This time, the system consists of 6 mol NO_2 and 1.5 mol N_2O_4 in a 3 L container.

We can determine whether this system is at equilibrium:

$$[NO_2] = \frac{n}{V} = \frac{6}{3} = 2 \text{ mol L}^{-1}$$

$$[N_2O_4] = \frac{n}{V} = \frac{1.5}{3} = 0.5 \text{ mol L}^{-1}$$

$$Q = \frac{[\text{products}]}{[\text{reactants}]} = \frac{[N_2O_4]}{[NO_2]^2} = \frac{0.5}{2^2} = \frac{0.5}{4} = 0.1 \quad \text{(1 significant figure)}$$

$Q > K_{eq}$ (0.1 > 0.0056), which means the system has too much product and too little reactant. It will shift to the left until equilibrium is established.

Reaction quotient and Le Chatelier's principle

Le Chatelier's principle states that when a system in equilibrium experiences a change, the system shifts to minimise the change.

We have already calculated equilibrium concentrations for the nitrogen dioxide–dinitrogen tetroxide reaction at 298 K as 3 mol L^{-1} NO_2 and 0.05 mol L^{-1} N_2O_4. If we make some changes to the system, does Le Chatelier's principle allow us to make the same predictions?

Halving the volume

Halving the volume will increase the concentrations of both the product and the reactant. The system is closed and we haven't changed the number of moles, so a decrease in volume will result in an increase in pressure (according to Boyle's law covered in the Year 11 course).

Applying Le Chatelier's principle, we can predict a shift to the right to counter the increase in pressure. This shift favours the side of the equation with the lower number of moles of gas (2 mol NO_2 compared with 1 mol N_2O_4).

Using the previous values, we can calculate the reaction quotient after the change:
2 L container, 0.2 mol N_2O_4 and 12 mol NO_2

$$[NO_2] = \frac{n}{V} = \frac{12}{2} = 6 \text{ mol L}^{-1}$$

$$[N_2O_4] = \frac{n}{V} = \frac{0.2}{2} = 0.1 \text{ mol L}^{-1}$$

$$Q = \frac{[\text{products}]}{[\text{reactants}]} = \frac{[N_2O_4]}{[NO_2]^2} = \frac{0.1}{6^2} = \frac{0.1}{36} = 0.003 \quad \text{(1 significant figure)}$$

In this case, $Q < K_{eq}$, so the reaction needs more product to reach equilibrium. This means the system will shift to the right, as we predicted using Le Chatelier's principle.

Adding 1 mol NO_2

Using Le Chatelier's principle, we can predict a shift to the right. This counters the increase of NO_2 by conversion of NO_2 into N_2O_4.

Using our previous values, we can calculate the reaction quotient after the change:
4 L container, 0.2 mol N_2O_4 and 12 + 1 = 13 mol NO_2

$$[NO_2] = \frac{n}{V} = \frac{13}{4} = 3.25 \text{ mol L}^{-1}$$

$$[N_2O_4] = \frac{n}{V} = \frac{0.2}{4} = 0.05 \text{ mol L}^{-1}$$

$$Q = \frac{[\text{products}]}{[\text{reactants}]} = \frac{[N_2O_4]}{[NO_2]^2} = \frac{0.05}{3.25^2} = 0.005 \quad \text{(1 significant figure)}$$

Once again, $Q < K_{eq}$, so the reaction needs more product to reach equilibrium. This means the system will shift to the right, as we predicted using Le Chatelier's principle.

1.3.3 Equilibrium constants and temperature

Changing temperature does not affect calculation of an equilibrium constant; remember that K_{eq} is specific to a reaction at a particular temperature. We can use energy values for a reaction to determine whether the forward reaction is exothermic or endothermic.

If the temperature of a system is increased, the endothermic reaction is favoured to minimise the increase in energy. If the temperature is decreased, the exothermic reaction is favoured to minimise the decrease in energy.

1.3.4 Determining an equilibrium constant by investigation

To determine an equilibrium constant experimentally, we need to consider the measurements we need to make. For the reactants, we can carefully prepare solutions with known concentrations. How can we determine the degree to which each reaction proceeds and when equilibrium is reached?

One method we can use is colourimetry. An electronic instrument called a colourimeter measures a specific wavelength of light as it enters and leaves a solution. The principle behind colourimetry is the Beer–Lambert law, which describes the relationship between **absorbance**, A, and light intensity, I:

$$A = \log_{10}\frac{I_0}{I}$$

The solution to be analysed sits in a small glass or plastic tube called a cuvet, represented in Figure 1.10. Light of a particular wavelength, and with an intensity I_0, enters the cuvet and leaves the cuvet with intensity I. If the solution in the cuvet has not absorbed any light, $I_0 = I$, and $A = 0$.

The absorbance depends on the length of the path through the cuvet, the concentration, c, of the solution and the **molar absorptivity**, ε ($\text{L mol}^{-1}\,\text{cm}^{-1}$). Molar absorptivity represents the ability of a particular solution to absorb light of a given wavelength. The Beer–Lambert law can also be written using these terms (provided on your formulae sheet):

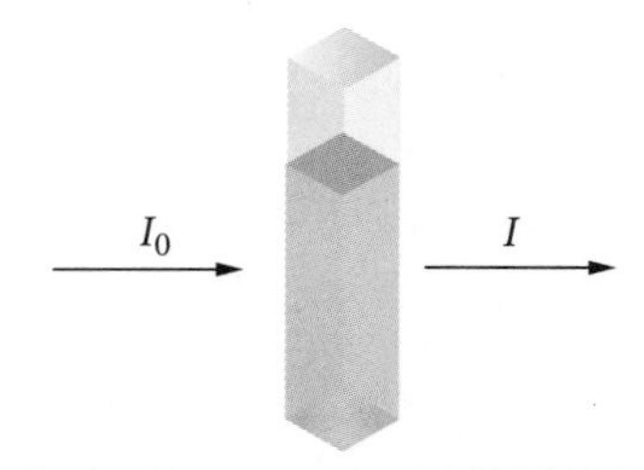

FIGURE 1.10 A cuvet containing a solution. Light enters from the left with an intensity of I_0 and exits to the right with an intensity of I.

$$A = \varepsilon l c$$

Assuming we use the same cuvets (so l remains constant) and the same substance in the cuvet (so ε remains constant), absorbance is in proportion to concentration:

$$A \propto c$$

To obtain an absorbance for a particular concentration, we construct a **calibration curve**. We can use a particular wavelength of light to measure the absorbance of solutions of known concentrations. These concentrations should cover a range within which the concentration of the unknown solution is expected to be. The relationship between A and c is proportional, so we can plot a straight-line graph of absorbance versus concentration. We then measure the absorbance of our sample and interpolate from the calibration curve, as shown in Figure 1.11 for iron(III) thiocyanate ($FeSCN^{2+}$). The absorbance of 0.35 indicates a sample concentration of $0.032\ \text{mol L}^{-1}$.

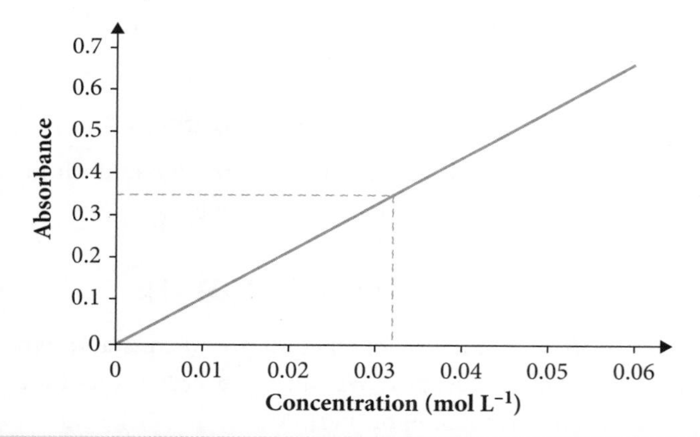

FIGURE 1.11 A colourimetry calibration curve for $FeSCN^{2+}$ shows the line of best fit and an interpolation for an unknown solution with a measured absorbance.

After determining the concentration of one ion in a reaction, the equation and initial values can be used to calculate the concentrations at equilibrium for all species, and from there the equilibrium constant. In practice, this may be an average of several values.

Consider the iron(III) thiocyanate equilibrium reaction. We can start with solutions of:

- 10 mL of 0.15 mol L^{-1} iron(III) nitrate (source of Fe^{3+})
- 10 mL of 0.10 mol L^{-1} potassium thiocyanate (source of SCN^-).

The product, iron(III) thiocyanate ($FeSCN^{2+}$), has a rich, blood-red colour, and its concentration can be determined by colourimetry.

The equation for the iron(III) thiocyanate equilibrium reaction is:

$$Fe^{3+}(aq) + SCN^-(aq) \rightleftharpoons FeSCN^{2+}(aq)$$

Using the concentration value of 0.032 mol L^{-1} from Figure 1.11, we can construct an ICE (Initial, Change and Equilibrium) table for our calculations (Table 1.4). Mole values can be easily converted into concentrations for calculation of the equilibrium constant (if the volume of the container is constant).

TABLE 1.4 An ICE table to calculate equilibrium concentrations for $Fe^{3+}(aq) + SCN^-(aq) \rightleftharpoons FeSCN^{2+}(aq)$

Equation	$Fe^{3+}(aq) + SCN^-(aq) \rightleftharpoons FeSCN^{2+}(aq)$		
Species	Fe^{3+}	SCN^-	$FeSCN^{2+}$
Mole ratio	1	1	1
I: Initial number of moles	$n = cV$ $= 0.15 \times 0.01$ $= 1.5 \times 10^{-3}$ mol	$n = cV$ $= 0.10 \times 0.01$ $= 1 \times 10^{-3}$ mol	0
C: Change in number of moles at equilibrium ($n_{reactants}$ decreased, $n_{products}$ increased)	mole ratio is 1 : 1 : 1 $\therefore n(Fe^{3+}) = n(FeSCN^{2+})$ $= 6.4 \times 10^{-4}$ mol	mole ratio is 1 : 1 : 1 $\therefore n(SCN^-) = n(FeSCN^{2+})$ $= 6.4 \times 10^{-4}$ mol	$n = 6.4 \times 10^{-4}$ mol
Number of moles at equilibrium (calculated from sample colourimetry for $FeSCN^{2+}$)	$n_{equilibrium} = n_{initial} -$ change $= 1.5 \times 10^{-3} - 6.4 \times 10^{-4}$ $= 8.6 \times 10^{-4}$ mol	$n_{equilibrium} = n_{initial} -$ change $= 1 \times 10^{-3} - 6.4 \times 10^{-4}$ $= 3.6 \times 10^{-4}$ mol	$c(FeSCN^{2+})$ $= 0.032$ mol L^{-1} (from Figure 1.11) $n = cV$ $= 0.032 \times 0.02$ $= 6.4 \times 10^{-4}$ mol
E: Concentration at equilibrium	$c = \frac{n}{V}$ $= \frac{8.6 \times 10^{-4}}{0.02}$ $= 0.043$ mol L^{-1}	$c = \frac{n}{V}$ $= \frac{3.6 \times 10^{-4}}{0.02}$ $= 0.018$ mol L^{-1}	$c = \frac{n}{V}$ $= \frac{6.4 \times 10^{-4}}{0.02}$ $= 0.032$ mol L^{-1}

To calculate the equilibrium constant for the iron(III) thiocyanate equilibrium reaction:

$$K_{eq} = \frac{[\text{products}]}{[\text{reactants}]}$$
$$= \frac{[FeSCN^{2+}]}{[Fe^{3+}][SCN^-]}$$
$$= \frac{0.032}{0.043 \times 0.018}$$
$$= 41 \quad \text{(2 significant figures)}$$

We need to compare any experimental result with a theoretical result and account for any discrepancies. These might include:

- *reliability*: Is there consistency between repetitions? It is not sufficient to repeat the experiment; we need to ensure that each repetition yields a consistent set of results
- *validity*: Did the method match the aim? Was there a valid range of concentrations to construct a useful calibration curve? Was the choice of wavelength of light appropriate for the solution being studied? Were there any variables we did not adequately control? Were all experiments conducted at the same temperature?
- *accuracy*: How accurately did we measure absorbance? How accurate were our original solutions? Did they correspond to the values used in the calculation?

Has there been human error? Were any values inadvertently transposed? These are the sorts of questions you should consider during any experimental procedure and should form the basis of your discussion in your practical report.

1.3.5 Equilibria for dissociation of ionic solutions

What happens when an ionic salt is added to water? The simple answer is that it dissolves. It separates into its positive and negative ions, which is called **dissociation**. We know that not all ionic salts will dissolve in water, or at least not much of some salts dissolve. We can use our knowledge of equilibrium to determine the degree to which an ionic salt will dissolve in water.

An ionic salt has a lattice structure: this is a regular series of cations and anions held together in three dimensions by ionic bonds. When salt is added to water, the water molecules, which are polar, are attracted to the ions in the solid. If the force between a polar water molecule and an ion is greater than the forces within the ionic crystal that hold the ion in place, the ion will move away from the crystal structure (Figure 1.12).

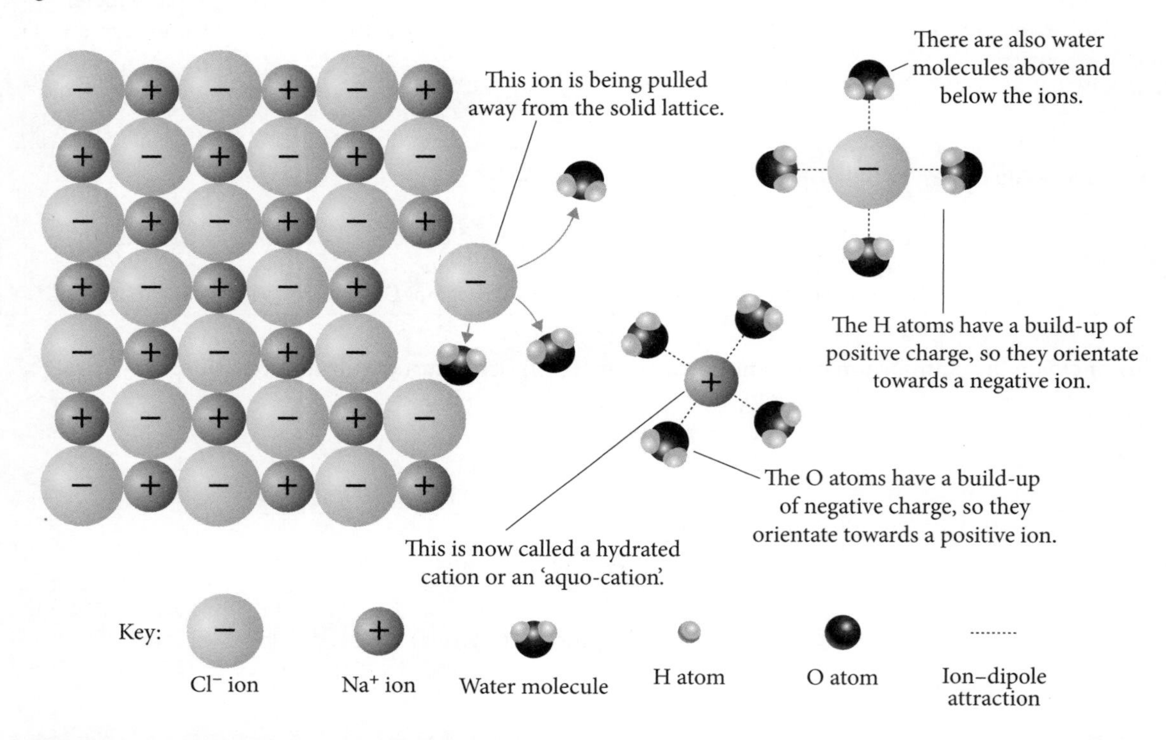

FIGURE 1.12 This model represents the dissolution of sodium chloride in water.

When an ion leaves the lattice, it may be surrounded by water molecules, keeping it away from other ions. However, as more and more ions of the crystal lattice are removed by water molecules, the concentration of ions in the solution increases. This can lead to an interaction between ions in solution and the ions in the crystal lattice, which may 'reclaim' some of the lost ions. This is another example of an equilibrium system.

The following equation shows an ionic solid (cation C^+ and anion A^-) in an equilibrium system involving its ions in solution:

$$CA(s) \rightleftharpoons C^+(aq) + A^-(aq)$$

We can write an equilibrium expression based on the equilibrium constant calculation for this system:

$$K_{eq} = \frac{[\text{products}]}{[\text{reactants}]}$$
$$= \frac{[C^+][A^-]}{[CA]}$$

This is an important example of a **heterogeneous equilibrium** system. The two product ions are in solution, but the solid is not part of the solution. It may have formed a **precipitate**. As a solid, its concentration is constant. If we have twice the mass of the solid (assuming no further dissolution) then it would take up twice the volume. So, if the concentration of CA does not change, we do not include it in our equilibrium constant calculation.

Multiplying both sides of the equation by the concentration value of CA:

$$K_{eq} \times [CA] = \frac{[C^+][A^-]}{[CA]} \times [CA]$$
$$= [C^+][A^-]$$

This new equilibrium constant is the **solubility product**, K_{sp}.

$$K_{sp} = [C^+][A^-]$$

This will be explored in more detail in section 1.4.

1.3.6 Equilibria for dissociation of acids and bases

Module 6 explores the properties and reactions of acids and bases in detail. In relation to equilibria, we need to consider the different strengths of acids and bases – the degree to which they ionise (dissociate) in water.

A strong acid ionises completely in water. Because this is in effect a completion reaction, it will not reach an equilibrium. An example is sulfuric acid (H_2SO_4).

A weak acid does not fully ionise in water. An equilibrium is reached between its molecular form and its ions. An example is ethanoic (acetic) acid (CH_3COOH), the acid in vinegar.

A strong base dissociates completely in water. Again, this is in effect a completion reaction, so it will not reach an equilibrium. An example is sodium hydroxide (NaOH).

A weak base does not fully dissociate or ionise in water. This means there is an equilibrium between its molecular form and its ions. An example is ammonia (NH_3).

In this section we will introduce the concept of equilibrium as it applies to weak acids and weak bases.

Note

Revisit this section after you have covered the section in Module 6 on acid–base reactions to better understand what is actually happening in solution.

Equilibrium constants for weak acids

Acids ionise in water, releasing H^+ ions (usually in the form of H_3O^+) and an anion. The most common types of strong acids are hydrochloric, nitric and sulfuric acids:

$$HCl(aq) + H_2O(l) \rightarrow H_3O^+(aq) + Cl^-(aq)$$

$$HNO_3(aq) + H_2O(l) \rightarrow H_3O^+(aq) + NO_3^-(aq)$$

$$H_2SO_4(aq) + 2H_2O(l) \rightarrow 2H_3O^+(aq) + SO_4^{2-}(aq)$$

The equilibria for these ionisations are far to the right, to the extent that we can consider these acids to be completely ionised. Hence, a single arrow is used. All of these are strong acids.

Weak acids only partially ionise in water and hence an equilibrium is established when they are in aqueous solutions. Two examples of weak acids are ethanoic and carbonic acids:

$$CH_3COOH(aq) + H_2O(l) \rightleftharpoons H_3O^+(aq) + CH_3COO^-(aq)$$

$$H_2CO_3(aq) + 2H_2O(l) \rightleftharpoons 2H_3O^+(aq) + CO_3^{2-}(aq)$$

The acids are weak because of the low concentration of H^+ in solution.

In general, for an acid ionisation equilibrium:

$$HA \rightleftharpoons H^+ + A^-$$

where HA represents the acid, H^+ the hydrogen ions in solution and A^- the anions in solution.

We can calculate the equilibrium constant, which for acids and bases is referred to as the **acid ionisation constant**, K_a.

The general expression for the acid ionisation constant is:

$$K_a = \frac{[H^+][A^-]}{[HA]}$$

Water is a liquid in this equilibrium, so its concentration is not used in the calculation.

Consider a 0.1 mol L^{-1} solution of carbonic acid at room temperature.

It has a pH of 3.68 at 25°C.

Note
pH = $-\log_{10}[H^+]$
The formula for pH is provided on the formulae sheet.

A pH of 3.68 means $[H^+] = 10^{-pH}$ or $10^{-3.68}$
$= 2.09 \times 10^{-4}$ mol L^{-1}

The equation is:

$$H_2CO_3(aq) + 2H_2O(l) \rightleftharpoons 2H_3O^+(aq) + CO_3^{2-}(aq)$$

Assumptions:

$H^+ : CO_3^{2-}$ is 2 : 1, so $[CO_3^{2-}] = 0.5 \times [H^+]$
$= 0.5 \times 2.09 \times 10^{-4}$
$= 1.04 \times 10^{-4}$ mol L^{-1}

- This equilibrium involves a weak acid, so the degree of ionisation is very low. This means the final concentration of the acid molecules is very close to their initial concentration $(0.1 - 1.04 \times 10^{-4})$. It is so close that we can approximate it to 0.1 mol L^{-1}.
- Carbonic acid is a diprotic acid. This means 1 mole of acid releases 2 moles of H^+ ions. However, the chemical behaviour associated with the loss of the first hydrogen ion is different from that associated with the loss of the second hydrogen ion. For now, we will assume they occur as part of a single system.

To calculate the acid ionisation constant:

$$K_a = \frac{[H^+]^2[CO_3^{2-}]}{[H_2CO_3]}$$

$$= \frac{(2.09 \times 10^{-4})^2 \times 1.04 \times 10^{-4}}{0.1}$$

$$= 4.56 \times 10^{-11}$$

The smaller the K_a value, the lower the concentration of H^+ in the solution, and the weaker the acid.

Equilibrium constants for weak bases

In this context, the common bases can generally be regarded as ionic substances and treated in the same way. A common base is sodium hydroxide:

$$NaOH(s) \rightleftharpoons Na^+(aq) + OH^-(aq)$$

$$K_{eq} = \frac{[Na^+][OH^-]}{[NaOH]}$$

NaOH solid, if there is any, does not affect the equilibrium. The equilibrium expression can then be:

$$K_{eq} = [Na^+] \times [OH^-]$$

Some other substances act as bases in water and establish an equilibrium. For example, in sodium hydrogen carbonate ($NaHCO_3$), the sodium ion is a spectator ion and the hydrogen carbonate (bicarbonate) ion interacts with a water molecule:

$$HCO_3^-(aq) + H_2O(l) \rightleftharpoons H_2CO_3(aq) + OH^-(aq)$$

This time, the equilibrium expression is:

$$K_{eq} = \frac{[H_2CO_3][OH^-]}{[HCO_3^-][H_2O]}$$

Water is not needed in this equation because it is a liquid and the solvent for the other ions, so:

$$K_{eq} = \frac{[H_2CO_3][OH^-]}{[HCO_3^-]}$$

This now looks similar to the expression used to form the acid equilibrium expression. This time the calculation is for a base, so:

$$K_b = \frac{[H_2CO_3][OH^-]}{[HCO_3^-]}$$

1.4 Solution equilibria

We previously identified that solid salt added to water will dissociate. If enough salt is added, the solution becomes **saturated**. This means the solution has as many ions as it can hold at a particular temperature. Any additional salt settles to the bottom of the solution.

During a precipitation reaction, if enough ions are present, equilibrium is reached between a precipitate and ions in solution. These sorts of reactions are examples of **solution equilibria**.

1.4.1 Equilibria for ionic salt precipitates

Some ionic compounds readily dissolve in water. Under standard laboratory conditions (**SLC**), potassium chloride (KCl) is a solid with a crystalline structure. When added to liquid water, it dissociates to become K^+ and Cl^- ions. The KCl lattice is dissociated by water molecules, which pull K^+ and Cl^- ions away from the lattice. This can be explained in terms of electrostatic attraction.

> **Note**
> Standard laboratory conditions (SLC) are a temperature of 25°C (298 K) and pressure of 100 kPa (about 1 atm). Chemists often refer to **STP** (standard temperature and pressure), which is 0°C (273 K) and 100 kPa (1 atm). Always calculate answers with the values provided on your formulae sheet.

- Water is a polar molecule that contains a slightly negative end ($O^{\delta-}$) and two slightly positive ends ($H^{\delta+}$). This slight separation of charge between the ends is a dipole.
- The $H^{\delta+}$ end of the molecule is attracted to Cl^- ions, and the $O^{\delta-}$ end is attracted to K^+ ions. This force of attraction between an ion and the oppositely charged region of a water molecule is an **ion–dipole bond**.
- The electrostatic attractions between these particles are stronger than the ionic bonds between K^+ and Cl^-.
- These stronger attractions cause K^+ and Cl^- ions to attach to and be surrounded by the water molecules, and they break away from the lattice structure.

The K^+ and Cl^- ions in the aqueous solution are hydrated – they move freely in the aqueous solution. Not all ionic substances are soluble in water; this is because in those substances the electrostatic attraction between the ions is stronger than their electrostatic attraction to water molecules.

As the solid KCl continues to dissolve, the concentrations of K^+ and Cl^- ions in solution increase. Some of these ions will recombine and leave the solution as a KCl precipitate. Solution equilibria suggests that at equilibrium the rate of the precipitation reaction equals the rate of dissociation:

$$K^+(aq) + Cl^-(aq) \rightleftharpoons KCl(s)$$

Equilibrium in this solution can be disrupted by:

- *increasing the temperature*: this usually shifts the equilibrium to the left because water often holds more ions at higher temperatures
- *adding K^+ ions to the solution*: this shifts the equilibrium to the right
- *adding Cl^- ions to the solution*: this shifts the equilibrium to the right.

These points are consistent with Le Chatelier's principle: increasing the concentration of an ion already present in a solution will shift the equilibrium to increase the rate at which K^+ and Cl^- become KCl. This is the **common ion effect**. Adding a salt that has no ions in common with those already in solution (e.g. NaBr) has no effect on the original equilibrium.

1.4.2 Indigenous chemistry – detoxifying foods

Aboriginal and Torres Strait Islander Peoples have the oldest continuous culture on Earth. Their history and culture have been passed down orally and through rock art over many generations.

The identities of Australia's first peoples are closely connected to Country – the various geographical places of which Aboriginal and Torres Strait Islander Peoples are traditional custodians.

Over millennia, first peoples have used their traditional knowledge of the environment for food, shelter, clothing, storage, healing and hunting.

One of the ways in which they have applied their knowledge of the land and its flora is in the extraction of toxins from certain seeds so that they can be eaten safely. Such a range of potential foods requires a wide variety of techniques, as well as a lot of patience.

One such food is the seeds of the cycad, shown in Figure 1.13. A simple representation of the process is:

$$\text{toxin(s)} \rightleftharpoons \text{toxin(aq)}$$

Cycads may contain several toxins, including mercury, cycasin and methylazoxymethanol glucosides. These can cause gastrointestinal problems, neural difficulties and liver damage. Table 1.5 shows some of the techniques in a process traditionally used by Aboriginal people of tropical north Queensland to prepare cycad seeds for safe consumption, and the associated chemistry. Cycasin's high water **solubility** makes this method a very effective way to detoxify cycad seeds.

FIGURE 1.13 A cycad species found in northern Queensland. Cycads can contain toxins.

TABLE 1.5 Detoxifying cycad seeds: techniques and chemistry

Technique	Result	Associated chemistry
Roasting	Temperature increase Production of carbon	An increase in temperature increases reaction rates. There is a chemical change as **dehydration** of organic toxins releases water and changes complex carbon compounds into carbon (charcoal).
Pounding/grinding	Squashed seeds with increased surface area	A higher surface area exposes more particles and creates the potential for more collisions. Reaction rate increases.
Washing	Dissolution of toxins in water	This is an open system so equilibrium cannot be reached. Any dissolved toxins are washed away with the running water.
Leaching (usually in woven baskets submerged in running water)	Dissolution of toxins in water	This is an open system so equilibrium cannot be reached. Any dissolved toxins are washed away as the water moves through the woven baskets. This technique is similar to washing, although the water contact is from the sides as well as from above, and potentially over a longer time frame.

9780170465281

1.4.3 Solubility rules

Some ionic salts can exist in very high concentrations in solution, whereas others cannot. The maximum amount of solute that can dissolve in a certain volume of a particular solvent indicates the solubility of a solute (in this context, an ionic salt). We are again referring to equilibria, so temperature also affects solubility.

Three terms are commonly used to describe solubility:

- *soluble*: a relatively large quantity of a solute (e.g. >0.1 mol) will dissolve in 1 L of water
- *slightly soluble*: a moderate amount of a solute will dissolve in 1 L of water
- *insoluble*: a very small amount of a solute (e.g. <0.01 mol) will dissolve in 1 L of water.

The Year 11 course includes a range of precipitation reactions – two solutions reacting to produce an insoluble solid (the precipitate). How can we identify the precipitate produced if it does not have a distinct colour? How can we predict which combinations of solutions will generate a precipitate?

We could carry out a series of precipitation reactions in small containers and record the presence and colour of any precipitates. However, this method is only qualitative, and it can be difficult at low ion concentration to observe the formation of a precipitate.

A better way is to use our knowledge of equilibrium to determine the solubility of a specific salt. This enables us to measure ion concentration quantitatively.

The solubility rules in Table 1.6 are based on experiments and can be applied in a range of contexts. The acronyms in bold in the second column may be helpful for remembering the rules.

Note
The HSC Chemistry data sheet contains a list of solubility constants at 25°C.

TABLE 1.6 The solubility rules (NAGSAG) for ionic compounds

Rule	Exceptions
N – nitrates All nitrates are soluble.	None
A – ammonium ions All ammonium salts are soluble.	None
G – group 1 alkali metals All group 1 alkali metals form soluble salts.	None
S – sulfates All sulfates are soluble.	Lead, mercury, silver (**LMS**); calcium, strontium, barium (**CaStroBear**)
A – ethanoates (acetates) All acetates are soluble.	None
G – group 17 elements (halogens) All group 17 salts are soluble.	Lead, mercury and silver (**LMS**) salts of group 17 elements

Note
Most common ions not in the above categories (carbonates, phosphates, hydroxides) are insoluble. However, there are exceptions to some of these (e.g. CaStroBear for hydroxides).

With some practical experience and the NAGSAG, LMS and CaStroBear rules, we can analyse a range of solutions to determine whether a precipitate will form. Each of the examples below can also be applied to laboratory investigations, so you can observe and describe each precipitate formed.

Potassium chloride and silver nitrate

The equation for the equilibrium reaction between potassium chloride and silver nitrate is:

$$KCl(aq) + AgNO_3(aq) \rightleftharpoons AgCl(?) + KNO_3(aq)$$

The ions in the initial solution are:

$$K^+(aq) \qquad Cl^-(aq) \qquad Ag^+(aq) \qquad NO_3^-(aq)$$

We want to focus on only those ions that might form a precipitate. We can rule out the spectators, if there are any, using NAGSAG:

- **N**(AG): all nitrates are soluble, so $NO_3^-(aq)$ is not involved in any precipitate that might form. It is a spectator ion.
- (NA)**G**: all group 1 alkali metals form soluble salts, so $K^+(aq)$ is not involved in any precipitate that might form. It is a spectator ion.
- (SA)**G**: all group 17 halides are soluble except LMS. Cl^- is a soluble halide, but Ag^+ is insoluble (S in LMS represents silver); hence, silver chloride will precipitate from this solution.

The equation can thus be written as:

$$Ag^+(aq) + Cl^-(aq) \rightleftharpoons AgCl(s)$$

$K_{sp} = [Ag^+][Cl^-]$. We will look at the reason for this later.

Potassium iodide and lead nitrate

The equation for the equilibrium reaction between potassium iodide and lead nitrate is:

$$2KI(aq) + Pb(NO_3)_2(aq) \rightleftharpoons PbI_2(?) + 2KNO_3(aq)$$

The ions in the initial solution are:

$$2K^+(aq) \qquad 2I^-(aq) \qquad Pb^{2+}(aq) \qquad 2NO_3^-(aq)$$

Once again, we can rule out the spectators, if there are any, using NAGSAG:

- **N**(AG): all nitrates are soluble, so $NO_3^-(aq)$ is not involved in any precipitate that might form. It is a spectator ion.
- (NA)**G**: all group 1 alkali metals form soluble salts, so $K^+(aq)$ is not involved in any precipitate that might form. It is a spectator ion.
- (SA)**G**: all group 17 halides are soluble except LMS. I^- is a soluble halide, but Pb^{2+} is insoluble (L in LMS represents lead); hence, lead iodide will precipitate from this solution.

The equation can thus be written as:

$$Pb^{2+}(aq) + 2I^-(aq) \rightleftharpoons PbI_2(s)$$

$$K_{sp} = [Pb^{2+}][I^-]^2$$

Sodium sulfate and barium nitrate

The equation for the equilibrium reaction between sodium sulfate and barium nitrate is:

$$Na_2SO_4(aq) + Ba(NO_3)_2(aq) \rightleftharpoons BaSO_4(?) + 2NaNO_3(aq)$$

The ions in the initial solution are:

$$2Na^+(aq) \qquad SO_4^{2-}(aq) \qquad Ba^{2+}(aq) \qquad 2NO_3^-(aq)$$

Once again, we can rule out the spectators, if there are any, using NAGSAG:

- **N**(AG): all nitrates are soluble, so $NO_3^-(aq)$ is not involved in any precipitate that might form. It is a spectator ion.
- (NA)**G**: all group 1 alkali metals form soluble salts, so $Na^+(aq)$ is not involved in any precipitate that might form. It is a spectator ion.
- **S**(AG): all sulfates are soluble except LMS and CaStroBear. Lead, mercury and silver are not present in the solution. Ba^{2+} is present and insoluble (Bear in CaStroBear represents barium); hence, barium sulfate will precipitate from this solution.

The equation can thus be written as:

$$Ba^{2+}(aq) + SO_4^{2-}(aq) \rightleftharpoons BaSO_4(s)$$

$$K_{sp} = [Ba^{2+}][SO_4^{2-}]$$

This procedure can be applied for mixtures of any two solutions, as long as we recognise the ions present and the rules explaining which combinations of ions are likely to produce a precipitate.

1.4.4 Saturated solutions

For a soluble solute in small quantities, all of the solute will dissolve in the solvent. However, we know that each solute has a unique solubility: a particular amount that will dissolve in a given volume of water at a given temperature.

- Sodium hydroxide (NaOH) has a solubility of 80 g/100 mL of water at 25°C.
- Sodium chloride (NaCl) has a solubility of 36 g/100 mL of water at 25°C.
- Copper sulfate ($CuSO_4$) has a solubility of 14 g/100 mL of water at 25°C.
- Calcium hydroxide ($Ca(OH)_2$) has a solubility of 0.12 g/100 mL of water at 25°C.

When no further substance dissolves, any extra substance settles to the bottom of the solution or, if very fine, remains in suspension. The **saturation point** for that substance at 25°C has been reached. A saturated solution is one in which no more solute will dissolve in the solvent.

Consider copper sulfate from the previous list; any mass up to and including 14 g will dissolve in 100 mL of water at room temperature. If we add 14.5 g, then 14.0 g will dissolve and 0.5 g will remain in solid form and sink to the bottom of the container, or remain in suspension.

Although we cannot see it with the naked eye, the rate that ions are added to the solution from the solid state is the same as the rate at which ions are combining to form the solid precipitate. For example:

$$PbCl_2(s) \rightleftharpoons Pb^{2+}(aq) + 2Cl^-(aq)$$

The equilibrium expression for saturated solutions of soluble ionic solids is, for example:

$$K_{eq} = \frac{[Pb^{2+}][Cl^-]^2}{[PbCl_2]}$$

In general, the more insoluble the substance, the smaller the value of the equilibrium constant (K_{eq}). This value is affected by the number of ions dissociated for each reactant and product. We can use a specific type of equilibrium expression for the solubility product constant (K_{sp}). The solubility product expression indicates that the concentration of the solid remains constant, so it is written as:

$$K_{sp} = [Pb^{2+}][Cl^-]^2$$

Like the equilibrium constant, the solubility product constant changes with temperature. For example, the solubility of lead chloride in water is 1.0 g/100 mL at 25°C and 3.3 g/100 mL at 100°C.

Values for the solubility product constant can only be compared if they have the same units, i.e. the same number of ions for each reactant and product.

> **Note**
> Relevant K_{sp} values are provided on the HSC Chemistry data sheet.

Consider barium carbonate ($BaCO_3$), which has a K_{sp} of 2.58×10^{-9}. How can we determine the concentration of barium ions at the saturation point and how much solid barium carbonate would be needed in 100 mL of water to reach this point?

First, write the equation:

$$BaCO_3(s) \rightleftharpoons Ba^{2+}(aq) + CO_3^{2-}(aq)$$

The mole ratio is:

$$1:1:1$$

The solubility product expression is:

$$K_{sp} = [Ba^{2+}][CO_3^{2-}]$$

We know the K_{sp} for $BaCO_3$ is 2.58×10^{-9} and that the concentrations of Ba^{2+} and CO_3^{2-} must be equal. Let's call them x.

$$K_{sp} = x \text{ times } x \text{ or } x^2$$

$$2.58 \times 10^{-9} = x^2$$

$$x = 5.08 \times 10^{-5}\ \text{mol L}^{-1}$$

So the concentration of barium ions at the saturation point is 5.08×10^{-5} mol L^{-1}.
In 100 mL of water:

$$\begin{aligned} n &= cV \\ &= 5.08 \times 10^{-5} \times 0.1 \\ &= 5.08 \times 10^{-6} \text{ mol} \end{aligned}$$

The mole ratio is 1 : 1 : 1, so 5.08×10^{-6} mol $BaCO_3$ was dissolved in 100 mL of water.
To find the mass, multiply number of moles by molar mass:

$$MM \text{ of } BaCO_3 \text{ is } (137.34 + 12.01 + 3 \times 16.00) = 197.35$$

$$\begin{aligned} m &= n \times MM \\ &= 5.08 \times 10^{-6} \times 197.35 \\ &= 1.00 \times 10^{-3} \text{ g} \\ &= 1.00 \text{ mg} \end{aligned}$$

So we need to add 1.00 mg of barium carbonate to 100 mL of water to make a saturated solution.

1.4.5 Predicting solubility of ionic salts

The solubility product constant is relevant in a range of precipitation reactions. In section 1.3.2, we looked at the application of the equilibrium constant expression in determining whether a system had reached equilibrium. Similarly, we can determine whether a precipitate will form when two solutions are mixed together and establish a heterogeneous equilibrium.

When determining whether a system has reached equilibrium and a precipitate will form, we calculate the reaction product and compare the value calculated with the K_{sp} value.

The ionic product (Q_{sp}) is calculated in the same way as the solubility product but uses the values of the system.

The general equation for a solubility equilibrium between CA and its ions, C^+ and A^-, is written as:

$$CA(s) \rightleftharpoons C^+(aq) + A^-(aq)$$

The ionic product expression is:

$$Q_{sp} = [C^+][A^-]$$

We can then compare the value of Q_{sp} with the solubility product constant, K_{sp}, for this reaction at the stated temperature. If:

- $Q_{sp} = K_{sp}$, the system has reached saturation
- $Q_{sp} < K_{sp}$, the system has not reached saturation and there will be no precipitate
- $Q_{sp} > K_{sp}$, the system has reached saturation; additional ions will form a precipitate.

You may wish to use a table to assist with your calculations.

Example 1: 25 mL of 0.1 mol L^{-1} sodium sulfate and 15 mL of 0.01 mol L^{-1} barium nitrate at 25°C

The equation for this reaction is:

$$Na_2SO_4(aq) + Ba(NO_3)_2(aq) \rightleftharpoons BaSO_4(s) + 2NaNO_3(aq)$$

As an ionic equation:

$$2Na^+(aq) + SO_4^{2-}(aq) + Ba^{2+}(aq) + 2NO_3^-(aq) \rightleftharpoons BaSO_4(s) + 2Na^+(aq) + 2NO_3^-(aq)$$

Cancelling the spectator ions (Na^+ and NO_3^-) from both sides gives the net ionic equation:

$$SO_4^{2-}(aq) + Ba^{2+}(aq) \rightleftharpoons BaSO_4(s)$$

The balanced net ionic equation shows a mole ratio of:

$$1 : 1 : 1$$

We can use a table like Table 1.7 to predict whether a precipitate will form.

TABLE 1.7 Determining equilibrium concentrations of barium and sulfate ions in solution for $SO_4^{2-}(aq) + Ba^{2+}(aq) \rightleftharpoons BaSO_4(s)$

Equation	$SO_4^{2-}(aq) + Ba^{2+}(aq) \rightleftharpoons BaSO_4(s)$	
Species	Ba^{2+}	SO_4^{2-}
Initial number of moles	$n = cV$ $= 0.01 \times 0.015$ $= 1.50 \times 10^{-4}$ mol	$n = cV$ $= 0.01 \times 0.025$ $= 2.50 \times 10^{-3}$ mol
Concentration at equilibrium	$c = \frac{n}{V}$ $= \frac{1.50 \times 10^{-4}}{0.04}$ $= 3.75 \times 10^{-3}$ mol L^{-1}	$c = \frac{n}{V}$ $= \frac{2.50 \times 10^{-3}}{0.04}$ $= 0.0625$ mol L^{-1}

$$\begin{aligned} Q_{sp} &= [C^+] \times [A^-] \\ &= [Ba^{2+}] \times [SO_4^{2-}] \\ &= 3.75 \times 10^{-3} \times 0.0625 \\ &= 2.34 \times 10^{-4} \end{aligned}$$

From the data sheet, K_{sp} of $BaSO_4$ is 1.08×10^{-10}.
So $Q_{sp} > K_{sp}$ and hence a precipitate will form.

Example 2: 5 mL of 0.01 mol L^{-1} sodium sulfate and 5 mL of 0.02 mol L^{-1} silver nitrate at 25°C

The equation for this reaction is:

$$Na_2SO_4(aq) + 2AgNO_3(aq) \rightleftharpoons Ag_2SO_4(s) + 2NaNO_3(aq)$$

As an ionic equation:

$$2Na^+(aq) + SO_4^{2-}(aq) + 2Ag^+(aq) + 2NO_3^-(aq) \rightleftharpoons Ag_2SO_4(s) + 2Na^+(aq) + 2NO_3^-(aq)$$

Cancelling the spectator ions (Na^+ and NO_3^- ions) from both sides gives the net ionic equation:

$$SO_4^{2-}(aq) + 2Ag^+(aq) \rightleftharpoons Ag_2SO_4(s)$$

The balanced net ionic equation shows a mole ratio of:

1 : 2 : 1

We can use a table like Table 1.8 to predict whether a precipitate will form.

TABLE 1.8 Determining equilibrium concentrations of silver and sulfate ions in solution for $SO_4^{2-}(aq) + 2Ag^+(aq) \rightleftharpoons Ag_2SO_4(s)$

Equation	$SO_4^{2-}(aq) + 2Ag^+(aq) \rightleftharpoons Ag_2SO_4(s)$	
Species	Ag^{2+}	SO_4^{2-}
Initial number of moles	$n = cV$ $= 0.02 \times 0.005$ $= 1.00 \times 10^{-4}$ mol	$n = cV$ $= 0.01 \times 0.005$ $= 5.00 \times 10^{-5}$ mol
Concentration at equilibrium	$c = \frac{n}{V}$ $= \frac{1 \times 10^{-4}}{0.01}$ $= 1.00 \times 10^{-2}$ mol L^{-1}	$c = \frac{n}{V}$ $= \frac{5 \times 10^{-5}}{0.01}$ $= 5.00 \times 10^{-3}$ mol L^{-1}

$$Q_{sp} = [C^+] \times [A^-]$$
$$= [Ag^+]^2 \times [SO_4^{2-}]$$
$$= (1.00 \times 10^{-2})^2 \times 5.00 \times 10^{-3}$$
$$= 5.00 \times 10^{-7}$$

From the data sheet, K_{sp} of Ag_2SO_4 is 1.20×10^{-5}.
So $Q_{sp} < K_{sp}$ and hence a precipitate will not form.

Example 3: 2 g of magnesium hydroxide in 100 mL of water at 2°C

The equation for this reaction is:

$$Mg(OH)_2(aq) \rightleftharpoons Mg^{2+}(aq) + 2OH^-(aq)$$

The balanced net ionic equation shows a mole ratio of:

1 : 1 : 2

We can use a table like Table 1.9 to predict whether a precipitate will form.

TABLE 1.9 Determining equilibrium concentrations of magnesium and hydroxide ions in solution for $Mg(OH)_2(aq) \rightleftharpoons Mg^{2+}(aq) + 2OH^-(aq)$

Equation	$Mg(OH)_2(aq) \rightleftharpoons Mg^{2+}(aq) + 2OH^-(aq)$		
Species	$Mg(OH)_2$	Mg^{2+}	OH^-
Mole ratio	1	1	2
Initial number of moles	$n = \frac{m}{MM}$ $MM = 24.31 + 16.00 \times 2 + 1.01 \times 2$ $= 58.33$ g mol L^{-1} $\frac{2}{58.33} = 0.034$ mol		
Number of moles at equilibrium		0.034 mol	0.069 mol
Concentration at equilibrium		$c = \frac{n}{V}$ $= \frac{0.034}{0.1}$ $= 0.34$ mol L^{-1}	$c = \frac{n}{V}$ $= \frac{0.069}{0.1}$ $= 0.69$ mol L^{-1}

$$Q_{sp} = [C^+] \times [A^-]$$
$$= [Mg^{2+}] \times [OH^-]^2$$
$$= 0.34 \times 0.69^2$$
$$= 0.16$$

From the data sheet, K_{sp} of $Mg(OH)_2$ is 5.61×10^{-12}.
So $Q_{sp} >> K_{sp}$ and hence a precipitate will form. In fact, most of this solid will not dissolve.

Example 4: Common ion effect – 1 g magnesium hydroxide solid added to sodium hydroxide solution (100 mL at 25°C)

You may occasionally encounter questions that are examples of the common ion effect or may involve a precipitation titration (see Module 8).

For example, what is the equilibrium concentration of magnesium ions if a small quantity of solid magnesium hydroxide is added to 100 mL sodium hydroxide solution at 25°C?

We can reasonably assume complete dissolution of NaOH and thus calculate the concentration of ions in the original NaOH solution:

$$NaOH(aq) \rightarrow Na^{+}(aq) + OH^{-}(aq)$$

$$\begin{aligned} MM(NaOH) &= 22.99 + 16 + 1.008 \\ &= 39.998\ g\ mol^{-1} \end{aligned}$$

$$\begin{aligned} n &= \frac{m}{MM} \\ &= \frac{1}{39.998} \\ &= 0.025\ mol \\ V &= 100\ mL\ (0.100\ L) \\ c &= \frac{n}{V} \\ &= \frac{0.025}{0.100} \\ &= 0.250\ mol\ L^{-1} \end{aligned}$$

Therefore, the concentration of Na^{+} and OH^{-} ions in the solution is 0.250 mol L^{-1}.

When magnesium hydroxide is added to water:

$$Mg(OH)_2(aq) \rightleftharpoons Mg^{2+}(aq) + 2OH^{-}(aq)$$

Mole ratio is:

$$1:1:2$$

So x mol $Mg(OH)_2$ dissociates into x mol Mg^{2+} and $2x$ mol OH^{-}.

At saturation, K_{sp} of $Mg(OH)_2$ is 5.61×10^{-12}.

To calculate the change, we can use the ICE method (Table 1.10).

TABLE 1.10 The common ion effect: determining equilibrium concentrations of magnesium and hydroxide ions in solution

Equation	$Mg(OH)_2(aq) \rightleftharpoons Mg^{2+}(aq) + 2OH^{-}(aq)$		
Species	$Mg(OH)_2(aq)$	Mg^{2+}	OH^{-}
Mole ratio	1	1	2
I: Initial number of moles		0	0.025
C: Change in number of moles at equilibrium		x mol	$2x$ mol
Number of moles at equilibrium		x mol	$(0.025 + 2x)$ mol
E: Concentration at equilibrium		$c = \frac{n}{V}$ $= \frac{x}{0.1}$ mol L^{-1}	$c = \frac{n}{V}$ $= \frac{0.025 + 2x}{0.1}$ mol L^{-1}

Substituting into the K_{sp} expression:

$$\begin{aligned} K_{sp} &= [C^+][A^-] \\ &= [Mg^{2+}][OH^-]^2 \\ &= 5.61 \times 10^{-12} \\ &= \left(\frac{x}{0.1}\right) \times \left(\frac{0.025 + 2x}{0.1}\right)^2 \end{aligned}$$

If your mathematics is strong, you may be able to expand the expression and solve for x.

Alternatively, if x is very small, we can assume that $0.025 + 2x \approx 0.025$. This makes the equation easier to solve:

$$\begin{aligned} 5.61 \times 10^{-12} &= \left(\frac{x}{0.1}\right) \times \left(\frac{0.025 + 2x}{0.1}\right)^2 \\ \frac{x}{0.1} &= \frac{5.61 \times 10^{-12}}{0.0625} \\ x &= 8.98 \times 10^{-12} \text{ mol L}^{-1} \end{aligned}$$

So the concentration of Mg^{2+} ions is 8.98×10^{-12} mol L^{-1}.

This is a very small value; hence, our assumption about x holds true.

We can compare the concentration of Mg^{2+} ions with what we might expect if we had added magnesium hydroxide directly to the water:

$$\begin{aligned} Mg(OH)_2(aq) &\rightleftharpoons Mg^{2+}(aq) + 2OH^-(aq) \\ x\text{ mol} &\rightleftharpoons x\text{ mol} \qquad 2x\text{ mol} \end{aligned}$$

Volume is a common factor, so we can solve by substituting the mole ratios in the K_{sp} equation:

$$\begin{aligned} K_{sp} &= [C^+][A^-] \\ &= [Mg^{2+}][OH^-]^2 \\ &= x \times (2x)^2 \\ &= 4x^3 \end{aligned}$$

$$\begin{aligned} 5.61 \times 10^{-12} &= 4x^3 \\ x &= \sqrt[3]{\frac{5.61 \ x \ 10^{-12}}{4}} \\ &= 1.12 \times 10^{-4} \text{ mol L}^{-1} \end{aligned}$$

So the equilibrium concentration for Mg^{2+} is 1.12×10^{-4} mol L^{-1}.

Comparing this concentration with the one calculated previously, we can see a much lower concentration of magnesium ions in the first solution. The common hydroxide ion has driven the equilibrium to the left and reduced the concentration of magnesium ions.

Note

Be sure to consider all of the ions in a solution, regardless of their origin, when calculating a solubility constant.

Glossary

absorbance A measure of the amount of light absorbed by a sample in UV–vis spectrophotometry

anhydrous Without water. Some crystals have water molecules within their structure, but an anhydrous form has no water in the crystals

acid ionisation constant The equilibrium constant expression applied to the release of a hydrogen ion when an acid is ionised in water

calibration curve A graph developed from a series of readings of absorbance of a particular wavelength of electromagnetic radiation at different, known concentrations for a particular cation

closed system A system in which matter cannot be exchanged between the system and the surroundings, e.g. if a solid precipitate forms in a chemical reaction, it may sink to the bottom of the solution in an open test tube but it does not leave the reaction system, and the system is closed

common ion effect Adding a soluble ionic salt to a solution containing one of the same ions results in a decrease in the solubility of a precipitate, e.g. adding soluble sodium hydroxide to a solution of copper hydroxide (which has a much lower solubility)

dehydration The removal of water

dissociation The separation of cations and anions when an ionic salt is dissolved in water

dynamic equilibrium When a system has reached equilibrium, even if no macroscopic changes can be seen (e.g. colour changes), the reaction is continuing at the microscopic level. Rate of forward reaction = rate of reverse reaction

equilibrium A chemical system that demonstrates reversibility, i.e. reactants form products and at the same time, products re-form reactants

enthalpy A measure of the energy in a system. Enthalpy is not an easy quantity to measure, so we usually refer to enthalpy change (the gain or loss of energy during a chemical reaction)

entropy A measure of the order in a system. Entropy is not an easy quantity to measure, so we usually refer to entropy change (the increase or decrease in disorder during a chemical reaction)

heterogeneous equilibrium An equilibrium system in which more than one state is present, e.g. a solid in equilibrium with its ions in an aqueous solution

hydrated Refers to the presence of water molecules in the crystal structure of a salt

hydrocarbon A compound containing hydrogen and carbon

ion–dipole bond A force of attraction between an ion and the oppositely charged pole of a dipole, e.g. a sodium cation and the negative pole of a water molecule

https://get.ga/aplus-hsc-chemistry-u34

A+ DIGITAL FLASHCARDS

Revise this topic's key terms and concepts by scanning the QR code or typing the URL into your browser.

isolated system A system in which both energy and matter are conserved

molar absorptivity The tendency of a particular concentration of a particular solution to absorb light of a particular wavelength

open system A system in which matter can be exchanged between the system and the surroundings, e.g. if the gas produced in a chemical reaction can escape through an open test tube, the system is open

precipitate An insoluble solid produced when two solutions are mixed together

qualitative Observations based on non-numerical descriptions, e.g. yellow precipitate, bubbles of a gas

quantitative Observations based on numerical descriptions, e.g. 20 seconds, 55°C

reaction quotient The application of the equilibrium constant expression to a system to determine whether it is in equilibrium; it is found by dividing the products by the reactants (with associated indices)

reversible reaction A type of chemical reaction that does not go to completion but in which the product particles will begin to re-form the reactant particles

saturated Refers to a solution in which no further solute particles will dissolve in a stated volume of water for a given solute

saturation point The molar concentration of the solute in a saturated solution

solubility The amount of solute that can dissolve in a given volume of solvent (usually water)

solubility product The product of the concentration of the cations and anions for a given solution

solution equilibria An equilibrium system established by the interaction between the ions in a solution and their precipitate in the same solution

static equilibrium An equilibrium in which there is no change at either a macroscopic or a microscopic level in a reaction system. Under all reasonable definitions of equilibrium, it cannot exist in a static state

SLC Standard laboratory conditions; assumed to mean a temperature of 25°C or 298 K, and a pressure of about 100 kPa (formerly 101.3 kPa) or 1 atm

STP Standard temperature and pressure; it is assumed to mean a temperature of 0°C or 273 K, and a pressure of about 100 kPa or 1 atm

Exam practice

Multiple-choice questions

Solutions start on page 181.

Question 1

Which of the following best describes an open system?

A A saturated solution of sodium chloride with a visible precipitate in a test tube

B A piece of magnesium metal reacting with hydrochloric acid in a beaker

C A combustion reaction taking place inside a bomb calorimeter

D An equilibrium established between iron(III) ions, thiocyanate ions and iron(III) thiocyanate ions in a sealed flask

Question 2

An example of an equilibrium system is:

$$2NO_2(g) \rightleftharpoons N_2O_4(g)\ \Delta H = -\text{ve}$$

Which of the following changes would you expect to increase the yield of N_2O_4?

A Increasing the temperature

B Adding a catalyst

C Removing half of the nitrogen dioxide

D Increasing the pressure

Question 3

The following equation represents an equilibrium system established between yellow chromate ions (CrO_4^{2-}) and orange dichromate ions ($Cr_2O_7^{2-}$).

$$2CrO_4^{2-}(aq) + 2H^+(aq) \rightleftharpoons Cr_2O_7^{2-}(aq) + H_2O(l)$$

Predict the effect that adding hydroxide ions (OH^-) would have on the colour of the solution.

A There would be no change in colour.

B The solution would become more orange.

C The solution would become more yellow.

D The solution would oscillate between yellow and orange.

Question 4

Which of the following mixtures would you expect to form a precipitate?

A Sodium chloride and potassium carbonate

B Ammonium sulfate and potassium iodide

C Barium nitrate and lithium sulfate

D Lead ethanoate and copper nitrate

Question 5

In the following chemical reaction, the ΔH value is negative and the sign of ΔS can be determined from the change in state.

$$NaOH(s) \rightleftharpoons Na^+(aq) + OH^-(aq)$$

From this, we can conclude the ΔG value is

A positive and the reaction is spontaneous.

B positive and the reaction is non-spontaneous.

C negative and the reaction is spontaneous.

D negative and the reaction is non-spontaneous.

Question 6

The best explanation involving collision theory for the shift in an equilibrium after adding heat to a system is that increasing the temperature

A favours the endothermic reaction.

B favours the exothermic reaction.

C raises the average kinetic energy of all particles but significantly affects those of the endothermic reaction more, increasing the endothermic reaction rate more than the exothermic reaction rate.

D raises the average kinetic energy of all particles but significantly affects those of the exothermic reaction more, increasing the exothermic reaction rate more than the endothermic reaction rate.

Question 7

Which of the following is the most correct statement about an Australian Indigenous practice related to the detoxification of foods? Aboriginal Australians used

A leaching to effectively remove toxins through dissolution by ensuring that an equilibrium between the solid and the ions was not established.

B leaching to effectively remove toxins through dissolution by ensuring that an equilibrium between the solid and the ions was established.

C grinding and pounding of seeds to increase the surface area and increase the natural release of the toxins into the air.

D washing to produce an acidic solution that could neutralise the toxins.

Question 8

Using the value for K_{sp} from the data sheet, determine the concentration of silver ions in a saturated solution (in $mol\,L^{-1}$).

A 9.43×10^{-9}

B 2.83×10^{-8}

C 4.26×10^{-5}

D 1.28×10^{-4}

Question 9

A student made up a solution of calcium phosphate by adding 1 g of the solid to 1 L of water. Which of the following statements is true?

A $Q_{sp} > K_{sp}$, hence a precipitate would form.

B $Q_{sp} > K_{sp}$, hence a precipitate would not form.

C $Q_{sp} < K_{sp}$, hence a precipitate would form.

D $Q_{sp} < K_{sp}$, hence a precipitate would not form.

Question 10

A chemist made a solution of calcium sulfate by adding 0.5 g of the solid to 1 L of water. Later, the chemist's assistant added 0.5 g of potassium sulfate to the solution. There was no significant change to the volume. Which of the following statements is correct?

A The chemist would not know there had been a change because there was no visible change to the solution.

B The chemist would know there had been a change because the solution would change colour.

C The chemist would know there had been a change because a precipitate of calcium sulfate would have formed.

D The chemist would not know there had been a change because there is no way to calculate the change.

9780170465281

Short-answer questions

Solutions start on page 184.

Question 11 (3 marks)

Soft drinks contain dissolved carbon dioxide, which bubbles out of solution when the pressure is reduced, according to the equation:

$$CO_2(aq) \rightleftharpoons CO_2(g)$$

Apply your understanding of equilibrium to explain what happens when a soft-drink bottle is opened and then resealed.

Question 12 (4 marks)

The following equation represents a system in equilibrium:

$$\underset{\text{pink}}{[Co(H_2O)_6]^{2+}(aq)} + 4Cl^-(aq) \rightleftharpoons \underset{\text{blue}}{[CoCl_4]^{2-}(aq)} + 6H_2O(l)$$

Explain the effect of adding potassium chloride to this system. Include any expected change in colour.

Question 13 (3 marks)

The following graph represents an equilibrium system involving the ions Fe^{3+}, SCN^- (thiocyanate) and $FeSCN^{2+}$. At time 0, the iron(III) ions and thiocyanate ions were mixed and allowed to reach equilibrium. At time *t*, a small quantity of iron(III) chloride was added to the mixture. Show this change on the graph and continue the lines to show how the system would adjust to this change. Absolute values are not necessary.

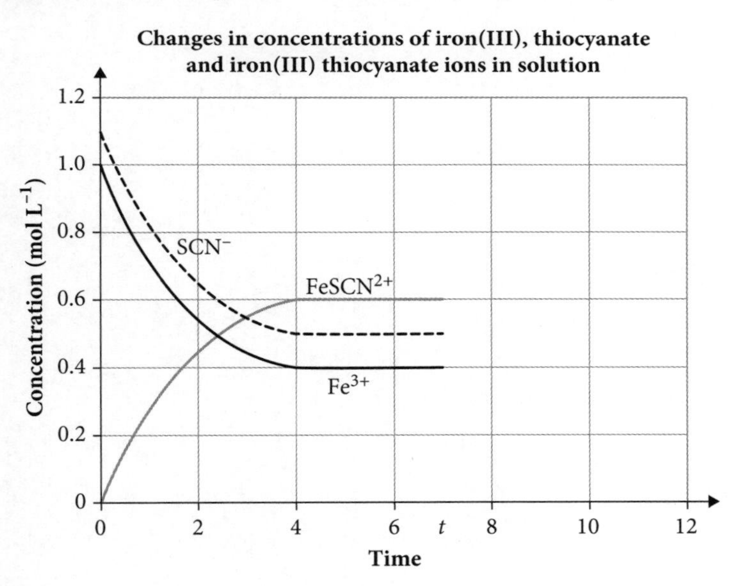

Question 14 (5 marks)

An equilibrium may be established when an ionic solid is dissolved in water. Choose a specific example of an ionic salt and explain how an equilibrium could be established between the solid salt and its ions. Show how the equilibrium expression for your example could be derived (no actual numbers are required).

9780170465281

Question 15 (8 marks)

3.0 moles of nitrogen dioxide and 2.0 moles of dinitrogen tetroxide were introduced into a 3.0 L sealed gas canister at SLC and allowed to come to equilibrium. The K_{eq} at this temperature for this reaction is 215.5 (when NO_2 is the reactant).

a Calculate the concentration of each species at equilibrium. 4 marks

b An additional 2.0 moles of dinitrogen tetroxide were introduced into the container. There was no change in temperature. Calculate the equilibrium constant and the concentrations of both species in these new conditions. 4 marks

Question 16 (2 marks) ©NESA 2019 SII Q29a

Stormwater from a mine site has been found to be contaminated with copper(II) and lead(II) ions. The required discharge limit is 1.0 mg L^{-1} for each metal ion. Treatment of the stormwater with $Ca(OH)_2$ solid to remove the metal ions is recommended.

Explain the recommended treatment with reference to solubility. Include a relevant chemical equation.

CHAPTER 2 MODULE 6: ACID/BASE REACTIONS

Chapter 2
Module 6: Acid/base reactions

Module summary

Acidic and basic solutions exist throughout natural and human environments. Acids are sour and change the colour of blue litmus paper to red. They are vital in body systems and processes such as digestion. Acids are found naturally in many foods, they are added to soft drinks and foods for flavour and preservation, and they are a key electrolyte in many car batteries. Basic solutions are common in cleaning products and are responsible for the bitter taste of some foods. They change the colour of red litmus paper to blue. We use both acids and bases to maintain swimming pools, gardens and agricultural soils.

Pure water is neither acidic nor basic. Acids and bases dissolve in water, interacting with many other substances.

In this module, we explore several ideas about acids and bases.

- What is the difference between acidic substances and basic substances?
- Why is water so important in the context of acids and bases?
- How are acidic and basic solutions analysed?

We will explore a range of aqueous solutions, how and why they affect the colour of indicators, and why classification systems for acids and bases have changed over time. We will also look at some of the quantitative methods used to analyse different types of acid and base solutions.

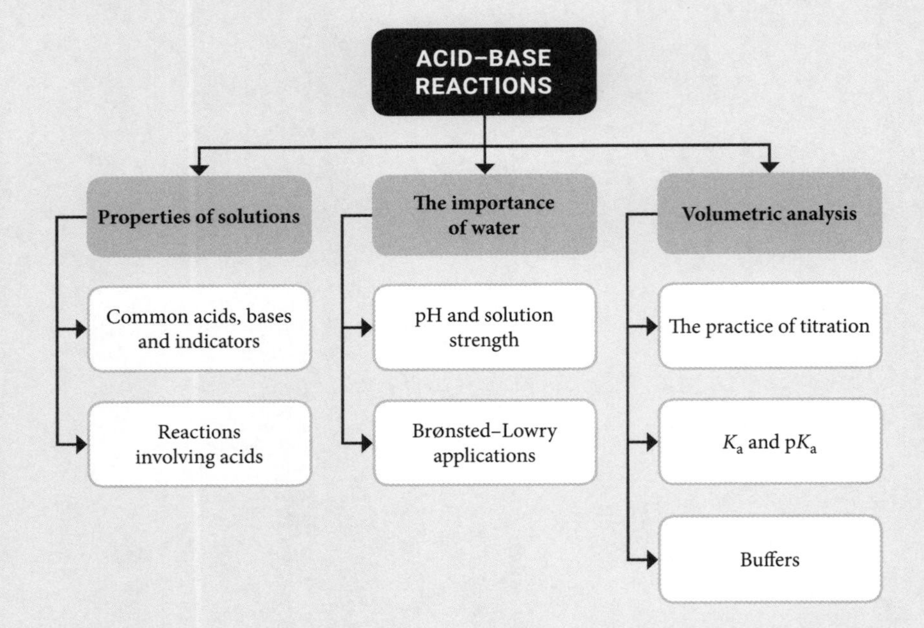

Outcomes

On completing this module, you should be able to:

- describe, explain and quantitatively analyse acids and bases using contemporary models

Working Scientifically skills

In this module, you are required to demonstrate the following Working Scientifically skills:

- develop and evaluate questions and hypotheses for scientific investigation
- design and evaluate investigations in order to obtain primary and secondary data and information
- conduct investigations to collect valid and reliable primary and secondary data and information
- analyse and evaluate primary and secondary data and information

2.1 Properties of acids and bases

French chemist Antoine Lavoisier suggested that an **acid** is a substance that contains oxygen (Figure 2.1). The definition has changed over time, largely because of the observed properties of acids.

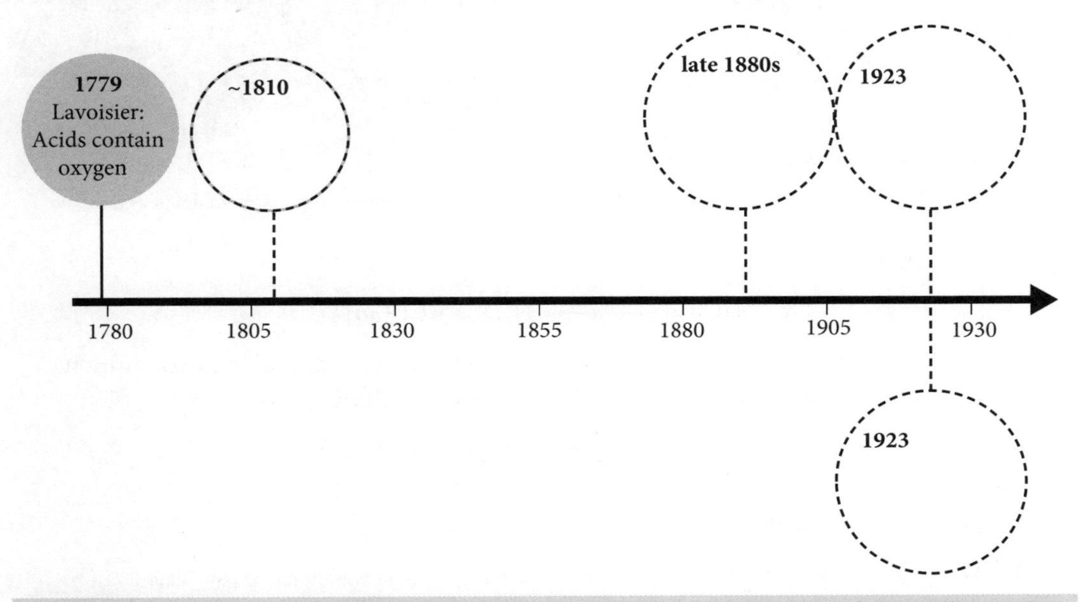

FIGURE 2.1 Alternatives to Lavoisier's definition of acids were developed in the 19th and early 20th centuries.

The most commonly known and used acids in the 18th century were sulfuric acid (H_2SO_4), carbonic acid (H_2CO_3), phosphoric acid (H_3PO_4) and possibly nitric acid (HNO_3), so the assumption of the presence of oxygen in acidic substances was logical.

Lavoisier's conclusions were based on his observations about the behaviour of non-metallic oxides in water. He noted that when substances such as an oxide of sulfur (e.g. sulfur trioxide, SO_3), carbon (e.g. carbon dioxide, CO_2) or phosphorus (e.g. phosphorus pentoxide, P_2O_5) were mixed with water, many of these oxides partly or fully dissolved in water to create acidic solutions. The reactions he observed were:

$$SO_3(g) + H_2O(l) \rightarrow H_2SO_4(aq)$$

$$CO_2(g) + H_2O(l) \rightarrow H_2CO_3(aq)$$

You may have noticed that Lavoisier's definition is too narrow. When metallic oxides are mixed with water, many of them dissolve and produce a basic solution that contains oxygen:

$$Na_2O(s) + H_2O(l) \rightarrow 2NaOH(aq)$$

$$CaO(s) + H_2O(l) \rightarrow Ca(OH)_2(aq)$$

Lavoisier's definition was only part of the story; the definitions of acids and bases based on chemical and physical properties would later be further refined.

2.1.1 Common inorganic acids and bases

Inorganic acids and **bases** are often derived from inorganic compounds or minerals. Many of these ionise in solution. Hydrochloric acid and sodium hydroxide (a base) are two examples, although there are a range of organic and inorganic acids and bases in and around our homes (Figure 2.2).

Science Photo Library / Alamy Stock Photo

FIGURE 2.2 Some common household acids and bases

> **Note**
> We generally refer to substances that contain carbon as organic substances. This would suggest that substances that do not contain carbon are inorganic compounds. This is a useful working definition, but there are some exceptions.

Hydrochloric acid is hydrogen chloride (HCl) dissolved in water. In water, it forms positive hydrogen ions and chloride anions, and is an acidic solution.

$$HCl(g) \xrightarrow{\text{in water}} H^+(aq) + Cl^-(aq)$$

Sodium hydroxide (NaOH) in water dissociates to form sodium cations and hydroxide anions. The product is a basic solution, also called an alkaline solution or **alkali**.

$$NaOH(s) \xrightarrow{\text{in water}} Na^+(aq) + OH^-(aq)$$

IUPAC nomenclature for acids

The relevant system of nomenclature used for the naming of inorganic acids and bases in the HSC Chemistry course follows the IUPAC system. IUPAC (International Union of Pure and Applied Chemistry) is the world authority on chemical naming (**nomenclature**) and terminology. IUPAC is also responsible for naming new elements to be included in the periodic table and for standardised methods of measurement.

Many inorganic acids are more often referred to by their non-systematic names; for example, 'hydrochloric acid' and 'sulfuric acid'. Memorisation is the best tool for these acids. The systematic names are derivative and beyond the scope of the HSC Chemistry course.

Table 2.1 shows some of the most common inorganic acids and their chemical formulae.

TABLE 2.1 Some common inorganic acids

Non-systematic name	Chemical formula
Bromic acid	$HBrO_3$
Carbonic acid	H_2CO_3
Chloric acid	$HClO_3$
Hydrochloric acid	HCl
Nitric acid	HNO_3
Nitrous acid	HNO_2
Perchloric acid	$HClO_4$
Phosphoric acid	H_3PO_4
Sulfuric acid	H_2SO_4
Sulfurous acid	H_2SO_3

IUPAC nomenclature for bases

When naming inorganic bases, follow the rules for naming inorganic salts.

1 Name the cation. This is usually a metal whose entire name is included in the base name. Another common example is the cation NH_4^+, which has the non-systematic but acceptable name 'ammonium'.

2 Name the anion. If the anion:
- is a single element, change the end of the element name to 'ide' (e.g. 'chloride')
- contains two or more elements, one of which is oxygen, include 'ate' in the element name (e.g. 'carbonate')
- fits the previous category and contains hydrogen, add 'hydrogen' before the element name or use the prefix 'bi' (e.g. 'hydrogen carbonate'/'bicarbonate')
- needs to be distinguished from compounds with similar names (e.g. 'carbon monoxide' and 'carbon dioxide'), use a prefix such as 'mono'.

As for acids, there are a number of acceptable non-systematic names for many of the most common bases. Recall too, a basic substance that is soluble in water is known as an alkali. Table 2.2 highlights the most common types of inorganic bases and their chemical formulae.

TABLE 2.2 Some common inorganic bases

Non-systematic name	Chemical formula
Ammonia	NH_3
Ammonium hydroxide	NH_4OH
Barium hydroxide	$Ba(OH)_2$
Calcium hydroxide	$Ca(OH)_2$
Sodium carbonate	Na_2CO_3
Sodium hydrogen carbonate (sodium bicarbonate)	$NaHCO_3$
Sodium hydroxide	$NaOH$

Properties of acids and bases

In your Year 11 course, you would have discussed the properties of matter. You may recall that chemical substances can have both physical and chemical properties. This is also the case with acids and bases (Table 2.3). Physical properties are often related to the type and arrangement of atoms in a substance, the types of bonds holding these atoms together and the way electromagnetic radiation behaves as it passes through the substance. Chemical properties relate to the energy in the bonds and how easy it is to break the chemical bonds and create new bonds.

TABLE 2.3 Physical and chemical properties of acids and bases

	Acids	Bases
Physical properties	• Sour (in dilute form) • Conduct electricity in **aqueous** solutions	• Bitter (in dilute form) • Conduct electricity in aqueous solutions • Slippery or soapy
Chemical properties	• Corrosive (e.g. to metals or to skin) in **concentrated** form • Change colour of: – litmus indicator to red – methyl orange indicator to red – phenolphthalein to colourless • Neutralisation reaction with bases to form ionic salts and water • Used for cleaning, in fire extinguishers, for etching, and to produce fertilisers, explosives and dyes	• Frequently corrosive (e.g. to skin or some metals) in concentrated form • Change colour of: – litmus indicator to blue – methyl orange indicator to yellow – phenolphthalein to wine red or magenta • Neutralisation reaction with acids to form ionic salts and water • Used to produce soaps

2.1.2 Indicators

Acid–base **indicators** change colour in response to the degree of acidity or basicity of a solution. Most indicators are weak acids whose corresponding base product is a different colour. The acid and the base are in equilibrium with each other, meaning the equilibrium can be shifted to favour one colour by adding hydrogen ions or hydroxide ions:

$$\underset{\text{red}}{HY(aq)} \rightleftharpoons H^+(aq) + \underset{\text{blue}}{Y^-(aq)}$$

You may also see HInd and Ind^- used to represent the equilibrium mixture for an indicator solution:

$$\underset{\text{red}}{HInd(aq)} \rightleftharpoons H^+(aq) + \underset{\text{blue}}{Ind^-(aq)}$$

As we saw in Module 5, an equilibrium is established when a reaction does not go to completion. Using Le Chatelier's principle, we can predict how a system will adjust to counter a change.

A range of indicators

Indicators are often a combination of plant-based dyes, and the indicator we select for an experiment depends on what we want to measure.

One indicator can be prepared by boiling red cabbage leaves to extract the coloured dye. The dye is purple. Figure 2.3 shows the indicator in different acidic and basic solutions.

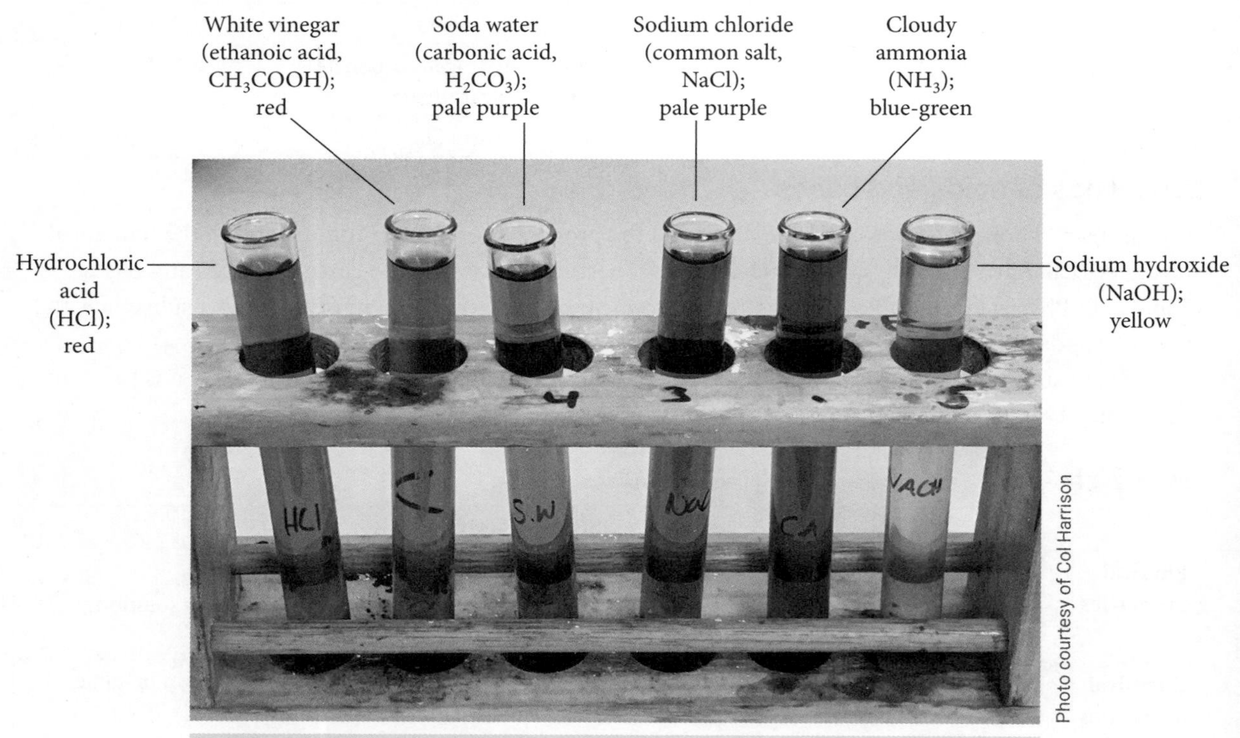

FIGURE 2.3 Red cabbage indicator in some common acidic, neutral and basic solutions

In Figure 2.3, we can see that the indicator colour:

- changes from purple to red in a **strong** acid solution (HCl)
- changes from purple to red in a **weak** acid solution (CH_3COOH)
- does not change colour in soda water or in a neutral salt solution (NaCl)
- changes from purple to blue-green in a weak base solution (NH_3)
- changes from purple to yellow in a strong base solution (NaOH).

Using red cabbage indicator, we can easily see colour differences between acidic, neutral and basic solutions, and even some colour change to show the degree of basicity (but not as evident for different strengths of acid).

Like red cabbage indicator, universal indicator has a range of different colours indicating how acidic or basic a solution is, as well as acidic or basic **strength**. This makes it an excellent indicator for testing when we don't know if a solution is an acid, a base or neutral. It is accurate enough to be a broad indicator of pH.

Universal indicator is not as good for analytical techniques such as titration. The combination of dyes makes it much more difficult to identify exactly where a particular point, such as neutralisation, may occur. For these techniques, it is better to use more simple indicators.

Litmus

Litmus indicator is available as a solution and as paper treated with litmus solution. There are several types, and choice depends on the solution being tested. Adding a few drops of indicator to a solution (or dipping the paper into the solution) indicates whether the solution is an acid, a base or neutral.

- Acids change blue litmus to red. Red litmus is not changed by acids.
- Bases change red litmus to blue. Blue litmus is not changed by bases.
- Neutral substances do not change the colour of either red or blue litmus.

Methyl orange

Methyl orange is an indicator solution that changes colour from red through orange to yellow. Strong acids change the colour to red, and colour changes through orange to yellow all happen in the acidic range. This means that we cannot use methyl orange to tell the difference between a neutral solution and a basic solution.

Phenolphthalein

Phenolphthalein is an indicator solution that changes from colourless to magenta. Strong bases change the colour to a deep magenta, but colour changes happen in the basic range. This means that we cannot use phenolphthalein to tell the difference between a neutral solution and an acidic solution.

Combining indicators

At times we may need to combine indicators to get a sense of the acidity or basicity of a solution.

- If we combine methyl orange (changes between pH 3.1 and 4.4) and phenolphthalein (changes between pH 8.3 and 10), we can distinguish between a strong and weak acid and a neutral substance and a base, but not between a weak acid and a neutral substance.
- If we now add bromothymol blue, we can distinguish between the weak acid and the neutral substance.

This shows why universal indicator, which is made of a number of dyes that change colour at different pH values, can be used to identify acidity and basicity for a wide range of solutions.

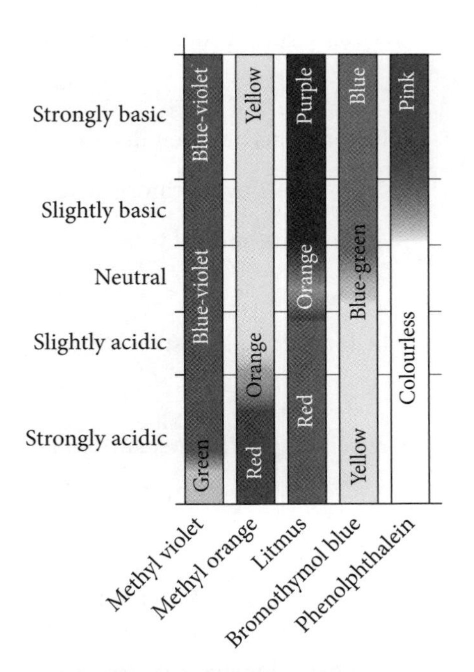

FIGURE 2.4 Colour change ranges for different indicators

Acids, bases and equilibria

Let's look at the equilibrium for an indicator solution we discussed earlier:

$$\underset{\text{red}}{HY(aq)} \rightleftharpoons H^+(aq) + \underset{\text{blue}}{Y^-(aq)}$$

What happens to this equilibrium if we add some acid?

An acid contains hydrogen ions, so we expect additional hydrogen ions in the solution. If so, the concentration of hydrogen ions increases. This change disrupts the equilibrium, and the system adjusts by shifting to the left. Excess hydrogen ions react with the Y^- ion and produce more HY. This deepens the red colour of the solution.

What happens to this equilibrium if we add some base?

A base often contains hydroxide ions, so we expect additional hydroxide ions in the solution. If so, the concentration of hydrogen ions decreases because the hydroxide ions neutralise the hydrogen ions, decreasing their concentration. This change in the system disrupts the equilibrium, and the system responds by shifting to the right. Excess hydrogen ions are produced to replace the reacted H^+ ions, which increases the concentration of Y^- and decreases the concentration of HY. This deepens the blue colour of the solution.

When a reaction equilibrium favours one dye, that colour dominates (e.g. red litmus in acids). When the equilibrium shifts so that there are roughly equal quantities of dyes of two colours, we may see an intermediate colour. Litmus solution may look purple in neutral solutions because it has equal amounts of the red and blue dyes; methyl orange appears orange because it has equal amounts of the red and yellow dyes.

2.1.3 Reactions involving acids

Looking at some of the general reactions involving acids gives us further clues to the chemical properties of acids. It is also helps us to expand and refine our definition of acids.

Acid and metal reactions

Reactions involving metals and acids depend on the activity of the metal and the concentration of the acid. We will focus here on the reactions of **dilute** acids with active metals.

We can write a general word equation for these reactions:

dilute acid + active metal → salt + hydrogen gas

The reactions between dilute acids and active metals are displacement reactions. For example:

hydrochloric acid + magnesium → magnesium chloride + hydrogen gas

We can write and balance the chemical equation:

$$2HCl(aq) + Mg(s) \rightarrow MgCl_2(aq) + H_2(g)$$

The ionic equation is:

$$2H^+(aq) + 2Cl^-(aq) + Mg(s) \rightarrow Mg^{2+}(aq) + 2Cl^-(aq) + H_2(g)$$

Cancelling out the spectator ions on both sides leaves the net ionic equation:

$$2H^+(aq) + \cancel{2Cl^-(aq)} + Mg(s) \rightarrow Mg^{2+}(aq) + \cancel{2Cl^-(aq)} + H_2(g)$$

$$2H^+(aq) + Mg(s) \rightarrow Mg^{2+}(aq) + H_2(g)$$

This is a single displacement reaction, although you may have also recognised it as a redox reaction. Two hydrogen ions have accepted electrons to form hydrogen gas (reduction), and magnesium atoms in the solid have donated electrons to form magnesium ions (oxidation).

This type of displacement reaction led English chemist Humphry Davy to improve on Lavoisier's definition of acids. He stated that an acid is a substance with a hydrogen that can be replaced by another cation, such as a metal (Figure 2.5). If you remember D for Davy and D for Displacement, you should be able to recall Davy's definition.

Davy's definition of acids was more accurate than Lavoisier's, but it still left a lot of room for development and improvement.

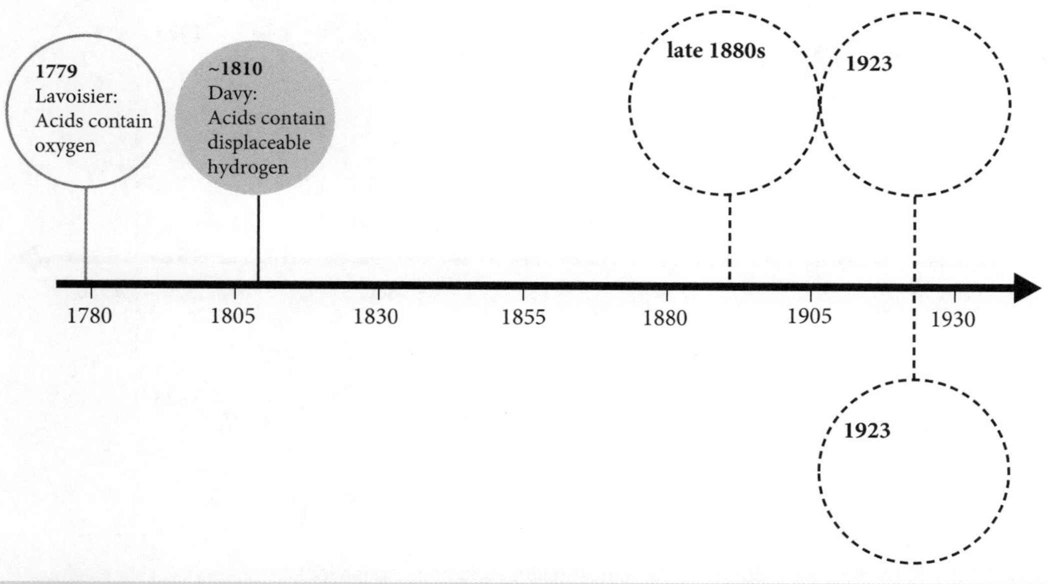

FIGURE 2.5 Davy's definition of acids describes displacement of atoms.

Acid and base reactions

We can write a general word equation for acid and base reactions:

$$\text{acid} + \text{base} \rightarrow \text{salt} + \text{water}$$

Acids and bases neutralise each other when they react, so the reactions are neutralisation reactions. For example:

$$\text{hydrochloric acid} + \text{sodium hydroxide} \rightarrow \text{sodium chloride} + \text{water}$$

We can write and balance the chemical equation:

$$HCl(aq) + NaOH(aq) \rightarrow NaCl(aq) + H_2O(l)$$

The ionic equation is:

$$H^+(aq) + Cl^-(aq) + Na^+(aq) + OH^-(aq) \rightarrow Na^+(aq) + Cl^-(aq) + H_2O(l)$$

Cancelling out the spectator ions on both sides leaves the net ionic equation:

$$H^+(aq) + \cancel{Cl^-(aq)} + \cancel{Na^+(aq)} + OH^-(aq) \rightarrow \cancel{Na^+(aq)} + \cancel{Cl^-(aq)} + H_2O(l)$$

$$H^+(aq) + OH^-(aq) \rightarrow H_2O(l)$$

Note

As many of these neutralisation reactions will take place in water, and produce water, the state is aqueous, '(aq)'.

The ionic equation shows the reaction between the hydrogen ions from the acid and hydroxide ions from the base to produce water. This reaction leads us to the third significant development in our understanding of acids and bases: Arrhenius' definitions. Svante Arrhenius was a Swedish chemist who came up with the following definitions for acids and bases.

- An acid is a substance that liberates *hydrogen* ions in water.
- A base is a substance that liberates *hydroxide* ions in water.

Arrhenius' definitions (Figure 2.6) are adequate when we are considering acids and bases in solution. These definitions also explain an important measure of acidity: pH.

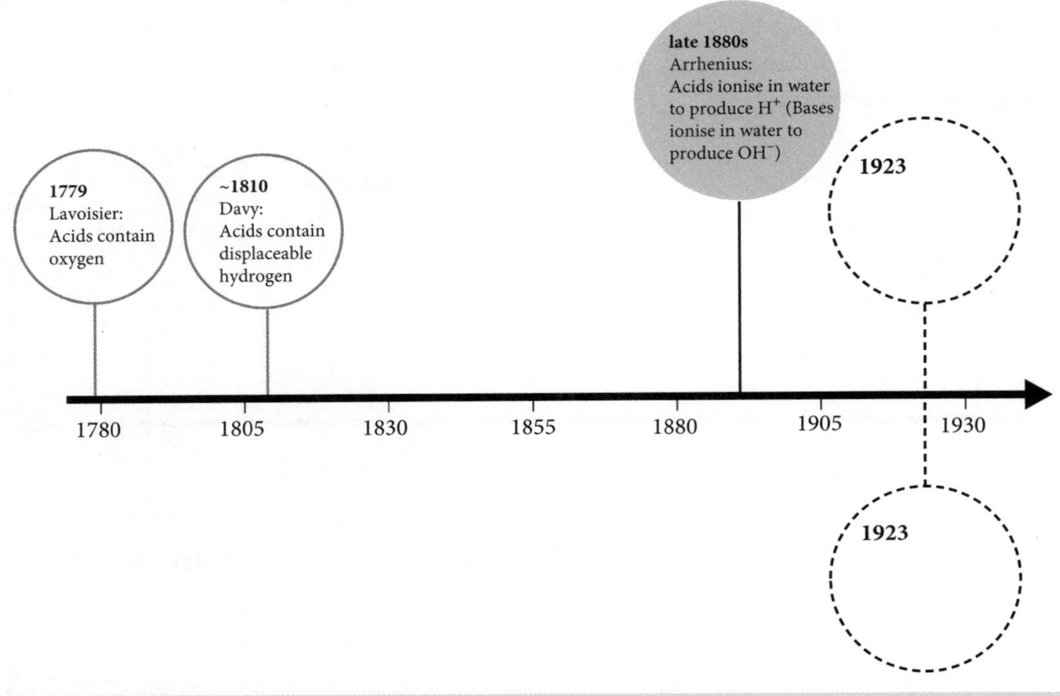

FIGURE 2.6 Arrhenius' definition of acids relates to liberation of hydrogen ions.

Acid and carbonate reactions

The general word equation for reactions between acids and carbonates is:

acid + carbonate → salt + water + carbon dioxide

These are neutralisation reactions in which an acid reacts with a metal carbonate (or ammonium carbonate). For example:

hydrochloric acid + sodium carbonate → sodium chloride + water + carbon dioxide

We can write and balance the chemical equation:

$$2HCl(aq) + Na_2CO_3(aq) \rightarrow 2NaCl(aq) + H_2O(l) + CO_2(g)$$

The ionic equation is:

$$2H^+(aq) + 2Cl^-(aq) + 2Na^+(aq) + CO_3^{2-}(aq) \rightarrow 2Na^+(aq) + 2Cl^-(aq) + H_2O(l) + CO_2(g)$$

Cancelling out the spectator ions on both sides leaves the net ionic equation:

$$2H^+(aq) + \cancel{2Cl^-(aq)} + \cancel{2Na^+(aq)} + CO_3^{2-}(aq) \rightarrow \cancel{2Na^+(aq)} + \cancel{2Cl^-(aq)} + H_2O(l) + CO_2(g)$$

$$2H^+(aq) + CO_3^{2-}(aq) \rightarrow H_2O(l) + CO_2(g)$$

We can see that this is a neutralisation reaction because water is a product. This time, carbon dioxide is also a product. Carbon dioxide is a gas at room temperature and will appear as bubbles. This observation allows you to identify if a neutralisation reaction involves a carbonate.

2.1.4 Common neutralisation reactions

Neutralisation reactions occur when an acidic substance reacts with a basic substance to form a salt and water. The hydrogen ion (proton) from the acid reacts with hydroxide ions from the base to form water. As we have seen, this can also involve the production of carbon dioxide if the base involved is a carbonate (or a bicarbonate). Either way, neutralisation reactions have several important applications.

Neutralisation in food and our bodies

Several organic acids, such as vinegar, citric acid and tartaric acid, are used in cooking and to preserve food. Sometimes we bake with them; for example, we use sodium bicarbonate to help cakes rise. The carbon dioxide bubbles in the cake produce a light and fluffy crumb.

Hydrochloric acid in our stomach helps to break down proteins in food during digestion. It is a strong acid, and sometimes it can return from the stomach into the oesophagus. This is acid reflux and it sometimes causes a burning feeling in the chest (which is why it is often called heartburn). If this is not treated, the acid can eventually damage the lining of the oesophagus.

Neutralising stomach acids with an antacid gives short-term relief for the symptoms of heartburn. Antacids like the one shown in Figure 2.7 may contain magnesium hydroxide ($Mg(OH)_2$) or aluminium hydroxide ($Al(OH)_3$). Some others contain a carbonate such as sodium carbonate (Na_2CO_3). If a carbonate is used, then carbon dioxide is produced during neutralisation. This is why we may burp after drinking an antacid solution.

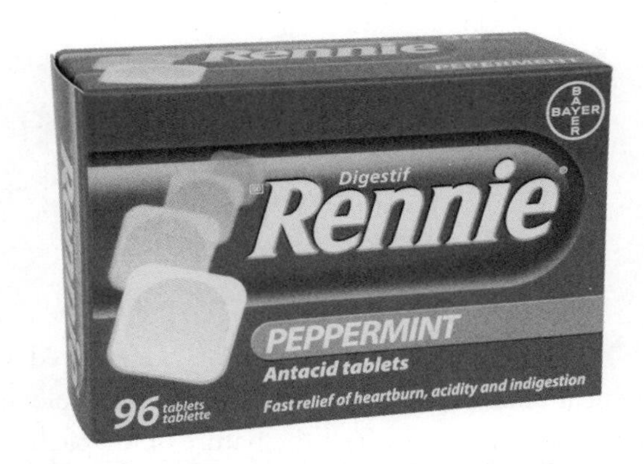

Neil Overy / Alamy Stock Photo

FIGURE 2.7 A common antacid contains several bases to neutralise stomach acids

Many of the enzymes (biological catalysts) in the body, including proteases in the stomach, only function within a certain pH range. If the pH of a body system is too high or too low, reactions in the body adjust to try to neutralise the excess acid or base and restore the system pH. This includes the pH of our blood, urine and saliva, especially after we have eaten.

Industrial neutralisation

Gardeners and farmers use their knowledge of neutralisation reactions to change the acidity of soils to best suit the plants they wish to grow. Slaked lime (calcium hydroxide) can be added to acidic soils, and gypsum or iron salts can be added to alkaline soils. Some plants grow better in acidic or neutral soils while others prefer alkaline soils. For example, hydrangeas are living indicators of soil pH: their flowers are pink in alkaline soils and blue in acidic soils.

Two common fertilisers – sulfate of ammonia ($(NH_4)_2SO_4$) and ammonium nitrate (NH_4NO_3) – are produced in neutralisation reactions. Other industrially important applications of neutralisation reactions include:

- manufacture of fabrics
- production of esters (in, for example, polymers and cosmetics)
- treatment of wastewater
- management of chemical spills.

2.1.5 Enthalpy of neutralisation

Enthalpy is a measure of heat energy; however, this is not easy to measure in a chemical reaction. Instead, we can measure the *change* in enthalpy, indicated by a change in the temperature of a system. Enthalpy of neutralisation is the enthalpy change during a neutralisation reaction.

The general reaction is the reaction between an acid and a base:

$$\text{acid} + \text{base} \rightarrow \text{salt} + \text{water}$$

The solutions we react together to demonstrate neutralisation are usually a strong acid, such as hydrochloric acid, and a strong base, such as sodium hydroxide.

The equation for the reaction of hydrochloric acid and sodium hydroxide is:

$$HCl(aq) + NaOH(aq) \rightarrow H_2O(l) + NaCl(aq)$$

The ionic equation is:

$$H^+(aq) + Cl^-(aq) + Na^+(aq) + OH^-(aq) \rightarrow H_2O(l) + Na^+(aq) + Cl^-(aq)$$

Cancelling out the spectator ions (Na^+ and Cl^-) on both sides leaves the net ionic equation:

$$H^+(aq) + OH^-(aq) \rightarrow H_2O(l)$$

The net ionic equation shows that neutralisation is a proton transfer reaction. A proton from the acid transfers to the hydroxide ion of the base. This is the same for any acid or base we choose. For example:

$$H_2SO_4(aq) + 2KOH(aq) \rightarrow 2H_2O(l) + K_2SO_4(aq)$$

The ionic equation is:

$$2H^+(aq) + SO_4^{2-}(aq) + 2K^+(aq) + 2OH^-(aq) \rightarrow 2H_2O(l) + 2K^+(aq) + SO_4^{2-}(aq)$$

Cancelling out the spectator ions (K^+ and SO_4^{2-}) on both sides leaves the net ionic equation:

$$2H^+(aq) + 2OH^-(aq) \rightarrow 2H_2O(l)$$

This is the same as:

$$H^+(aq) + OH^-(aq) \rightarrow H_2O(l)$$

When the heat of neutralisation is measured for a range of strong acids and strong bases, the amount of heat released is approximately 57 kJ per mole of water formed. This is because the particular acid and base pair used may affect the nature of the particular salt formed, but not the net reaction.

The net ionic equation and enthalpy change for the reaction between a strong acid and a strong base is:

$$H^+(aq) + OH^-(aq) \rightarrow H_2O(l) \qquad \Delta H = -57\text{ kJ mol}^{-1}$$

A study of the heat energy released for various (strong) acid–base reactions shows that the values are very close to one another. This suggests that the same type of reaction – production of salt and water – is occurring each time. More specifically, this is the transfer of a proton from a proton donor to a proton acceptor. This definition was stated independently at a very similar time by Danish chemist Johannes Brønsted and English chemist Thomas Lowry. It is known as the Brønsted–Lowry definition.

- An acid is a proton donor.
- A base is a proton acceptor.

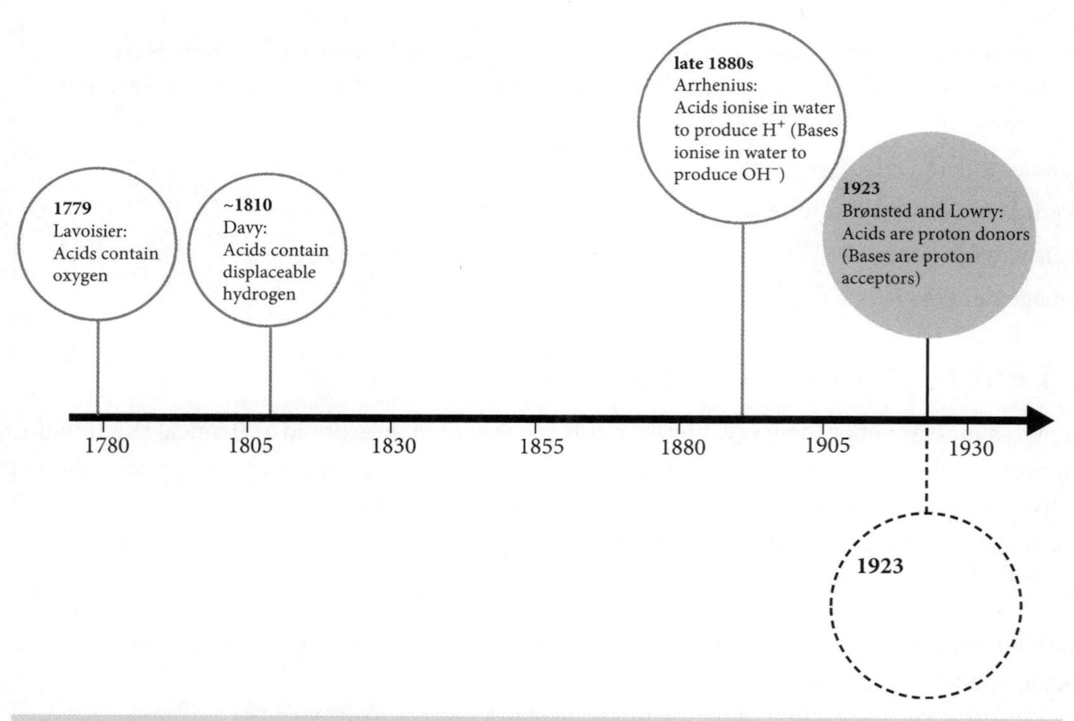

FIGURE 2.8 Brønsted and Lowry defined acids in terms of protons.

Measuring enthalpy change

We can calculate heat quantity, q, as a way to measure enthalpy (ΔH) of a solution. It is calculated from the mass of a substance, m, its specific heat capacity, c, and the change in temperature, ΔT:

$$q = mc\Delta T$$

Let's say we have 25 mL of a 0.5 mol L^{-1} HCl solution in one beaker and 25 mL of a 0.5 mol L^{-1} NaOH solution in a polystrene cup. When we measure the initial temperature of each solution, we should find that each one is at room temperature (e.g. 25°C).

Initial temperature: 25°C (298 K)

Now we add the HCl to the NaOH in the polystyrene cup. Energy is produced when the acid and base react in the cup. (We use a polystyrene cup because it is a better insulator than glass and may keep more of the reaction energy in the solution.) This is an exothermic reaction, so the temperature of the solution will increase and be detected by the thermometer.

Now we have a 50 mL solution (25 mL + 25 mL) in the cup. Let's say the temperature is now 28°C.

Final temperature: 28°C (301 K)

Change in temperature is the difference between the initial and final temperatures:

$$\Delta T = 301 - 298 = 3\,\text{K}$$

Assume the final solution will have the same specific heat (c) as pure water (this is a potential source of error). The density of water is 1 g mL^{-1}, so 50 mL of water weighs 50 g or 0.05 kg:

$$\begin{aligned} q &= mc\Delta T && \text{(equation from the formulae sheet)} \\ &= 0.05 \times 4.18 \times 10^3 \times 3 && \text{(value of } c \text{ from the formulae sheet)} \\ &= 627\,\text{J} \end{aligned}$$

To express this enthalpy change as a change per mole, we have to take into account the amount of water produced in the reaction. This can be calculated as shown in Table 2.4. We will assume a complete reaction with no substance in excess.

Note
Assuming no substance is in excess is another potential source of error. This depends on accurate measurement when preparing solutions.

TABLE 2.4 Calculating the heat of neutralisation per mole for $H^+(aq) + OH^-(aq) \rightleftharpoons H_2O(l)$

	$H^+(aq)$ +	$OH^-(aq)$ $\rightleftharpoons$	$H_2O(l)$
Mole ratio	1	1	1
V (L)	25×10^{-3}	25×10^{-3}	
c (mol L^{-1})	0.5	0.5	
Number of moles	$n = cV$ $= 0.5 \times 25 \times 10^{-3}$ $= 1.25 \times 10^{-2}$ mol	$n = cV$ $= 0.5 \times 25 \times 10^{-3}$ $= 1.25 \times 10^{-2}$ mol	$n = cV$ $= 2.5 \times 10^{-3}$ mol

Table 2.4 shows that 2.5×10^{-3} mol of water is produced in this reaction.
Using a simple ratio calculation:

$$2.5 \times 10^{-3} : 627$$
$$1 : x$$

$$\begin{aligned} x &= \frac{627}{2.5 \times 10^{-3}} \\ &= 250\,800\,\text{J mol}^{-1} \\ &= 250.8\,\text{kJ mol}^{-1} \end{aligned}$$

9780170465281

We can compare our experimental and theoretical values, and try to explain any differences. These may include:

- accuracy of the temperature meter (probe or thermometer)
- assumption about the specific heat capacity of pure water when we have a salt solution
- energy losses from the solution to the atmosphere or the cup
- actual concentrations of original solutions (and how we extracted the samples).

2.1.6 Defining acids and bases

Having looked at a range of chemical reactions involving acids and bases, we can now summarise our knowledge about the development of ideas about acids and bases (Figure 2.9).

Figure 2.9 includes the definitions of American chemist Gilbert Lewis. He extended the ideas of Brønsted and Lowry to some very specific reactions, resulting in the following definition.

- Acids are electron pair acceptors.
- Bases are electron pair donors.

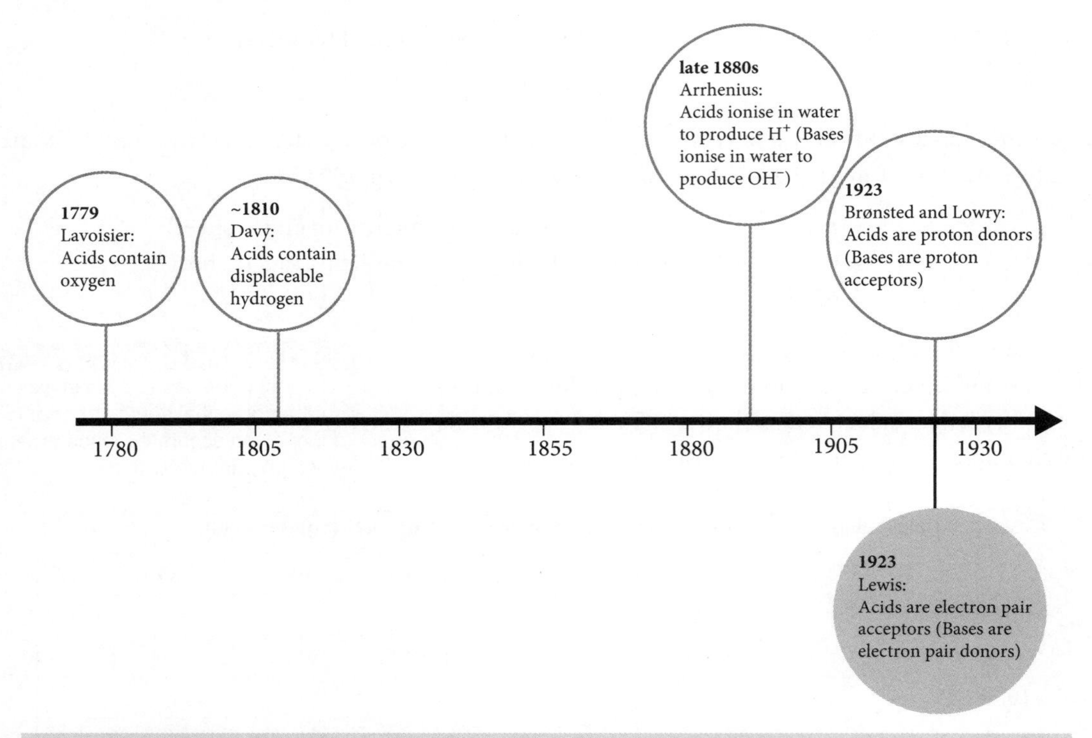

FIGURE 2.9 Addition of Lewis definition to our history of ideas about acids

The most important thing to notice about the development of these definitions is that each successive model expanded previous definitions to include other substances that behaved, or were thought to behave, like acids. For the HSC Chemistry course, we focus on the Brønsted–Lowry definition and its implications.

> **Note**
> It is unlikely that you would need to apply the Lewis definition for acids and bases; however, it is provided here to complete the story. Lewis acids and bases require neither protons nor an aqueous solvent.

Arrhenius acids and bases

A substance that produces H^+ ions in aqueous solutions (or a compound with hydrogen that reacts water to produce hydronium ions) is an Arrhenius acid. The excess H^+ gives the solution the properties of an acid. An example is sulfuric acid:

$$H_2SO_4(aq) \rightarrow 2H^+(aq) + SO_4^{2-}(aq)$$

A substance that produces hydroxide ions in aqueous solutions is an Arrhenius base. The excess OH^- gives the solution the properties of a base. An example is calcium hydroxide:

$$Ca(OH)_2(aq) \rightarrow Ca^{2+}(aq) + 2OH^-(aq)$$

Arrhenius suggested that when an acid and base react to form salt and water, the H^+ ions in the acid neutralise the OH^- ions in the base to form H_2O:

$$H^+(aq) + OH^-(aq) \rightarrow H_2O(l)$$

The salts produced are ionic compounds formed from the union of the cation from the base and the anion from the acid. Salts are classified according to the nature of the parent acid.

Brønsted–Lowry acids and bases

The Arrhenius definition works for many acid–base reactions, such as those involving common acids and hydroxides, but it does not explain them all. This is why we use the Brønsted–Lowry definitions. The following examples use the Arrhenius definition of an acid or base. Try to work out why the Brønsted–Lowry definition is a better definition of acids and bases.

Example 1: $CuCO_3(s) + 2H^+ \rightarrow Cu^{2+} + CO_2(g) + H_2O(l)$

Copper carbonate ($CuCO_3$) neutralises the acid, producing water, so it fits the classification of a base. However, copper carbonate does not produce hydroxide ions in aqueous solution as per the Arrhenius definition.

Example 2: $Na_2CO_3(s) \rightarrow 2Na^+ + CO_3^{2-}$

Sodium carbonate dissolves in water to produce a solution that contains OH^- ions. Sodium carbonate solutions change the colour of red litmus to blue and neutralise acids, liberating CO_2 gas.

$$Na_2CO_3(s) \rightarrow 2Na^+ + CO_3^{2-}$$
$$CO_3^{2-} + H_2O \rightarrow HCO_3^- + OH^-$$
$$CO_3^{2-} + 2H^+ \rightarrow H_2O(l) + CO_2(g)$$

The hydroxide ions form as a result of the ionising reaction between CO_3^{2-} and H_2O. Reactions involving water that lead to changes in acidity or basicity are **hydrolysis** reactions. In the reactions in this example, the resulting solutions are said to be alkaline, owing to the presence of OH^- ions.

Example 3: $NH_3(g) + HCl(g) \rightarrow NH_4Cl(s)$

Mixing ammonia and hydrogen chloride gases produces a white smoke. These are particles of ammonium chloride (Figure 2.10). We know HCl dissolved in water forms hydrochloric acid. This strong acid can neutralise a solution of ammonia, but is the same thing happening when their gases are mixed?

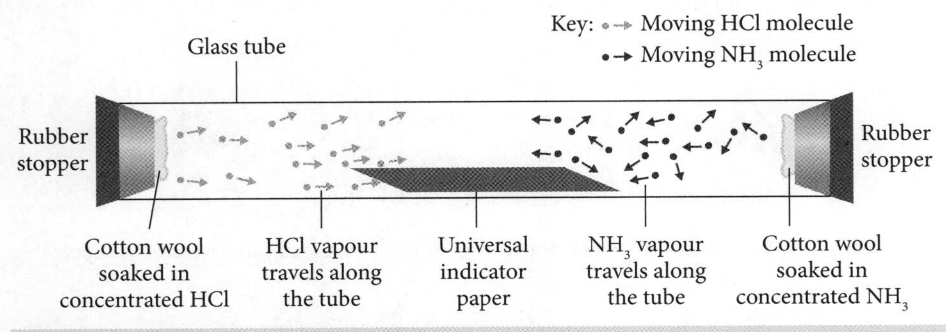

FIGURE 2.10 The reaction of ammonia and hydrogen chloride gases

We can better explain this behaviour using the Brønsted–Lowry definition of acids and bases, relating to the exchange of protons.

Protons are ionised hydrogen atoms. The most common isotope of hydrogen is hydrogen-1 (^1H). This has one proton in its nucleus and no neutrons. When the hydrogen atom donates its electron, it becomes a cation that is in this case a proton.

According to the Brønsted–Lowry definition, an acid–base reaction involves the transfer of protons from one species to another.

- A species capable of donating a proton (molecule or ion) is classified as an acid.
- A species capable of accepting a proton is classified as a base.

Acid–base reactions involve the transfer of a proton from an acid species to a base species. For example:

$$HCl(g) + NH_3(g) \rightarrow NH_4Cl(s)$$

A coordinate covalent bond forms due to the donation of two electrons by the nitrogen atom; the H^+ has no electrons to share. A **coordinate bond** is a covalent bond in which both of the shared electrons have come from the same atom, rather than one from each.

This also happens on the rare occasion when two water molecules ionise, as shown in Figure 2.11.

A proton (H^+ ion) is donated by one water molecule to the other water molecule. Water is acting as both an acid and a base. The proton bonds covalently to the other oxygen atom in the water molecule; the two electrons in the bond were both originally part of the oxygen atom. A hydronium ion (H_3O^+) and a hydroxide ion (OH^-) are formed.

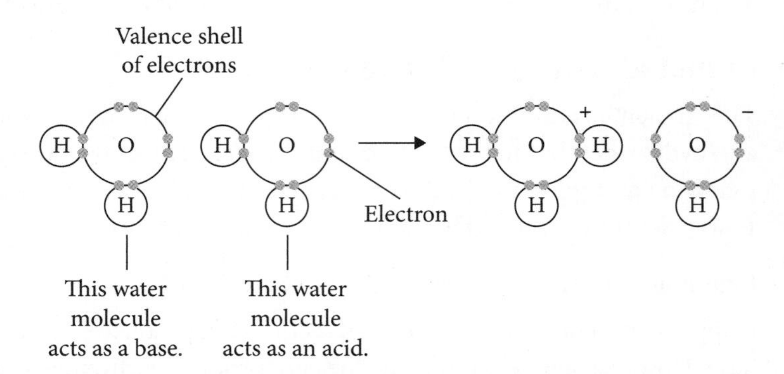

FIGURE 2.11 Water molecules acting as both an acid and a base and forming a coordinate bond to produce a hydronium ion (H_3O^+)

2.2 Using Brønsted–Lowry theory

Using the Brønsted–Lowry definition, we can more easily explain a range of observed behaviours for acids and bases. Ammonia is a weak base, but it has no hydroxide ions, so it does not easily fit the Arrhenius definition of a base.

The Brønsted–Lowry definition explains what is happening when ammonia is added to water, as shown in Figure 2.12.

A water molecule donates a proton (H^+), which produces a hydroxide ion (OH^-). This proton donation fits the Brønsted–Lowry definition of an acid. The ammonia molecule accepts a proton, forming the ammonium ion (NH_4^+). This proton acceptance fits the Brønsted–Lowry definition of a base.

If we think of water as a combination of H^+ and OH^-, we can more easily identify if substances are behaving as acids or bases when they interact with water.

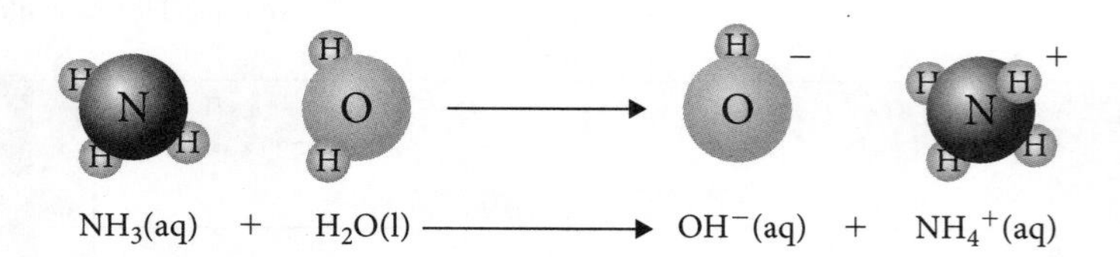

FIGURE 2.12 Acid–base interaction between an ammonia molecule and a water molecule

2.2.1 Measuring pH

In simple terms, pH is a measure of the proportion of hydrogen ions in a solution. This value is much higher in acidic solutions and lower in basic solutions. Table 2.5 shows a range of common substances, their relative acidity or basicity and their pH values.

Previously we looked at indicators, especially universal indicator, as a method for identifying the acidity of a solution. Universal indicator has also been calibrated so the colour of the dyes in different solutions also correspond to a particular pH value. Strong acids have a pH of 1–2, water and other neutral substances have a pH of 7, whereas strong bases have a pH of 12–14.

TABLE 2.5 The pH values of a range of common substances

Colour of universal indicator	$[H_3O^+]$	pH	Substance(s)	
Red	10	−1	Concentrated hydrochloric acid	ACID
	1	0	Car battery acid 1 mol L^{-1} hydrochloric acid	ACID
	10^{-1}	1	0.1 mol L^{-1} hydrochloric acid	ACID
	10^{-2}	2	Stomach acid	ACID
	10^{-3}	3	Vinegar Lemon juice	ACID
Orange	10^{-4}	4	Soft drinks Soda water	ACID
Orange-yellow	10^{-5}	5	Wine Black coffee	ACID
	10^{-6}	6	Rain water Milk Saliva	ACID
Yellow-green	10^{-7}	7	Very pure water	NEUTRAL
	10^{-8}	8	Blood Sea water	ALKALINE
Green	10^{-9}	9	Bore water Baking soda solution	ALKALINE
	10^{-10}	10	Hand soap	ALKALINE
Blue-green	10^{-11}	11	Laundry detergents	ALKALINE
	10^{-12}	12	Household ammonia Dishwashing machine powders	ALKALINE
Blue	10^{-13}	13	Chlorine bleach solutions	ALKALINE
	10^{-14}	14	Oven cleaners 1.0 mol L^{-1} sodium hydroxide	ALKALINE

pH meters and/or probes are a more accurate way to determine the acidity or basicity of a solution. pH probes can provide a single reading or they can be attached to a data logger to track pH change over time.

In practical investigations in which you measure pH, you should be able to identify the ways in which you measure pH as well as a range of common substances or solutions and their pH values. You should be able to recall which ones are acidic, which are neutral and which are basic, and the differences between the strong and weak acids, and the strong and weak bases.

2.2.2 pH, pOH and ions in solution

We can use the pH scale to compare not only acids and bases, but also relative strengths of different acids.

Water is a weak electrolyte that partly ionises in the following equilibrium reaction:

$$H_2O \rightleftharpoons H^+(aq) + OH^-(aq)$$

At 25°C, $[H^+] = [OH^-] = 10^{-7}\,mol\,L^{-1}$. This corresponds to a pH of 7.

In water, acids increase the concentration of protons (H^+). Protons are highly reactive and react with water molecules to produce hydronium ions (H_3O^+).

We can say that:

- for acidic solutions: $[H^+] > 10^{-7}\,mol\,L^{-1}$
- for basic solutions: $[H^+] < 10^{-7}\,mol\,L^{-1}$
- neutral solutions have a pH of 7.

An acid such as hydrochloric acid completely ionises in water to produce a very strong acid solution with a very low pH. A $0.1\,mol\,L^{-1}$ HCl solution has a pH of 1. The same is not true for a weaker acid such as carbonic acid (H_2CO_3). A $0.1\,mol\,L^{-1}$ solution of carbonic acid does not have a pH of 1.

pH and the Brønsted–Lowry definition of acids and bases help explain why acids of similar concentrations have solutions with different pH values.

pH and pOH

Table 2.6 shows pH values of acids for a range of H^+ concentrations.

TABLE 2.6 Acids: relationship between hydrogen ion concentration and pH

$[H^+]$ ($mol\,L^{-1}$)	1.0	0.1	0.01	0.001	0.0001	0.000 01	0.000 001
pH	0	1	2	3	4	5	6

The equation that relates pH to H^+ concentration is:

$$pH = -\log_{10}[H^+]$$

Table 2.7 shows pH values of bases for a range of OH^- concentrations.

TABLE 2.7 Bases: relationship between hydroxide ion concentration and pH

$[OH^-]$ ($mol\,L^{-1}$)	1.0	0.1	0.01	0.001	0.0001	0.000 01	0.000 001
pH	14	13	12	11	10	9	8

The equation that relates pH to OH^- concentration is:

$$pOH = -\log_{10}[OH^-]$$

An understanding of pH and pOH allows us to incorporate two other very important mathematical relationships:

$$pH + pOH = 14$$
$$[H^+] \times [OH^-] = 10^{-14}$$

At a pH of 4: $[H^+] = 10^{-pH} = 10^{-4} = 0.0001\,mol\,L^{-1}$.

However, at a pH of 12: $[OH^-] = 0.01\,mol\,L^{-1}$.

We can use either of two methods to help understand this relationship:

$$\begin{aligned} pOH &= -\log_{10}[OH^-] \\ &= -\log_{10}(0.01) \\ &= 2 \end{aligned}$$

$$\begin{aligned} pH + pOH &= 14 \\ pH &= 14 - pOH \\ &= 14 - 2 \\ &= 12 \end{aligned}$$

or

$$\begin{aligned} [H^+] \times [OH^-] &= 10^{-14} \\ [H^+] &= \frac{10^{-14}}{[OH^-]} \\ &= \frac{10^{-14}}{10^{-2}} \\ &= 10^{-12} \end{aligned}$$

$$\begin{aligned} pH &= -\log_{10}[H^+] \\ &= -\log_{10}(10^{-12}) \\ &= 12 \end{aligned}$$

Example: Calculate the pH of 0.1 mol L^{-1} NaOH solution

The equation for this reaction is:

$$NaOH(s) \rightarrow Na^+(aq) + OH^-(aq)$$

Sodium hydroxide is a strong base, so we assume that it completely dissociates in water. This means that a 0.1 mol L^{-1} solution of NaOH contains 0.1 mol L^{-1} OH^-.

$$\begin{aligned} pOH &= -\log_{10}[OH^-] \\ &= -\log_{10}(0.1) \\ &= 1 \end{aligned}$$

$$\begin{aligned} pH + pOH &= 14 \\ \therefore pH &= 14 - pOH \\ pH &= 14 - 1 \\ &= 13 \end{aligned}$$

2.2.3 Solution strength

Acids can be compared using measures of their strength and concentration. Sometimes we use the word 'strong' when we actually mean concentrated (e.g. a 'strong' cordial).

The strength of an acid relates to its ability to ionise in an aqueous solution. A strong acid ionises completely, whereas a weak acid only partially ionises.

Figure 2.13a shows that hydrochloric acid (HCl) is a strong acid. There are no HCl molecules, only H^+ ions and Cl^- ions. Figure 2.13b shows that hydrofluoric acid (HF) is a weak acid. The solution contains mostly HF molecules.

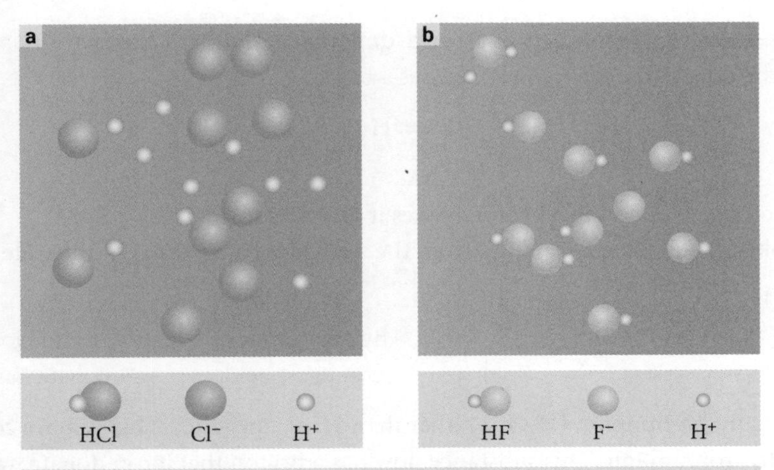

FIGURE 2.13 a HCl(aq) and b HF(aq). HCl is fully ionised, whereas very few HF molecules are ionised. HCl is a stronger acid than HF.

The concentration of an acid is equivalent to its molarity. It is a measure of the relative ratio of acid to water.

- A strong acid (or base) is fully ionised (or dissociated).
- A weak acid (or base) is partially ionised (or dissociated).
- A concentrated solution contains a lot of solute.
- A dilute solution contains less solute.

Note
Do not confuse the terms 'weak' and 'strong' with 'dilute' and 'concentrated'.

This section is linked to an investigation you may have carried out. To collect meaningful results, the pH of a range of both acids and bases are measured, and both the type of substance and the concentration of the substance must be varied, as in the examples in Table 2.8.

In this way, you can compare the same acid (or base) at different concentrations and different acids (or bases) at the same concentrations.

- Different concentrations of the same acid should give different pH values because of the different ratios of acid to water in the solutions.
- Different acids with the same concentration may give different pH values because of their different strengths. Weaker acids release fewer hydrogen ions in solution; hence, they will have higher pH values.

TABLE 2.8 A range of acid solutions and their pH values

Type of acid	Characteristics	Example concentration	pH at this concentration
Hydrochloric acid (HCl)	Strong, monoprotic	0.1	1
Hydrochloric acid (HCl)	Strong, monoprotic	0.01	2
Ethanoic acid (CH_3COOH)	Weak, monoprotic	0.1	2.9
Hydrochloric acid (HCl)	Strong, monoprotic	0.001	3

Note
The acids in Table 2.8 are monoprotic, which means they can donate one hydrogen ion per molecule. Some acids, such as sulfuric acid, can release two hydrogen ions per molecule – they are **diprotic**. Calculations of the ratio of concentration to pH must account for this.

2.2.4 Conjugate pairs

The Brønsted–Lowry definition treats reacting acids and bases as pairs. The pairs are related by the transfer of the hydrogen ion.

In this example, a water molecule acts as an acid, donating a proton to another water molecule. The other accepts the donated proton, thus acting as a base.

$$\underset{\text{acid}}{H_2O(l)} + \underset{\text{base}}{H_2O(l)} \rightleftharpoons H_3O^+(aq) + OH^-(aq)$$

Water's ability to donate or accept a proton makes it **amphiprotic.**

We can also look at the products resulting from the transfer of a proton. For the water reaction:

$$H_2O(l) + H_2O(l) \rightleftharpoons \underset{\substack{\text{hydronium ion} \\ \text{(conjugate acid)}}}{H_3O^+(aq)} + \underset{\substack{\text{hydroxide ion} \\ \text{(conjugate base)}}}{OH^-(aq)}$$

In this reaction, a hydronium ion (H_3O^+) rather than H^+ is a product. This is more common in water and it is called the hydronium ion. This substance now has a proton that it can donate, which makes it an acid. Likewise, the hydroxide ion was produced when the first water molecule donated its proton. It can now behave as a base.

Conjugate pairs

The Brønsted–Lowry definition allows us to look at the relationship between different species in an acid–base reaction in terms of hydrogen ions (protons).

Conjugate pairs are species involved in the transfer of protons (Figure 2.14). An acid is defined by its ability to donate a proton. After the acid donates a proton, the product is a **conjugate base**. A base is defined by its ability to accept a proton. After a base accepts a proton, the product is a **conjugate acid**.

In the previous example involving water molecules, the hydronium ion produced from water has a proton to donate, and is a conjugate acid. Likewise, the hydroxide ion produced when another water molecule donated its proton is a conjugate base.

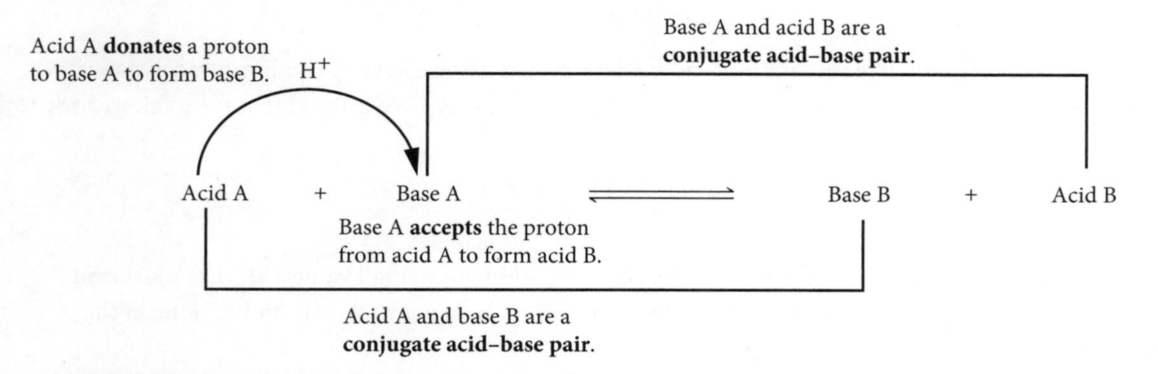

FIGURE 2.14 Acids and bases take part in a proton transfer to produce conjugate bases and conjugate acids.

For the reaction of ammonia with water:

$$NH_3(aq) + H_2O(l) \rightleftharpoons NH_4^+(aq) + OH^-(aq)$$

The conjugate pairs are:

- NH_3/NH_4^+: NH_3 is the base and NH_4^+ is the conjugate acid
- H_2O/OH^-: H_2O is the acid and OH^- is the conjugate base.

The equilibrium position of an acid–base system is determined by the relative strengths of the bases involved.

- An equilibrium to the *right* means the *stronger base* is on the *left-hand side* of the equilibrium arrow, because it has a greater ability to accept a proton.
- An equilibrium to the *left* means the *stronger base* is on the *right-hand side* of the equilibrium arrow.

In the reaction of ammonia with water, OH^- is a stronger base than NH_3, hence the equilibrium in this equation is to the left.

We know that the strength of an acid is a measure of its ability to fully ionise in water. Strong acids have weak conjugate bases and weak acids have strong conjugate bases.

For example, HCl is a strong acid; hence, its conjugate base, Cl^-, is a weak base. In water:

$$\underset{\text{acid}}{HCl(aq)} + H_2O(l) \rightleftharpoons H_3O^+(aq) + \underset{\text{conjugate base}}{Cl^-(aq)}$$

CH_3COOH is a weak acid; hence, its conjugate base, CH_3COO^-, is a relatively strong base:

$$\underset{\text{acid}}{CH_3COOH(aq)} + H_2O(l) \rightleftharpoons H_3O^+(aq) + \underset{\text{conjugate base}}{CH_3COO^-(aq)}$$

Similarly, strong bases have weak conjugate acids and weak bases have strong conjugate acids.

For example, NaOH is a strong base. In water, Na^+ remains in solution, so the OH^- accepts the proton to form water. In this case, water is the conjugate acid – a very weak one:

$$\underset{\text{base}}{OH^-(aq)} + H^+(aq) \rightleftharpoons \underset{\text{conjugate acid}}{HOH(l)}$$

NH_3 is a weak base; hence, its conjugate base, the ammonium ion (NH_4^+), is a relatively strong acid:

$$NH_3(aq) + H^+(aq) \rightleftharpoons NH_4^+(aq)$$
base conjugate acid

Acid and base dissociation

The stronger an acid, the more readily it is ionised. The weaker an acid, the less it is ionised.

An equilibrium exists between the molecules in a weak acid and its constituent ions:

$$\text{acid} \rightleftharpoons H^+ + \text{conjugate base}$$

$$HA(aq) \rightleftharpoons H^+(aq) + A^-(aq)$$

The extent to which acids are ionised can be determined using a special equilibrium constant introduced in Module 5, known as the acid ionisation constant, K_a. We learned that for weak **monoprotic** acids, which undergo partial ionisation:

$$K_a = \frac{[H^+] \times [A^-]}{[HA]}$$

The smaller the value of K_a, the weaker the acid. An acid with a small value of K_a has most acid molecules intact rather than ionised. Hence the concentration values of the H^+ and A^- ions in the numerator is low.

The strength of a base can be determined by the degree of dissociation. Many bases already exist as ions, which dissociate in water:

$$\text{base} + H^+ \rightleftharpoons \text{conjugate acid}$$

$$A^- + H^+ \rightleftharpoons HA$$

An example is $NaOH(aq) \rightleftharpoons Na^+(aq) + OH^-(aq)$.

When sodium carbonate reacts with water:

$$Na_2CO_3(aq) + H_2O(l) \rightleftharpoons 2Na^+(aq) + HCO_3^-(aq) + OH^-(aq)$$

The equilibrium constant for dissociation of bases is K_b, which we will consider later.

We can now re-examine an earlier example of acid dissociation, from section 1.3.6 (page 23).

Example: K_a of 0.1 mol L^{-1} carbonic acid solution with a pH of 3.68 at 25°C

> **Note**
> pH = $-\log_{10}[H^+]$ (provided on the HSC Chemistry formulae sheet)

A pH of 3.68 means $[H^+] = 10^{-pH}$ or $10^{-3.68}$

$= 2.09 \times 10^{-4}$ mol L^{-1}

The equation is:

$$H_2CO_3(aq) + 2H_2O(l) \rightleftharpoons 2H_3O^+(aq) + CO_3^{2-}(aq)$$

Assuming that the ratio of H^+ to CO_3^{2-} is 2:1:

$$[CO_3^{2-}] = \frac{1}{2} \times [H^+]$$

$$= \frac{1}{2} \times 2.09 \times 10^{-4}$$

$$= 1.04 \times 10^{-4} \text{ mol L}^{-1}$$

This is a diprotic acid, so it will have two separate ionisation reactions, each with a separate K_a value:

$$H_2CO_3(aq) + H_2O(l) \rightleftharpoons H_3O^+(aq) + HCO_3^-(aq) \quad K_a = 2.5 \times 10^{-4}$$

$$HCO_3^-(aq) + H_2O(l) \rightleftharpoons H_3O^+(aq) + CO_3^{2-}(aq) \quad K_a = 5.6 \times 10^{-11}$$

This is an equilibrium involving a weak acid, so we can assume the degree of ionisation is very low. This means the final concentration of the acid molecules is very close to their initial concentration $(0.1 - 1.04 \times 10^{-4})$. In fact, it is so close we can approximate it to 0.1 mol L^{-1}.

We know the specific K_a for the dissociation of carbonic acid, so we can calculate the concentrations for each of these species (Table 2.9). Our assumptions still hold, so we can simplify the equation and avoid solving a full quadratic equation.

TABLE 2.9 Calculating K_a for $H_2CO_3(aq) + H_2O(l) \rightleftharpoons H_3O^+(aq) + HCO_3^-(aq)$

Equation	$H_2CO_3(aq) + H_2O(l) \rightleftharpoons H_3O^+(aq) + HCO_3^-(aq)$		
Species	H_2CO_3	H_3O^+	HCO_3^-
Initial (mol)	0.1	0	0
Change (mol)	x	x	x
Equilibrium (mol)	$0.1 - x$ ≈ 0.1	x	x

$$K_a = \frac{[H_3O^+][HCO_3^-]}{[H_2CO_3]}$$

$$2.5 \times 10^{-4} = \frac{x \times x}{0.1 - x} \quad \text{or} \quad 2.5 \times 10^{-4} \approx \frac{x \times x}{0.1}$$

You are not expected to create and solve a quadratic equation, so you can ignore the left-hand equation above. If the value for x is small, the equation can be solved as follows:

$$K_a = \frac{[H_3O^+][HCO_3^-]}{[H_2CO_3]}$$

$$2.5 \times 10^{-4} \approx \frac{x \times x}{0.1}$$

$$x^2 = 2.5 \times 10^{-5}$$

$$x = 0.005$$

Note

Because the value of x is much smaller than 0.1, our assumption is reasonable.

Carbonic acid is a diprotic acid. This means 1 mole of acid releases 2 moles of H^+ ions. The chemical behaviour when the first hydrogen ion is donated is different from when the second hydrogen ion is donated. For now, we will assume they occur as part of a single system.

We can use the values we calculated for the first ionisation to determine the values for the second ionisation, as shown in Table 2.10.

TABLE 2.10 Calculating K_a for $HCO_3^-(aq) + H_2O(l) \rightleftharpoons H_3O^+(aq) + CO_3^{2-}(aq)$

Equation	$HCO_3^-(aq) + H_2O(l) \rightleftharpoons H_3O^+(aq) + CO_3^{2-}(aq)$		
Species	HCO_3^-	H_3O^+	CO_3^{2-}
Initial (mol)	0.005	0.005	0
Change (mol)	x	x	x
Equilibrium (mol)	$0.005 - x$	$0.005 + x$	x

$$K_a = \frac{[H_3O^+][CO_3^{2-}]}{[HCO_3^-]}$$

This time, we will use the simplified values:

$$5.6 \times 10^{-11} \approx \frac{(0.005 + x) \times x}{0.005 - x}$$

So:

$$0.005 - x \approx 0.005 + x \approx 0.005$$

$$5.6 \times 10^{-11} \approx \frac{(0.005 + x) \times x}{0.005 - x}$$

$$5.6 \times 10^{-11} \approx \frac{0.005 \times x}{0.005}$$

$$x \approx 5.6 \times 10^{-11}$$

Note

If the value of x (representing degree of ionisation) is very small compared to the initial concentration, it does not make a significant difference to the answer. (Check your answer each time to verify this.) Hence, we can be confident that $n_i - x = n_i + x = n_i$ (where n_i is the initial number of moles of the acid or base and x is the number of moles ionising or dissociating). For constant volume, this will also apply to the concentrations: $c_i - x = c_i + x = c_i$.

We can now calculate the K_a value again:

$$K_a = \frac{[H^+]^2[HCO_3^-]}{[H_2CO_3]} = \frac{(0.005)^2 \times 5.6 \times 10^{-11}}{0.095} = 1.47 \times 10^{-14}$$

This value is significantly less than the one we calculated previously.

The major difference between the values is the concentration of the carbonate ion. Our first assumption caused a major error in the original calculation because we assumed a 2:1 ratio of hydrogen to carbonate ion. However, hydrogen carbonate ion is amphiprotic and did not ionise in water as much as expected. A higher proportion of the ion remained in the solution, and fewer carbonate ions were released.

Amphiprotic substances

Amphiprotic substances can behave as either an acid or a base. Depending on the conditions, they can donate a proton or accept a proton.

Sodium hydrogen carbonate

Sodium hydrogen carbonate ($NaHCO_3$) is amphiprotic. In a strong base such as sodium hydroxide ($NaOH$), it forms the carbonate ion and water. This proton transfer decreases the pH of the solution.

The sodium ion is a spectator ion.

As an acid:

$$\underset{\text{acid}}{HCO_3^-(aq)} + OH^-(aq) \rightleftharpoons \underset{\text{conjugate base}}{CO_3^{2-}(aq)} + H_2O(aq)$$

In water (which is also amphiprotic), the hydrogen carbonate ion acts as a base – it doesn't donate a proton to water; it accepts one.

As a base:

$$\underset{\text{base}}{HCO_3^-(aq)} + H_2O(l) \rightleftharpoons \underset{\text{conjugate acid}}{H_2CO_3(aq)} + OH^-(aq)$$

Potassium dihydrogen phosphate

Potassium dihydrogen phosphate (KH_2PO_4) is also amphiprotic. How do we know whether the dihydrogen phosphate ion ($H_2PO_4^-$) is more likely to act as an acid or a base?

We could measure the pH of a potassium dihydrogen phosphate solution, which would give a pH of slightly less than 7. It is a neutral to very slightly acidic substance. Another way is to look at its K_a or K_b values.

The dihydrogen phosphate ion is a **polyprotic** species, which means it can donate more than one proton. We can show how this and some related ions can behave as Brønsted–Lowry acids or bases.

1 Phosphoric acid (H_3PO_4) behaves as an acid in water. Here, the dihydrogen phosphate ion ($H_2PO_4^-$) is its conjugate base:

$$\underset{\text{acid}}{H_3PO_4(aq)} + H_2O(l) \rightleftharpoons H_3O^+(aq) + \underset{\text{conjugate base}}{H_2PO_4^-(aq)}$$

2 In this reaction, $H_2PO_4^-$, which was a conjugate base in step 1, behaves as an acid. Here, the hydrogen phosphate ion (HPO_4^{2-}) is its conjugate base:

$$\underset{\text{acid}}{H_2PO_4^-(aq)} + H_2O(l) \rightleftharpoons H_3O^+(aq) + \underset{\text{conjugate base}}{HPO_4^{2-}(aq)}$$

3 In this reaction, HPO_4^{2-}, which was a conjugate base in step 2, behaves as an acid. Here, the phosphate ion (PO_4^{3-}) is its conjugate base:

$$\underset{\text{acid}}{HPO_4^-(aq)} + H_2O(l) \rightleftharpoons H_3O^+(aq) + \underset{\text{conjugate base}}{PO_4^{3-}(aq)}$$

Overall: If we add them together, the two intermediate ions cancel out:

$$H_3PO_4(aq) + 3H_2O(l) \rightleftharpoons 3H_3O^+(aq) + PO_4^{3-}(aq)$$

This net equation does not take into account that some of the intermediates will ionise more readily than others.

2.2.5 Modelling acid and base behaviour

As mentioned earlier, acids are classified according to their strength and concentration.

You can use a **model** to distinguish between the terms 'strong' and 'weak', and 'concentrated' and 'dilute', as they apply to acid and base solutions.

Models help us understand very small (e.g. atoms and molecules interacting with another) and very large (e.g. solar systems and galaxies) systems. In chemistry, models can be drawn (2D), constructed in 3D with molecular modelling kits or created in a computer animation. A 2D example is shown in Figure 2.15.

You should be able to evaluate the efficacy of your model. You can do this by asking some questions.

- What is my model trying to demonstrate?
- How well does my model approximate what is actually happening?
- Does my model have any predictive power?
- What are the limitations of my model?

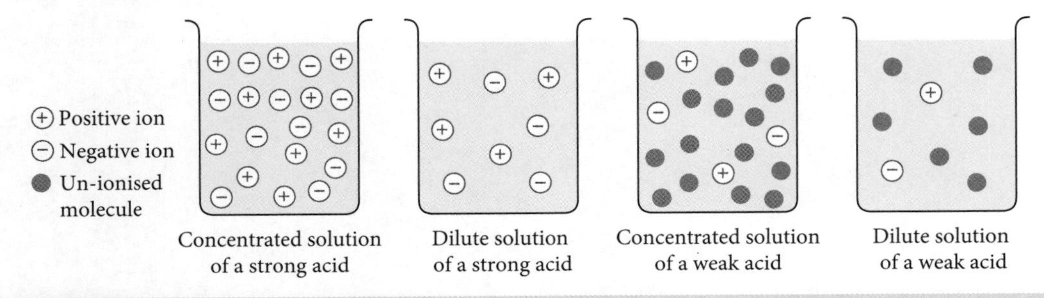

FIGURE 2.15 Contrasting the pairs 'weak' and 'strong', and 'concentrated' and 'dilute' using a model

Comparing strong and weak acids

The strength of an acid relates to its degree of ionisation. Strong acids are fully ionised, while weak acids are only partially ionised.

The photos in Figure 2.16 show the difference between hydrochloric acid and the weaker hydrofluoric acid. The white balls represent hydrogen, the darker balls are chlorine and fluorine.

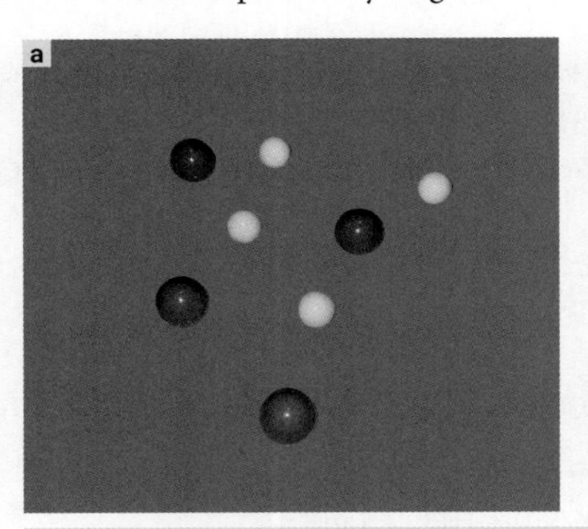

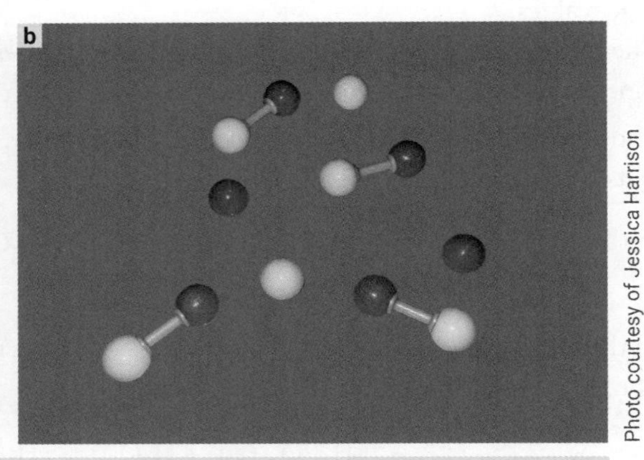

Photo courtesy of Jessica Harrison

FIGURE 2.16 **a** Modelling the strong acid full ionisation of HCl, and **b** modelling the weak acid partial ionisation of HF

The model in Figure 2.16a shows that all HCl molecules have been ionised into H^+ and Cl^-. The model in Figure 2.16b shows HF molecules, H^+ ions and F^- ions. Not all molecules have ionised, showing the behaviour of a weak acid.

Comparing concentrated and dilute acids

The concentration of an acid is equivalent to its molarity. It is a measure of the ratio of solute to solvent. Put simply, the more water there is for a given amount of solute, the more dilute the solution. Figure 2.17 shows hydrogen fluoride (HF) as a concentrated solution and a more dilute solution.

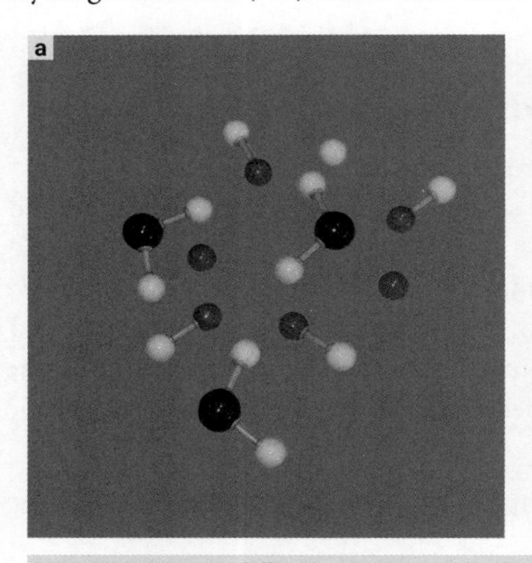

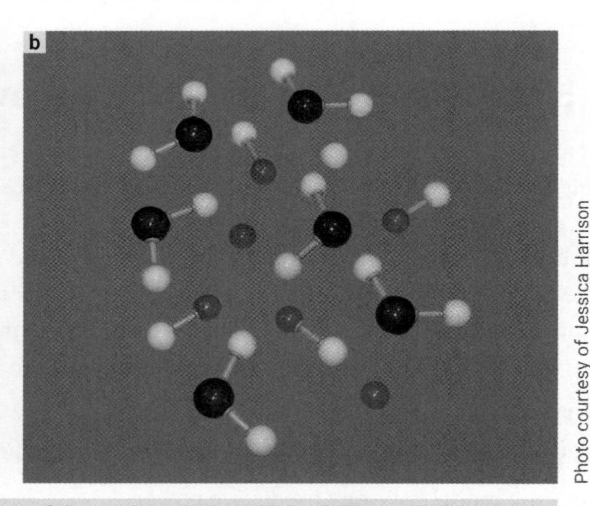

Photo courtesy of Jessica Harrison

FIGURE 2.17 **a** Modelling a concentrated solution of the weak acid HF, and **b** modelling a dilute solution (more water molecules) of the weak acid HF

2.2.6 Dilution and neutralisation

The pH of a solution changes when water is added to the solution or if there is a neutralisation reaction with either an acid or base.

Recall that:

$$pH = -\log_{10}[H^+]$$

So, in order to calculate pH, we need to determine the concentration of hydrogen ions. If the solution is a basic solution, it may be easier to determine $[OH^-]$, work out the pOH, and hence pH, using:

$$pOH = -\log_{10}[OH^-] \qquad \text{and} \qquad pH + pOH = 14$$

Dilutions and pH

Adding water to a solution changes the concentration of the solute in the solution; however, it does not change the number of moles of solute. Provided the only change we make is to increase the volume of the solution by adding water:

Moles of initial solution:

$$n_i = c_i \times V_i$$

Moles of final solution:

$$n_f = c_f \times V_f$$

Because we only add water, the number of moles of solute has not changed:

$$n_i = n_f$$
$$\therefore c_i V_i = c_f V_f$$

This is a simple way to calculate changes in concentration as a result of dilutions.

Example: pH of 25 mL of 0.5 mol L^{-1} sulfuric acid solution made up to 100 mL with distilled water

First, work out how the concentration of the sulfuric acid solution has changed.

Distilled water was used, so:

$$\begin{aligned} c_i V_i &= c_f V_f \\ 0.5 \times 0.025 &= c_f \times 0.1 \\ c_f &= \frac{0.5 \times 0.025}{0.1} \\ &= 0.125 \text{ mol L}^{-1} \end{aligned}$$

So the final concentration of sulfuric acid is 0.125 mol L^{-1}.

We need to write out the ionisation equation to calculate the pH:

$$H_2SO_4(aq) \rightarrow 2H^+(aq) + SO_4^{2-}(aq)$$

We will assume that the sulfuric acid fully ionises, although the second ionisation (of HSO_4^-) may not happen fully.

This means 1 mol SO_4^{2-} liberates 2 mol H^+.

$$\begin{aligned} [H^+] &= 2 \times 0.125 \\ &= 0.25 \text{ mol L}^{-1} \\ &= -\log_{10}[H^+] \\ &= -\log_{10}(0.25) \\ &= 0.6 \end{aligned}$$

Note

If you have the relevant K_a value for the second ionisation, you can do this calculation more accurately. If not, qualify your response by indicating an assumption of full ionisation or a value slightly higher than 0.6.

Neutralisation and pH

Neutralisation reactions will also change the pH, but sometimes by less than we might expect.

To calculate final concentration and pH for a neutralisation reaction, we need to write out the equation in full, work out the total moles of acid and base and determine whether either is in excess.

Example: pH of 25 mL of 0.2 mol L^{-1} HCl mixed with 25 mL of 0.25 mol L^{-1} NaOH

First, we construct a solution table as shown in Table 2.11.

TABLE 2.11 Calculating the final concentrations and any excess reactant for $HCl(aq) + NaOH(aq) \rightarrow NaCl(aq) + H_2O(l)$

Equation	$HCl(aq)$ +	$NaOH(aq)$ →	$NaCl(aq) + H_2O(l)$
Mole ratio	1	1	
V_i **(L)**	0.025	0.025	
c_i **(mol L^{-1})**	0.2	0.25	
n_i **(mol)**	$n = cV$ $= 0.2 \times 0.025$ $= 0.005$	$n = cV$ $= 0.25 \times 0.025$ $= 0.006\,25$	
n_f **(excess after reaction, mol)**	0 (limiting reagent)	$n = cV$ $= 0.006\,25 - 0.005$ $= 0.001\,25$	
V_f **(L)**	–	0.05	
c_f **(mol L^{-1})**		$c_f = \frac{n_f}{V_f}$ $= \frac{0.00125}{0.05}$ $= 0.025$	

This means we have a final concentration of 0.025 mol L^{-1} of the OH^- ions (which were in excess in this reaction):

$$\begin{aligned} pOH &= -\log_{10}[OH^-] \\ &= -\log_{10}(0.025) \\ &= 1.6 \end{aligned}$$

$$\begin{aligned} pH + pOH &= 14 \\ pH &= 14 - pOH \\ &= 14 - 1.6 \\ &= 12.4 \end{aligned}$$

This may seem like a very high pH value for a neutralisation reaction, but remember that we had more of the sodium hydroxide than was needed to neutralise the acid. The hydroxide ions remained unreacted in the solution and contributed to the final concentration.

2.3 Quantitative analysis

Volumetric analysis is an important quantitative tool in chemistry. In particular, **titration** is the technique used to measure the volume of a solution of known concentration that is required to neutralise a solution of unknown concentration.

During practical investigations using titration, you will determine the concentration of an unknown acid or base solution. You may be able to undertake a simple single-step titration to do this or you may need to standardise a solution first and then carry out a second titration to find the concentration of the unknown solution. It is important to be familiar with the specialised equipment and procedures associated with volumetric analysis. Mastering the technique of titration requires practice and precision.

2.3.1 Titration

Titration is the process used to accurately determine the volume (titre) of an acid or a base that is required to neutralise a solution whose concentration is not known.

Standard solutions and primary standards

For a titration, the solution of known concentration is the **standard solution**. It can be determined by titration against a **primary standard**, or it can be prepared by adding a known weight of a reagent in a certain volume of solution to make a primary standard solution.

A primary standard:

- has a high molecular mass
- is soluble in water
- is stable in solid and aqueous form
- is pure, and preferably anhydrous.

Some reagents used as primary standards in acid–base titrations are sodium carbonate and potassium hydrogen phthalate.

Not all substances are suitable as primary standards. NaOH is hygroscopic (it absorbs water from the air) and thus its mass may change, or its concentration as a solution may change. However, even a sodium hydroxide solution may be later standardised by reacting it with other known (standard) solutions.

Apparatus

Titrations are set up as shown in Figure 2.18. We must prepare the glassware in a particular way to avoid incorrect concentrations of the solutions:

- *bulb (graduated) pipette* – this is used to transfer a known volume of solution from a larger bottle or volumetric flask into a conical flask. Before use, it should be rinsed with the solution that is going to be added to it
- *burette* – this long, narrow piece of glassware is graduated and has a tap to allow very precise volumes of solutions to be added to the reaction vessel. Before use, it should be rinsed with the solution that will be added to it
- *conical (Erlenmeyer) flask* – the narrow neck reduces the chance of error due to splashing when the titration is happening. Before use, it should be rinsed with distilled water – *not* the solution
- *volumetric flask* – the line on the narrow neck allows preparation of a precise volume of solution. It should be clean and dry before use.

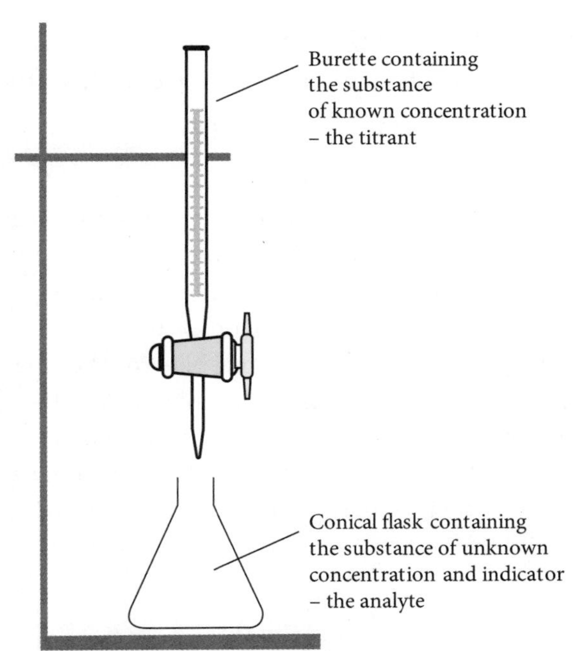

FIGURE 2.18 A burette and flask set up for a titration

9780170465281

2.3.2 Finding the equivalence point

The solutions in acid–base reactions usually remain colourless, so colour change cannot be used to determine when one solution has neutralised the other.

This point of neutralisation is called the **equivalence point**. It is when the number of moles of acid exactly balances the number of moles of base.

We need a chemical substance or electronic device to help us determine the equivalence point. This may be a pH probe (data logger) or meter, or a conductivity meter.

The pH probe measures changes in the pH as the reaction progresses. If we start with an acid (either weak or strong), the pH will be low. It will increase as the base is added. How quickly this happens will depend on the nature of the acid and base.

A conductivity meter measures the concentration of ions in a solution. A strong acid fully ionises, whereas a weak acid only partially ionises. As the base is added, it combines with H^+ ions to form water molecules. This will reduce the ion concentration of the strong acid and shift the equilibrium of the weak acid to release more ions (Le Chatelier's principle). When neutralisation happens, additional ions of the base are added and the conductivity increases. A conductivity curve for a titration like this could look like the one in Figure 2.19.

In most titrations at school, you will use an indicator in the solution in your conical flask to identify an **end point**. In neutralisation reactions, a very small excess amount of either the acid or the base can have a significant effect on the pH. The pH change is very rapid at the equivalence point, so choosing the right indicator is very important. The choice depends on the nature of the acid and/or base involved in the neutralisation.

You need to stop a titration as soon as the indicator changes to the desired colour. This is the end point and will usually be just after the equivalence point.

Acid–base titrations can be classified by strengths of the acid and base.

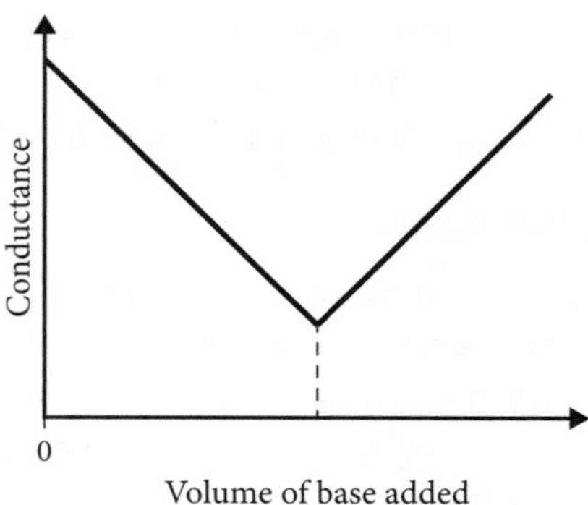

FIGURE 2.19 Change in conductance as a strong base is slowly added to a strong acid

Strong acid–strong base titrations

In these reactions, both species are completely dissociated:

$$H^+ + OH^- \rightarrow H_2O$$

The mole ratio here is 1 : 1.

At the equivalence point for a strong acid–strong base reaction, the pH is about 7.

We can plot pH against volume of acid added to give a titration curve like the one in Figure 2.20.

Phenolphthalein or bromothymol blue could be the indicator for this titration. Using phenolphthalein, the base would need to be in the burette. The end point is indicated by the faintest of pink tinges in the conical flask. Phenolphthalein is colourless in acids, making the faint colour visible.

> **Note**
> The mole ratio will be different for a diprotic acid or base, or for a triprotic acid. Make sure you write an equation for a reaction to determine the mole ratio before doing calculations.

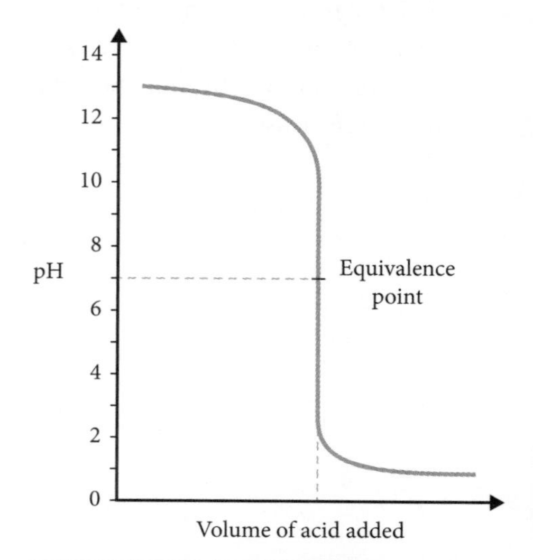

FIGURE 2.20 A titration curve for a strong acid–strong base titration

9780170465281

Weak acid–strong base titrations

Weak acids only partially ionise in solution, so much of the acid remains in molecular form during a weak acid–strong base titration.

Consider a titration of ethanoic acid (CH_3COOH) and sodium hydroxide (NaOH). The OH^- ions of the base react directly with the acid molecules. The equilibrium constant is very large; hence, the reaction proceeds to completion. The equation is:

$$CH_3COOH + OH^- \rightarrow CH_3COO^- + H_2O$$

Note
Not all weak acids are monoprotic (e.g. H_2CO_3 is diprotic), so write out an equation before doing calculations.

Again, the mole ratio is 1 : 1.

At the equivalence point for a weak acid–strong base reaction, the pH is about 9 and the titration curve looks like the one in Figure 2.21.

Phenolphthalein is again the best choice of indicator because the weak acid has a strong conjugate base. The indicator interacts with the water molecules to give a slightly basic solution (increased pH) at the equivalence point. Once again it is preferable to have the base in the burette, to progress from a colourless solution to one that will have a faint pink tinge. A magenta solution has a beautiful colour, but it is a clear indication that you have gone past the equivalence point.

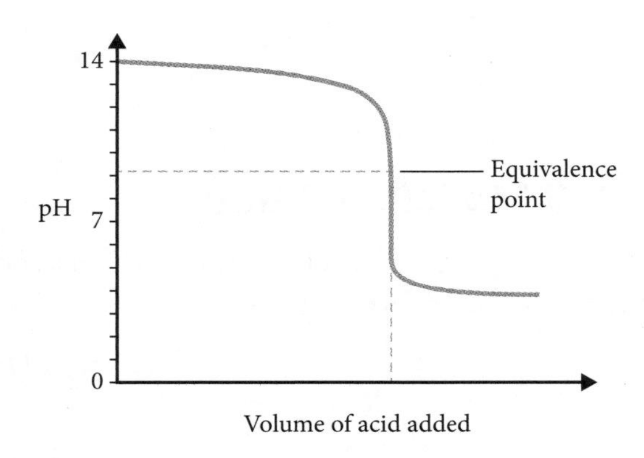

FIGURE 2.21 A titration curve for a weak acid–strong base titration

Strong acid–weak base titrations

For strong acid–weak base titrations, much of the base remains in molecular form. The H^+ from the acid reacts directly with the un-ionised base. The reaction again goes to completion. An example is the base ammonia:

$$NH_3(aq) + H^+(aq) \rightarrow NH_4^+(aq)$$

At the equivalence point for a strong acid–weak base reaction, the pH is about 5 and the titration curve looks like the one in Figure 2.22.

Methyl orange is the best choice of indicator for this reaction. Phenolphthalein is not a good choice of indicator because the weak base will have a strong conjugate acid. This interacts with the water molecules to produce a slightly acidic solution at the equivalence point and hence lower its pH. Phenolphthalein would remain colourless, so methyl orange is needed to identify this point. Methyl orange is red in strong acid but changes through orange to yellow as the reaction proceeds. Again, care must be taken because methyl orange remains yellow whether the solution has a pH of 5 or 12. For this reason it is advisable to place the acid in the burette.

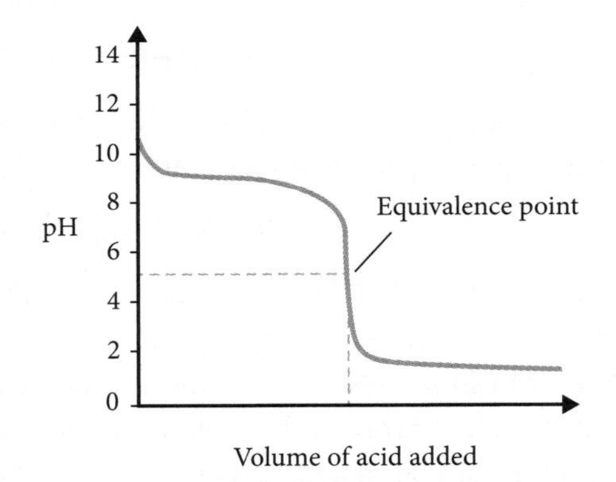

FIGURE 2.22 A titration curve for a strong acid–weak base titration

Weak acid–weak base titrations

Your titrations at school will only include the previous three acid–base combinations because titrating a weak acid and a weak base takes a very long time! Both equilibria need to shift, so the reaction rate is very slow and the equilibrium point (and hence the end point) can be difficult to identify (Figure 2.23).

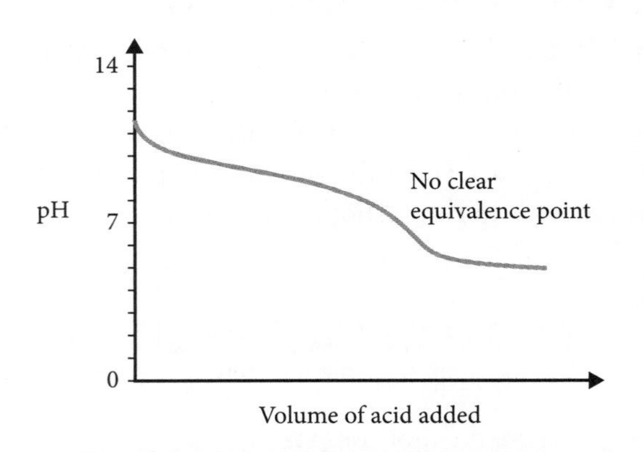

FIGURE 2.23 A titration curve for a weak acid–weak base titration

2.3.3 Modelling titration

We have considered a neutralisation reaction as one in which hydrogen (or hydronium) ions react with hydroxide ions to form water molecules:

$$H^+(aq) + OH^-(aq) \rightleftharpoons H_2O(l)$$

or

$$H_3O^+(aq) + OH^-(aq) \rightleftharpoons 2H_2O(l)$$

For reactions between strong acids and strong bases, the neutralisation reaction above is a good approximation and the pH is about 7 at the equivalence point.

However, reactions between weak acids and strong bases, and between strong acids and weak bases, do not have a pH of 7 at the equivalence point.

To understand different types of neutralisation reactions, we can model the reactions using graphs.

Weak acid–strong base

Consider the neutralisation reaction to produce sodium ethanoate (CH_3COONa):

$$CH_3COOH(aq) + NaOH(aq) \rightarrow CH_3COONa(aq) + H_2O(l)$$

The sodium will remain in solution, but the ethanoate ion will interact with water molecules, increasing the concentration of OH^-:

$$CH_3COO^-(aq) + H_2O(l) \rightleftharpoons CH_3COOH(aq) + OH^-(aq)$$

This increase in OH^- ions is the reason why the pH at the equivalence point is greater than 7.

We can show how the ion concentration changes as this reaction proceeds by looking at a conductivity graph for a similar reaction (Figure 2.24). We could also use a computer simulation to model this process.

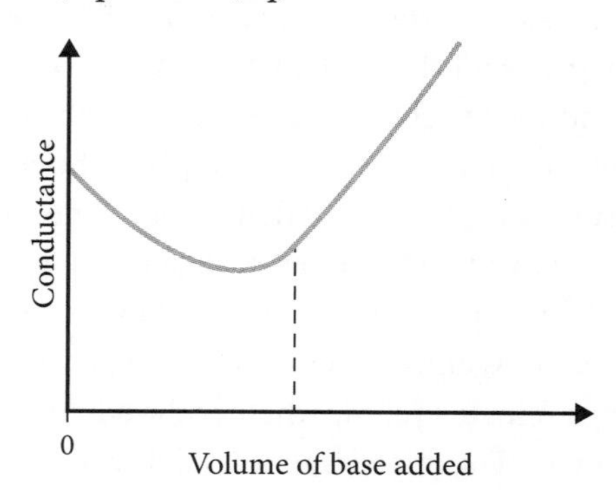

FIGURE 2.24 A conductivity graph for a weak acid–strong base titration

Strong acid–weak base

Consider the neutralisation reaction to produce ammonium chloride (NH_4Cl):

$$NH_3(aq) + HCl(aq) \rightarrow NH_4Cl(aq)$$

The chloride will remain in solution, but the ammonium ion will interact with water molecules, increasing the concentration of H_3O^+:

$$NH_4^+(aq) + H_2O(l) \rightleftharpoons NH_3(aq) + H_3O^+(aq)$$

This increase in H_3O^+ ions is the reason why the pH at the equivalence point is less than 7.

We can show how the ion concentration changes as this reaction proceeds by looking at a conductivity graph for a similar reaction (Figure 2.25). We could also use a computer simulation to model this process.

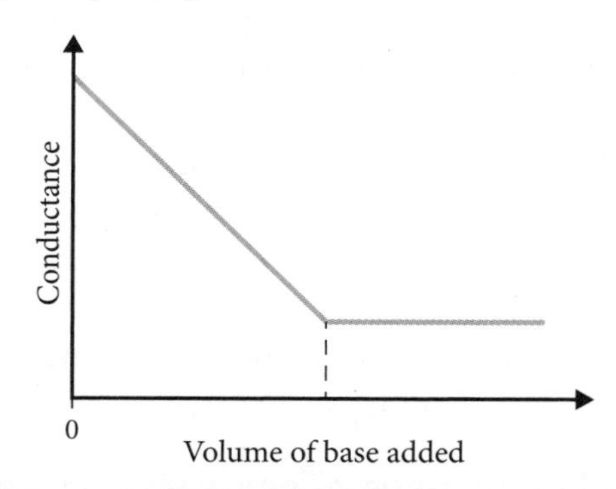

FIGURE 2.25 A conductivity graph for a strong acid–weak base titration

2.3.4 Indicators and the acid ionisation constant, K_a

We know that for a weak acid an equilibrium exists between the acid molecule and its ions:

$$HA(aq) \rightleftharpoons H^+(aq) + A^-(aq)$$

The extent to which acids are ionised can be determined using the acid ionisation constant, K_a. For weak, partially ionised monoprotic acids that undergo partial ionisation:

$$K_a = \frac{[H^+][A^-]}{[HA]}$$

Both acid strength and pH can be visualised on a log scale. Strong acids ionise completely, and weak acids only partially ionise; hence, we can use another indicator of acid strength, pK_a. Similarly to the calculation of pH, pK_a is calculated as:

$$pK_a = -\log_{10}(K_a)$$

The weaker the acid, the greater its pK_a.

For comparison:

- sulfuric acid is a strong acid with a K_a of 1.2×10^{-2} and a pK_a of 1.92
- carbonic acid is a weak acid with a K_a of 4.8×10^{-11} and a pK_a of 10.32.

The values for carbonic acid indicate that pK_a is a more useful measure where K_a values are extremely small.

pK_a is also an important feature of indicators. An interesting relationship is the one between an indicator and the pH associated with a colour change. We know that phenolphthalein is a good choice of indicator when a strong base is involved. Phenolphthalein has a pK_a of 9.3. This value is close to the pH at which we see the colour change in phenolphthalein; hence, the colour change will occur over a narrow range (8.3–10.3) – exactly as required for this type of titration.

2.3.5 Acid–base analyses

The techniques associated with acid–base reactions, volumetric analysis and the nature of salt production have wide applications. You should be familiar with applications in industry, Aboriginal and Torres Strait Islander traditional uses, and technology (using digital probes and instruments).

Acid–base analysis in industry

Acid–base analysis is essential in industries such as those for food and beverages, pharmaceuticals, viticulture (winemaking) and agriculture. Impurities, or additives such as nitrogen-based oxides (NO_x) and sulfur-based oxides (SO_x), can change the acidity of industrial mixtures, so they need to be monitored and possibly neutralised.

Under certain conditions, the ethanol in some alcoholic beverages can oxidise to form ethanoic acid. For example, this results in a vinegar taste in wines, making them unpleasant to drink. Adding sulfur dioxide to wines is a way to preserve them, reducing the chances of microbe contamination and oxidation damage as they age. Acid–base titrations can be used to monitor sulfur oxide concentration to check it is at the right level.

Although other techniques are replacing titration analysis in the mining industry, titrations were once a common way to analyse the composition of ores. Nitric acid is used for ore extraction in many transition metal ore bodies.

Many industries, including mining, must conduct regular environmental monitoring of the waste products they release into the atmosphere or into wastewater systems. Environment protection laws specify the quantities, concentrations and pH levels of certain chemicals that can be emitted by industries.

Traditional uses of acids and bases by Aboriginal and Torres Strait Islander Peoples

Aboriginal and Torres Strait Islander Peoples have traditional uses for a range of plants as food or medicines. Knowledge of these uses has been passed through generations.

One example involves a 'soap tree' from tropical north Queensland. A lather can be produced from this tree by crushing the leaves in water, as shown in Figure 2.26.

The lather can be used for cleaning or as an antiseptic. Other traditional uses for this plant are in treatments for an upset stomach, headaches or vision problems, muscular aches and toothache. Qualitative analysis indicates the presence of saponins (which are emulsifiers), organic acids and organic salts such as methyl salicylate (a component of modern aspirin).

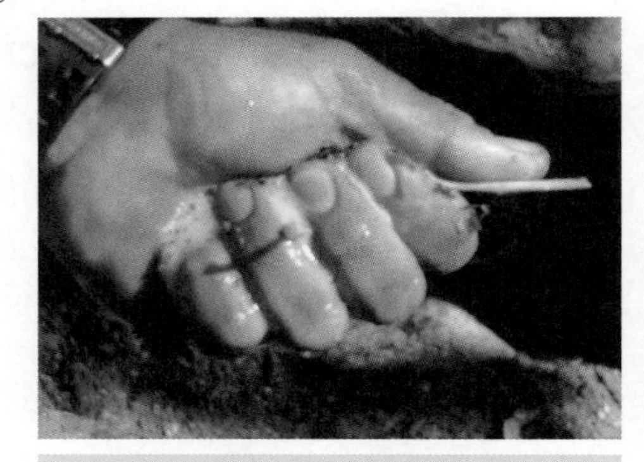

FIGURE 2.26 The lather produced by a soap tree can be used in personal cleaning.

Across Australia, Aboriginal and Torres Strait Islander Peoples have traditional knowledge about the best local plants to use to treat different stings and bites; for example, from insects and jellyfish.

It would be useful to be ready to discuss at least one specific example of these applications for an exam question.

Technological applications

We have already discussed some of the applications of technology in volumetric analysis. Computer-controlled procedures can improve accuracy and automate the process of calculating unknown concentrations, enabling quick analysis at a large scale.

Automated titrators can carry out industrial-scale titrations quickly and accurately, and conductivity probes can monitor the ion levels of solutions and identify potential equivalence points and end points (Figure 2.27).

You should come across at least one example of a digital probe or electronic logging instrument during your course. Universities and chemical laboratories have technological capabilities that allow quick and accurate volumetric analysis of solutions.

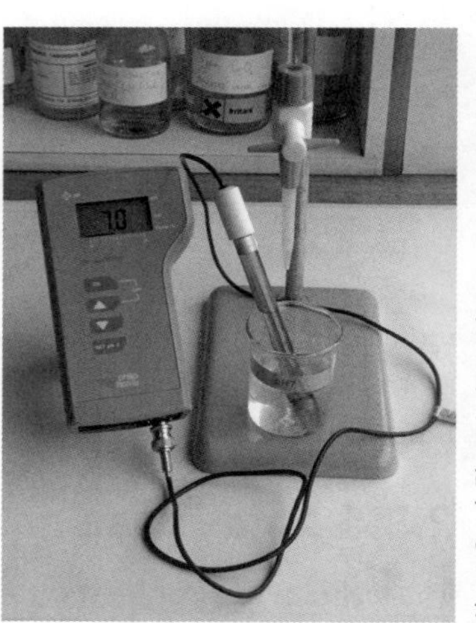

Alamy Stock Photo/sciencephotos

FIGURE 2.27 Using a pH probe to measure acidity

2.3.6 Analysing common household substances

In a practical investigation, you may have used titration to determine the concentration of an unknown acid or base solution (e.g. ethanoic acid concentration in vinegar). Many substances contain acids or bases in solution and hence may be analysed using titration. This section brings together our knowledge about titration techniques, strong and weak solutions and the appropriate selection of an indicator (particularly if an analyte is coloured). We need all of this information to use titration to analyse the acidity of solutions, including soft drinks, wines, juices and medicines.

As part of a practical investigation, you need to write your own method, select appropriate reagents and other materials and equipment, and consider aspects of safety, accuracy, reliability and validity. You also need to show your calculations and evaluate your final result.

Acid content of white wine

You should already be familiar with the procedure for a titration, so this example will focus on an analysis rather than the technique.

Titration notes

- 10 mL **aliquots** (subsamples) of the analyte sample (white wine) were used.
- The acid in the biggest quantity in white wine is tartaric acid ($C_4H_6O_6$). Tartaric acid is a diprotic acid and has a molar mass of $(4 \times 12.01 + 6 \times 1.008 + 6 \times 16) = 150.088\,g\,mol^{-1}$.
- Sodium hydroxide was the base, with a standardised concentration of $0.101\,mol\,L^{-1}$.
- The indicator was phenolphthalein.
- The average titre was 32.6 mL sodium hydroxide.

Analysis

Table 2.12 shows the calculations for this titration. Start at the shaded table cell and follow the arrows.

TABLE 2.12 Calculating the concentration of acid in wine

Equation	$C_4H_6O_6(aq)$ +	$2NaOH(aq)$	→ $Na_2C_4H_4O_6(aq) + 2H_2O(l)$
Mole ratio	1	2	
n	$n = CV$ $= \frac{1}{2} \times 3.29 \times 10^{-3}$ $= 1.65 \times 10^{-3}$	$n = cV$ $= 0.101 \times 0.0326$ $= 3.29 \times 10^{-3}$	
V_f (L)	0.01 (aliquot volume)	0.0326 (average titre for NaOH)	
c_f ($mol\,L^{-1}$)	$= \frac{1.65 \times 10^{-3}}{0.01}$ $= 0.165$	0.101 (standardised NaOH concentration)	

The concentration of tartaric acid in the white wine sample is $0.165\,mol\,L^{-1}$. This can also be calculated as a percentage of the total solution:

$c = 0.165\,mol\,L^{-1}$

$= 0.165$ mol of tartaric acid in 1000 mL of water

$= 0.165 \times 150.088$ ($m = n \times MM$) in 1000 g of water (density of water is $1\,g\,mL^{-1}$)

$= 24.8$ g of tartaric acid in 1000 g of water

$$= \frac{24.8}{1000} \times 100$$

$= 2.48\%$ tartaric acid (by mass or % m/m)

Acid content of aspirin

This analysis is best carried out as a **back titration.** This is a two-step titration (not including any standardisation steps) that involves adding an excess of one reagent and then titrating to determine the exact amount of the excess. By working backwards, we can then work out the concentration or mass of the initial substance.

Note
Not all of the aspirin is the active ingredient, assumed to be acetylsalicylic acid. There is often a binding agent, such as starch, which makes the aspirin not particularly water soluble. To carry out this procedure effectively, ethanol is often mixed with the crushed tablets to assist with their dissolution.

Titration notes

- 0.36 g samples of solid aspirin were mixed with distilled water and ethanol.
- The acid in aspirin is acetylsalicylic acid ($C_9H_8O_4$). Acetylsalicylic acid is a diprotic acid with a molar mass of $(9 \times 12.01 + 8 \times 1.008 + 4 \times 16) = 180.154\,\text{g mol}^{-1}$.
- Sodium hydroxide was the base, with a standardised concentration of $0.11\,\text{mol L}^{-1}$.
- 50 mL of the base was added to each aspirin solution. This ensured complete neutralisation of the acid in aspirin, and an excess of sodium hydroxide after the reaction had reached completion.
- Excess sodium hydroxide was titrated against standardised $0.080\,\text{mol L}^{-1}$ HCl solution.
- The indicator was phenolphthalein.
- The average titre was 21.6 mL HCl.

Analysis

Table 2.13 shows the calculations for this titration. Start at the shaded table cell and follow the arrows.

TABLE 2.13 Calculating the number of moles of excess NaOH for $HCl(aq) + NaOH(aq) \rightarrow NaCl(aq) + H_2O(l)$

Equation	$HCl(aq)$ +	$NaOH(aq)$ →	$NaCl(aq) + H_2O(l)$
Mole ratio	1	1	
n	$n = cV$ $= 0.080 \times 0.0216$ $= 1.73 \times 10^{-3}$	$n = cV$ $= 1.73 \times 10^{-3}$	
V (L)	0.0216	(average titre for HCl)	
c (mol L^{-1})	0.080		

Note
The key to a back titration is to start at the end and work backwards. If you start at the beginning, you will not have enough values to solve the problem.

Now we have the number of moles of NaOH that were left over after all of the acid from the aspirin tablets reacted with the original sodium hydroxide solution.

For the original sodium hydroxide solution:

$$c = 0.11\,\text{mol L}^{-1}$$
$$V = 0.05\,\text{L}$$

$$n = cV = 0.11 \times 0.05 = 5.5 \times 10^{-3}\,\text{mol}$$

This means 5.5×10^{-3} mol NaOH were added to the aspirin and 1.73×10^{-3} mol were left over to react with the HCl.

So the number of moles of NaOH neutralised the acid in the aspirin tablet was:

$$n = 5.50 \times 10^{-3} - 1.73 \times 10^{-3} = 3.77 \times 10^{-3}\,\text{mol}$$

Now we can complete the initial reaction table, as shown in Table 2.14.

TABLE 2.14 Calculating the concentration of acid in aspirin

Equation	$C_9H_8O_4(aq)$ +	$2NaOH(aq)$ →	$Na_2C_9H_6O_4(aq) + 2H_2O(l)$
Mole ratio	1	2	
n	$n = cV$ $= \frac{1}{2} \times 3.77 \times 10^{-3}$ $= 1.886 \times 10^{-3}$	$n = cV$ $= 3.77 \times 10^{-3}$	
MM **($g\,mol^{-1}$)**	180.154		
m **(g)**	$n = MM$ $= 1.886 \times 10^{-3} \times 180.154$ $= 0.340$		

The average mass of acetylsalicylic acid in an aspirin tablet is 0.34 g. This could also be calculated as a percentage of the total mass:

$$\text{Purity of tablet} = \frac{0.34}{0.36} \times 100$$
$$= 94.3 \text{ (94\% to 2 significant figures)}$$

Note
Always check the number of significant figures in the data in the question to inform how you express your final answer. Keep each step of the calculation in your calculator rather than round answers at each step.

2.3.7 Buffer solutions

A **buffer** solution can resist changes in the pH of a solution caused by small amounts of added acid or base. It is usually a mixture of a weak acid and its conjugate base at equal concentrations. A weak base does not readily ionise – much of it remains in molecular form in a solution. To produce a buffer, a salt containing the anion associated with the weak acid is required to raise the concentration of the conjugate base to an equivalent level.

An example of a buffer solution is carbonic acid and its conjugate base, hydrogen carbonate. The equation for this buffer solution is:

$$H_2CO_3(aq) \rightleftharpoons H^+(aq) + HCO_3^-(aq)$$

If an acid (H^+) is added to a solution containing buffer, the buffer equilibrium shifts to the left. The buffer equation is:

$$H_2CO_3(aq) \leftharpoondown\!\!\rightharpoonup H^+(aq) + HCO_3^-(aq)$$

The equilibrium shift removes additional H^+ by converting the ions to H_2CO_3.

If a base (OH^-) is added to a solution containing buffer, the hydrogen ions in the buffer solution begin to neutralise the hydroxide ions:

$$H^+(aq) + OH^-(aq) \rightleftharpoons H_2O(l)$$

As the hydroxide ions are used up, the H^+ concentration decreases and the buffer equilibrium shifts to the right to produce H^+, to maintain the pH. The buffer equation is:

$$H_2CO_3(aq) \rightharpoonup\!\!\leftharpoondown H^+(aq) + HCO_3^-(aq)$$

9780170465281

Titration curves can represent the behaviour of buffers during the early part of a weak acid or weak base titration. Figure 2.28 represents the progress of a titration where the analyte is a weak base. As the reaction begins, the hydroxide ions are neutralised by the hydrogen ions in the added acid. The buffer equilibrium shifts to the right to ionise more of the base and increase the hydroxide concentration. Eventually, so much acid is added that there is insufficient base to restore the reacted hydroxide ions, and so the pH decreases steeply, dropping vertically across the equivalence point.

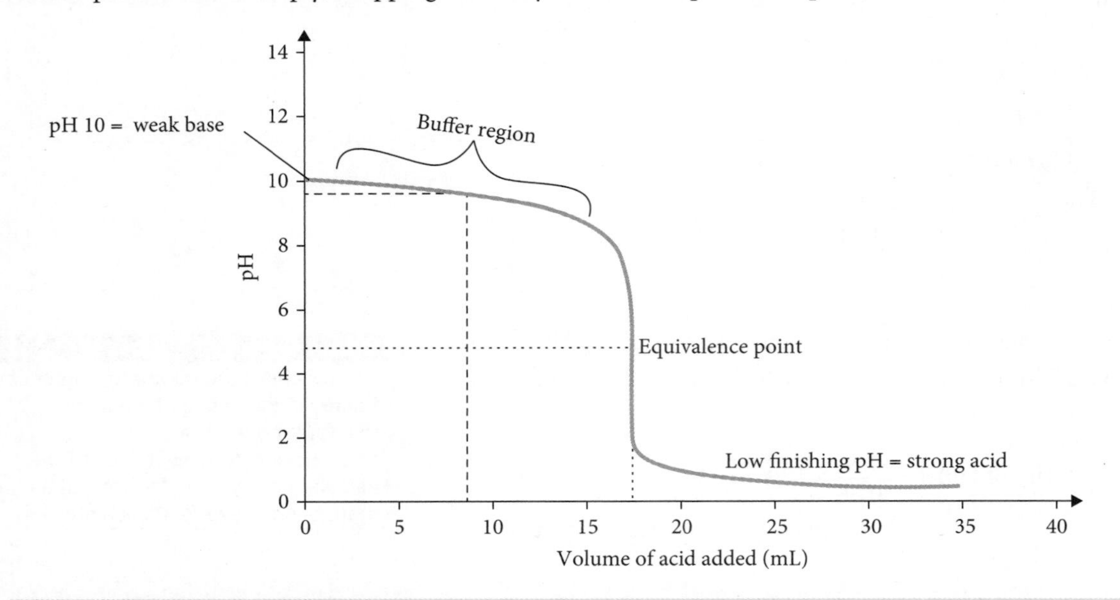

FIGURE 2.28 A titration curve showing the effect of the buffer at the start of the reaction

You should be able to conduct an investigation in which you make a buffer solution and test your buffer by adding both acid and base. To make a buffer solution, you add the salt form of the conjugate base to its weak acid; for example, sodium ethanoate and ethanoic acid. You may wish to calculate exactly how much salt you need to add to the acid solution, given the concentration of the acid, the volume and its pH. The pH will give you a clue about the degree of ionisation of the acid. It is possible to do this by trial and error, although this is not recommended.

2.3.8 Buffers in natural systems

Buffers in blood

A number of natural systems involve buffer solutions. We have already looked briefly at the buffering system in blood. The pH of blood must be maintained for body systems to function. Two important pairs of chemicals in blood are CO_3^{2-}/HCO_3^- and $H_2PO_4^-/HPO_4^{2-}$.

A buffer resists a change in pH by having roughly equal concentrations of an acid and its conjugate base. The two equations for the blood buffers are:

$$HCO_3^-(aq) \rightleftharpoons H^+(aq) + CO_3^{2-}(aq)$$

$$H_2PO_4^-(aq) \rightleftharpoons H^+(aq) + HPO_4^{2-}(aq)$$

Buffers in swimming pools

We need to add chemicals to pools to keep the water hygienic for swimming. However, these chemicals can damage pool equipment and make the water damaging to skin and eyes. Buffers are used to maintain the pH in swimming pools within safe limits. The following equilibrium represents a buffer that could be used in pool water maintenance:

$$OCl^-(aq) + H_2O(l) \rightleftharpoons HOCl(aq) + OH^-(aq)$$

Buffers in oceans

Rising carbon dioxide levels in the atmosphere are having an impact on ocean ecosystems. The important buffer system in the oceans is being tested by rising atmospheric carbon dioxide levels and the subsequent decrease in carbonate levels.

When CO_2 dissolves in the ocean, it combines with water to form hydrogen (H^+) and bicarbonate (HCO_3^-) ions:

$$CO_2(g) + H_2O(l) \rightleftharpoons H^+(aq) + HCO_3^-(aq)$$

Some of the hydrogen ions combine with carbonate ($CO_3^{2-}(aq)$) ions to form additional bicarbonate ions, resulting in a decrease in the carbonate ions and an increase in bicarbonate ions:

$$H^+(aq) + CO_3^{2-}(aq) \rightleftharpoons HCO_3^-(aq)$$

The net effect of these changes is to increase concentrations of H^+, CO_2 and HCO_3^- and decrease the concentration of CO_3^{2-}.

As is the case with all buffers, there is a point beyond which continual addition of acid can no longer be buffered, and the pH will start to decrease.

The carbonate ion is a major component of corals and the shells of many marine organisms. Decreasing the CO_3^{2-} concentration affects the calcification of exoskeletons for corals, marine molluscs and other marine life (Figure 2.29).

FIGURE 2.29 Buffer solutions are critically important in natural systems, including marine environments like coral reefs.

Glossary

acid A substance that can donate a proton (or accept an electron pair – Lewis definition)

acid–base indicator A substance, usually a plant dye, that changes colour in the presence of an acid or base

aliquot A standard volume of a sample used in a titration; this represents a part of a larger volume, e.g. 25 mL aliquot taken from a bottle of orange juice

alkali A base that has been dissolved in water to form a solution

amphiprotic A substance that can donate or accept a proton

aqueous Present in water

back titration A titration in which a solution containing an unknown amount of a substance is neutralised with a known number of moles of reactant; a second titration is used to work out how much of the reactant remains and how much was used for neutralisation

base A substance that can accept a proton (or donate an electron pair – Lewis definition)

buffer A solution that resists changes to its pH (often a solution of an acid and its conjugate base in approximately equal concentrations)

concentrated Having a high ratio of solute to solvent

conjugate acid The species formed when a base accepts a proton

conjugate base The species formed when an acid loses a proton

coordinate bond A covalent bond in which both of the shared electrons have come from the same atom, rather than one from each

dilute Having a high ratio of solvent to solute

diprotic An acid that can donate two protons

end point The point in a titration at which we stop adding the titrant; this is usually indicated by a specific colour change

https://get.ga/aplus-hsc-chemistry-u34

A+ DIGITAL FLASHCARDS
Revise this topic's key terms and concepts by scanning the QR code or typing the URL into your browser.

equivalence point The point at which the number of moles of acid exactly balances the number of moles of base, and neutralisation is achieved; ideally, it is the point at which we want to stop the titration, but it is often passed and a suitable indicator is needed to provide an end point

hydrolysis The decomposition of a compound by water; it may refer to any chemical reaction that is facilitated by the addition of water

model A 2D or 3D representation of a chemical principle or concept

monoprotic An acid that can donate only one proton

neutralisation A chemical reaction between an acid and a base that produces water

nomenclature A set of rules for naming chemical compounds

polyprotic An acid that can donate more than one proton

primary standard A very pure, stable substance that is soluble in water and has a relatively high molecular mass; used in titrations

standard solution A solution of known concentration

strength A measure of the relative ionising or dissociating ability of a substance, usually in water

strong Refers to a substance that readily ionises or dissociates (usually in water)

titration A method of volumetric analysis commonly used to find the concentration of an unknown solution of an acid or base

weak Refers to a substance that does not readily ionise or dissociate (usually in water)

Exam practice

Multiple-choice questions

Solutions start on page 188.

Question 1

A small quantity of tartaric acid was dissolved in water to give a final hydrogen ion concentration of 2.3×10^{-6} mol L^{-1}. The pH of this solution was

A −5.6.
B −2.3.
C 2.3.
D 5.6.

Question 2

Which of the following would introduce an error to a titration?

A Rinsing the burette with the solution you are going to add to it
B Rinsing the reaction flask with distilled water prior to adding the aliquot
C Rinsing the reaction flask with the solution you are planning to add
D Rinsing the graduated pipette with the solution you are planning to add to it

Question 3

Which of the following is the *best* reason for not selecting sodium hydroxide as a primary standard?

A Its molar mass is too low.
B It is an alkali, not an acid.
C It is only slightly soluble in water.
D It absorbs chemicals from the atmosphere.

Question 4

Which of the following statements is consistent with the Lavoisier definition of an acid?

A Acids have a displaceable hydrogen ion.
B Acids release hydrogen ions in water.
C Sulfuric acid is an example of an acid.
D Sodium hydroxide is an example of a base.

Question 5

Which of the following is a correct statement about acids?

A A small K_a and a small pK_a indicate a weak acid.
B A small K_a and a small pK_a indicate a strong acid.
C A small K_a and a large pK_a indicate a weak acid.
D A large K_a and a small pK_a indicate a weak acid.

Question 6

Which of the following equations shows the dihydrogen phosphate ion acting as an acid?

A $H_2PO_4^-(aq) + H_2O(l) \rightleftharpoons H_3PO_4(aq) + OH^-(aq)$
B $H_2PO_4^-(aq) + H_2O(l) \rightleftharpoons HPO_4^{2-}(aq) + H_3O^+(aq)$
C $HPO_4^{2-}(aq) + H_2O(l) \rightleftharpoons H_2PO_4^-(aq) + OH^-(aq)$
D $HPO_4^{2-}(aq) + H_2O(l) \rightleftharpoons PO_4^{3-}(aq) + H_3O^+(aq)$

Question 7

The best choice of indicator for a strong acid–weak base titration is

A phenolphthalein.
B universal indicator.
C bromothymol blue.
D methyl orange.

Question 8

A 0.02 M solution of sulfuric acid is diluted by a factor of 10. Its new pH is closest to

A 0.7.

B 1.4.

C 1.7.

D 2.4.

Question 9

A student carried out a titration on an unknown monoprotic acid solution. She titrated a 10 mL aliquot of the acid against a 0.21 M sodium hydroxide solution. She repeated her procedure several times and calculated an average titre of 18.95 mL. The concentration (M) of the unknown acid is closest to

A 0.11.

B 0.22.

C 0.40.

D 0.80.

Question 10

The major improvement on the Arrhenius definition that was addressed by the Brønsted–Lowry definition of acids was

A acid–base reactions that did not occur in aqueous solutions.

B the presence of hydrogen ions in aqueous acid solutions.

C the displacement of hydrogen from a solution.

D the fact that electron pairs can be accepted by acid species through coordinate bonds.

Short-answer questions

Solutions start on page 189.

Question 11 (7 marks)

Two of the key ingredients in the chewable tablets of the antacid Mylanta are aluminium hydroxide (400 mg per tablet) and magnesium hydroxide (400 mg per tablet).

a Write an equation to show how aluminium hydroxide can be used to neutralise the hydrochloric acid in the stomach. 1 mark

b Use the information given to determine the number of moles of stomach acid that could be neutralised by two of these chewable tablets. 4 marks

c If the stomach held a volume of 2 L, what was the initial concentration of the hydrochloric acid in the stomach? 2 marks

Question 12 (3 marks)

Evaluate the accuracy of universal indicator as a tool for determining the pH of a solution.

Question 13 (3 marks)

Use an example and at least *two* chemical equations to demonstrate your understanding of amphiprotism.

Question 14 (4 marks)

25 mL of 0.20 M H_2SO_4 was mixed with 25 mL of 0.25 M NaOH. Calculate the pH of the final solution.

Question 15 (12 marks)

A group of students carried out a titration to calculate the quantity of base in chewable antacids. They checked the label and noticed that the major ingredient in the antacid tablet was sodium bicarbonate. They decided to conduct a back titration by adding an excess of HCl to dissolve all of the antacid and then back titrate the excess acid against a standardised NaOH solution.

a Describe one procedure they should follow to minimise any errors in this procedure and suggest how this procedure would minimise the error you identified. 2 marks

b The students added 42 mL of a 0.3 M HCl solution to 0.8 g of crushed antacid tablet that had been dissolved in distilled water. Suggest a reason why it is preferable to crush the tablets rather than react them whole. 1 mark

c The second reaction involved a titration of excess HCl against a 0.14 M NaOH standard solution. Justify your choice of indicator for this titration. 2 marks

d If the average titre for the reaction described in part **c** was 27.2 mL, calculate the number of moles of acid that were used in the neutralisation of the antacid. 4 marks

e Work out the percentage of sodium hydrogen carbonate in the antacid tablets. 3 marks

Question 16 (5 marks)

During your study of Chemistry, you have observed the effect of adding a strong, dilute acid such as 0.122 M HCl to an active metal, e.g. zinc.

A strong, concentrated acid may react with a low activity metal. One example of this is the reaction between copper and nitric acid to produce the distinctive but irritating brown nitrogen dioxide gas. The equation is:

$$Cu(s) + 4HNO_3(aq) \rightarrow Cu(NO_3)_2(aq) + 2H_2O(l) + 2NO_2(g)$$

a Contrast the products of the reaction between zinc and hydrochloric acid with the products of the reaction between copper and concentrated nitric acid. 3 marks

b Write the net ionic equation for the reaction between copper and concentrated nitric acid. Include all states. 2 marks

CHAPTER 3 MODULE 7: ORGANIC CHEMISTRY

Chapter 3
Module 7: Organic chemistry

Module summary

Organic chemistry is the chemistry of carbon. The unique bonding of carbon has produced Earth's hardest mineral (diamond), a widely used natural resource (fossil fuels), and life itself. All life on Earth relies on carbon in the form of molecules such as sugars (carbohydrates), proteins, fats (lipids) and nucleic acids. Most organic compounds come from coal, oil, or plant and animal products.

Human societies are reliant, many would say over-reliant, on carbon-based fuel sources, yet our early forays into biofuels have not been widely successful. We have also discovered the usefulness of plastics and can now produce them in vast numbers, often to the detriment of both terrestrial and marine life. Organic chemistry has yielded a range of beverages, solvents, soaps and detergents, perfumes and flavourings.

Module 7 is an introduction to the diverse compounds derived from carbon. In this module, we will:

- identify the unique characteristics of different types of organic compounds
- identify and classify the structures of different organic compounds
- identify the different type of reactions that occur between organic compounds
- look at how the structures of different organic compounds affect their properties and uses.

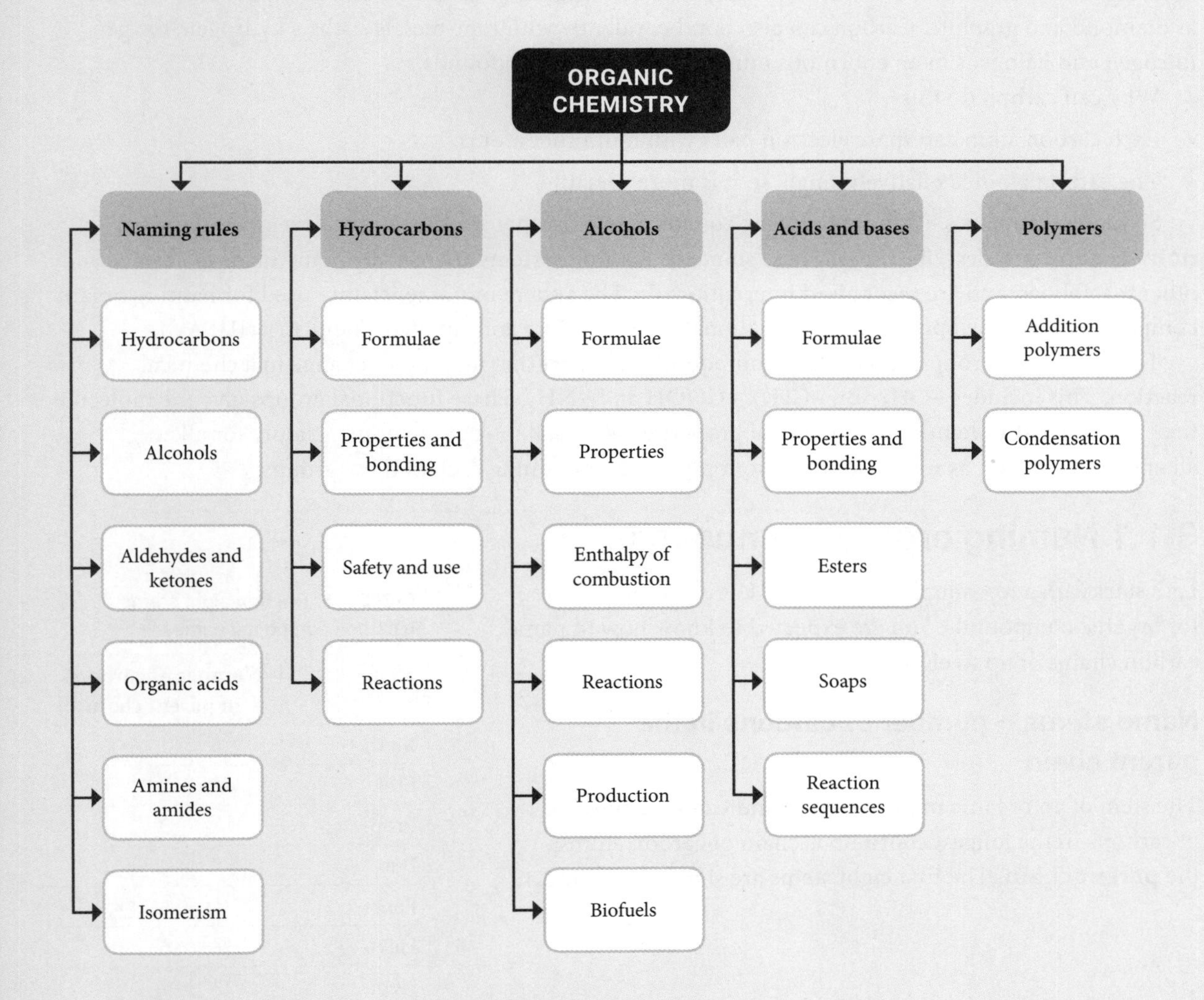

Outcomes

On completing this module, you should be able to:

- analyse the structure of, and predict reactions involving, carbon compounds

Working Scientifically skills

In this module, you are required to demonstrate the following Working Scientifically skills:

- analyse and evaluate primary and secondary data and information
- solve scientific problems using primary and secondary data, critical thinking skills and scientific processes
- communicate scientific understanding using suitable language and terminology for a specific audience or purpose

3.1 Nomenclature

Organic compounds are compounds containing carbon. Carbon atoms can form stable chains, networks and rings by means of carbon–carbon bonds. This bonding happens in substances of pure carbon, such as diamond and graphite. Carbon can also bond covalently with non-metals such as hydrogen, oxygen, nitrogen and halogens in an enormous number of different compounds.

Why can carbon do this?

- Each carbon atom can share electron pairs with four other atoms.
- The carbon atom is relatively small, so it is more versatile.

So many substances, including organic compounds, exist that scientists have developed standard **nomenclature** to describe them. These standard naming systems are used by many fields of science and other disciplines, and are recognised internationally. The system of nomenclature used for naming organic compounds was developed by the International Union of Pure and Applied Chemistry (IUPAC).

The functional group of an organic compound is the part that undergoes change in a chemical reaction. This includes –OH, –Br, –CHO, –COOH and $-NH_2$. These functional groups give the molecule their characteristic chemical and physical properties. We shall look at the nomenclature for alkanes, alkenes and alkynes, as well as for organic compounds containing each of these groups.

3.1.1 Naming organic compounds

Let's start with a few simple rules of the IUPAC nomenclature for organic compounds. You are expected to know how to name carbon chains of up to eight carbons.

Name stems – number of carbons in the parent chain

The stem of an organic molecule name indicates the number of carbons in the longest continuous chain of carbon atoms: the **parent chain**. The first eight stems are shown in Table 3.1.

TABLE 3.1 The first eight stems in carbon compound names

Stem	Number of carbons in parent chain
Meth-	1
Eth-	2
Prop-	3
But-	4
Pent-	5
Hex-	6
Hept-	7
Oct-	8

Name endings

The name ending indicates the family to which a compound belongs (Table 3.2); for example, alkanes or alcohols. In organic chemistry, compounds with the same general formula are part of the same family, referred to as a **homologous series**.

Atoms or groups of atoms substituted for hydrogen on the parent chain are called **substituents**. If they contain carbon, they are called alkyl groups (with prefixes methyl-, ethyl-, propyl- etc.). Some, but not all, substituents are functional groups.

TABLE 3.2 The name endings for carbon compounds and their homologous series

Name ending	Description	Homologous series	Example
-ane	Carbon compound with C–C bonds (no C=C or C≡C)	Alkanes	Butane
-ene	Carbon compound with at least one C=C bond	Alkenes	Ethene
-yne	Carbon compound with at least one C≡C bond	Alkynes	Propyne
-ol	Carbon compound with hydroxyl functional group (–OH)	Alcohols	Methanol
-al	Carbon compound with carbonyl functional group (C=O) at end of parent chain	Aldehydes	Ethanal
-one	Carbon compound with carbonyl functional group (C=O) not at end of parent chain	Ketones	Propanone
-oic acid	Carbon compound with carboxyl functional group (–COOH)	Carboxylic acids	Butanoic acid
-amine	Carbon compound with amino functional group ($-NH_2$ or $-NR_2$, where R can be H or another carbon chain; see page 113)	Amines	Pentanamine
-amide	Hydroxyl part (–OH) of –COOH in a carboxylic acid substituted with an amino functional group ($-NH_2$)	Amides	Hexanamide

General naming rules

To name an organic compound, we count the number of carbons in the parent chain to determine the name stem (Table 3.1). The name ending depends on whether the parent chain contains single, double or triple bonds (Table 3.2). We then number the carbons from the end that will give the lowest numbering for carbons that have side chains. The side chains must be identified by substituent and by the number of the carbon to which they are attached. When more than one number is used to identify substituents, use commas between them, and use a dash between a number and a letter.

The easiest way to learn the rules is to apply them to a number of different compounds.

Naming alkanes

An **alkane** contains only single carbon bonds. Consider the carbon compound represented in Figure 3.1. What is its molecular formula?

```
     H   H   H   H   H   H
     |   |   |   |   |   |
 H — C — C — C — C — C — C — H
     |   |   |   |   |   |
     H   H   H   H   H   H
```

FIGURE 3.1 What is the name of the alkane with this structural formula?

Step	Answer
1 Count the number of carbons in the parent chain.	6
2 Write the name stem for this number of carbons.	6 is hex-
3 Check the parent chain for single, double or triple bonds between carbon atoms.	(The parent chain of an alkane contains only carbon single bonds.)
4 Write the name ending to indicate this type of carbon bonding.	Carbon single bonds: -ane
5 Check for any substituents.	None
6 Put the parts of the name together.	Compound name is hexane

9780170465281

Naming alkenes

An **alkene** contains at least one carbon double bond. This makes the naming system for alkenes slightly different from the one used for alkanes. We need to know how many carbon double bonds there are and where they are located.

Consider the carbon compound represented in Figure 3.2. What is its molecular formula?

FIGURE 3.2 What is the name of the alkene with this structural formula?

Step	Answer
1 Count the number of carbon atoms in the parent chain.	4
2 Write the stem name for this number of carbon atoms.	4 is but-
3 Check the parent chain for single, double or triple bonds between carbon atoms.	(The parent chain of an alkene contains one or more carbon double bonds.) One carbon bond is C=C
4 Write the name ending to indicate this type of carbon bonding.	Carbon double bond: -ene
5 Determine the position of the carbon double bond.	Between carbons 1 and 2 (counting from right) or between carbons 3 and 4 (counting from left) Convention is to use the smaller number
6 Check for any substituents.	None
7 Put the parts of the name together.	Compound name is but-1-ene

Naming alkynes

An **alkyne** contains at least one carbon triple bond. Let's apply our rules to the carbon compound represented in Figure 3.3. What is its molecular formula?

FIGURE 3.3 What is the name of the alkene with this structural formula?

Step	Answer
1 Count the number of carbon atoms in the parent chain.	5
2 Write the stem name for this number of carbon atoms.	5 is pent-
3 Check the parent chain for single, double or triple bonds between carbon atoms.	(The parent chain of an alkyne contains one or more carbon triple bonds.) One carbon bond is C≡C
4 Write the name ending to indicate this type of carbon bonding.	Carbon triple bond: -yne
5 Determine the position of the carbon triple bond.	Between carbons 2 and 3 (counting from left) or between carbons 3 and 4 (counting from right) Convention is to use the smaller number
6 Check for any substituents.	None
7 Put the parts of the name together.	Compound name is pent-2-yne

Naming alcohols

An **alcohol** is identified by the presence of a hydroxyl group (–OH) attached to at least one carbon atom. The number of carbons to which that carbon is also bonded helps us classify the three different groups of alcohols (Table 3.3).

TABLE 3.3 The structures of primary, secondary and tertiary alcohols

Alcohol group	Structure	Example
Primary	–OH group is attached to a carbon atom at end of parent chain	Butan-1-ol
Secondary	–OH group is attached to a carbon atom that is also bonded to two other carbon atoms	Butan-2-ol
Tertiary	–OH group is attached to a carbon atom that is also bonded to three other carbon atoms	2-Methylpentan-2-ol

The most common alcohol is ethanol, a colourless liquid with a faint, sharp odour. It is the key ingredient in alcoholic beverages. Ethanol is a primary alcohol, and its structural formula is shown in Figure 3.4.

```
    H   H
    |   |
H — C — C — OH
    |   |
    H   H
```

FIGURE 3.4 The structural formula of ethanol

To name a secondary alcohol, we can use the rules we have been following, as well as considering the presence and location of substituents.

Consider the alcohol represented in Figure 3.5. What is its molecular formula?

```
    H   H   H
    |   |   |
H — C — C — C — H
    |   |   |
    H   OH  H
```

FIGURE 3.5 What is the name of the secondary alcohol with this structural formula?

Step	Answer
1 Count the number of carbon atoms in the parent chain.	3
2 Write the stem name for this number of carbon atoms.	3 is prop-
3 Check the parent chain for single, double or triple bonds between carbon atoms.	Only carbon single bonds
4 Write the name ending for this type of carbon bonding.	Carbon single bonds: -ane
5 Check for any substituents.	–OH group on carbon 2 (same number in either direction)
6 Adjust the name to indicate the substituent and position.	Adjust -ane to -an-2-ol
7 Put the parts of the name together.	Compound name is propan-2-ol

Naming aldehydes and ketones

An oxygen atom can be attached to a carbon atom by a double bond with a **terminal carbon** (a carbon atom at the end of a parent chain) or with a carbon atom attached to two other atoms (i.e. a non-terminal carbon atom). The location of the bond affects a compound's name, so there are two different homologous series: **aldehydes** and **ketones**.

Aldehydes

Aldehydes (also known as alkanals) have an oxygen atom attached to a terminal carbon by a double bond. The simplest example is methanal (also known as formaldehyde). Its structural formula is shown in Figure 3.6.

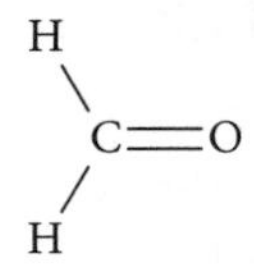

FIGURE 3.6 The structural formula of methanal

Consider the carbon compound represented in Figure 3.7. What is its molecular formula?

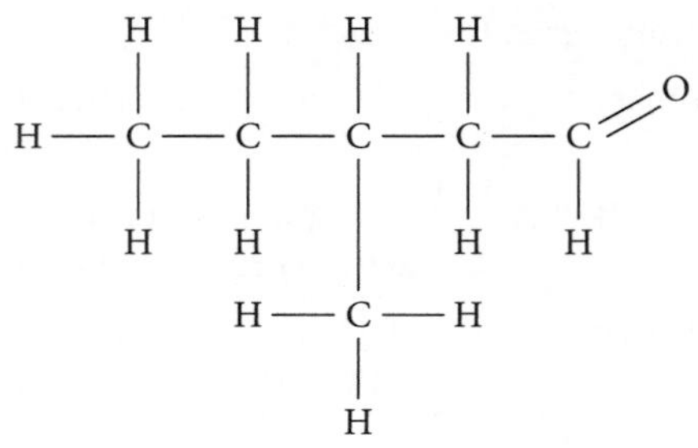

FIGURE 3.7 What is the name of the aldehyde with this structural formula?

Step	Answer
1 Count the number of carbon atoms in the parent chain.	5
2 Write the stem name for this number of carbons.	5 is pent-
3 Check the parent chain for single, double or triple bonds between carbon atoms.	Only carbon single bonds
4 Write the name ending for this type of carbon bonding.	Carbon single bonds: -ane
5 Check for any substituents.	C=O group at end of parent chain $-CH_3$ group attached to middle carbon atom. Numbering the C=O carbon as 1, the methyl group is attached to carbon 3
6 Adjust the name to indicate the substituents and positions.	Adjust -ane to -al Use a prefix to indicate methyl group: 3-methyl
7 Put the parts of the name together.	Compound name is 3-methylpentanal

Ketones

Ketones (also known as alkanones) have a C=O bond, like the aldehydes, but the carbon atom is bonded to two other carbon atoms (not with a terminal carbon). The simplest example is propanone (also known as acetone), represented in Figure 3.8.

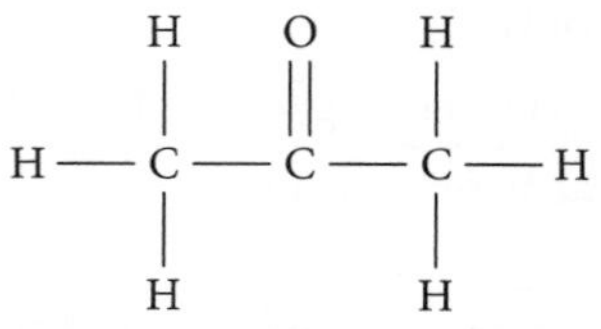

Why don't we call this 2-propanone or propan-2-one? The numbers in front of substituents are placed to avoid ambiguity. If the C=O group was on either terminal carbon, the compound would be called propanal, not propanone. For it to be propanone, it must be on a non-terminal carbon; this is the only possible arrangement for propanone, so we do not need the numbers.

FIGURE 3.8 The structural formula of propanone

Another example is shown in Figure 3.9. What is its molecular formula?

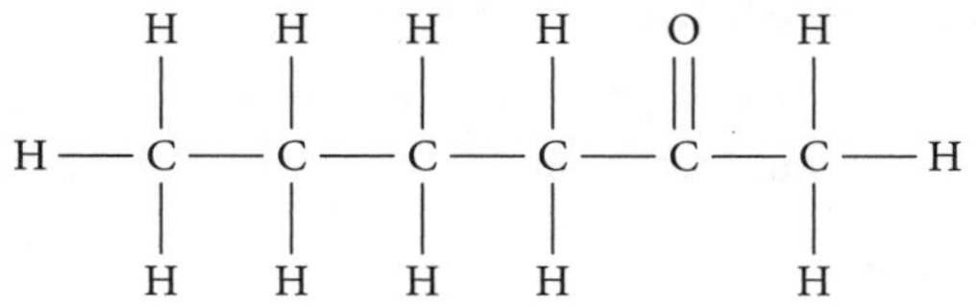

FIGURE 3.9 What is the name of the ketone with this structural formula?

Step	Answer
1 Count the number of carbon atoms in the parent chain.	6
2 Write the stem name for this number of carbon atoms.	6 is hex-
3 Check the parent chain for single, double or triple bonds between carbon atoms.	Only carbon single bonds
4 Write the name ending for this type of carbon bonding.	Carbon single bonds: -ane
5 Check for any substituents.	C=O group at non-terminal carbon atom: carbon 2 (numbering from right gives smaller number)
6 Adjust the name to indicate the substituent and position.	Adjust -ane to -an-2-one
7 Put the parts of the name together.	Compound name is hexan-2-one

Naming carboxylic acids

Organic acids, more properly named **carboxylic acids**, have the –COOH functional group. A terminal carbon shares a double bond with an oxygen atom and a single bond with a hydroxyl group. An example is ethanoic acid, represented in Figure 3.10.

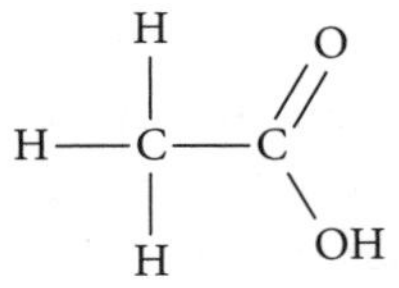

FIGURE 3.10 The structural formula of ethanoic acid

Another example is shown in Figure 3.11. What is its molecular formula?

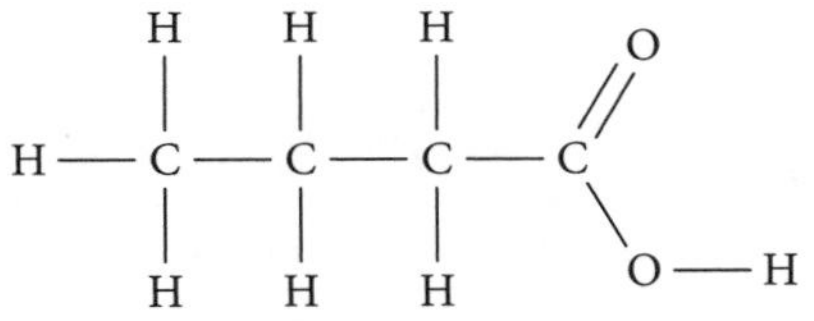

FIGURE 3.11 What is the name of the carboxylic acid with this structural formula?

Step	Answer
1 Count the number of carbon atoms in the parent chain.	4
2 Write the stem name for this number of carbon atoms.	4 is but-
3 Check the parent chain for single, double or triple bonds between carbon atoms.	Only carbon single bonds
4 Write the name ending for this type of carbon bonding.	Carbon single bonds: -ane
5 Check for any substituents.	–COOH group at terminal carbon atom
6 Adjust the name to identify the substituent and position.	Adjust -ane to -oic acid
7 Put the parts of the name together.	Compound name is butanoic acid (also known as butyric acid)

Naming amines and amides

There are two important homologous series you need to know that involve nitrogen: **amines** and **amides**.

Amines

Amines have an amino functional group. They can be primary, secondary or tertiary, depending on how many carbons (one, two or three) the nitrogen atom is bonded to.

An example is shown in Figure 3.12. What is its molecular formula?

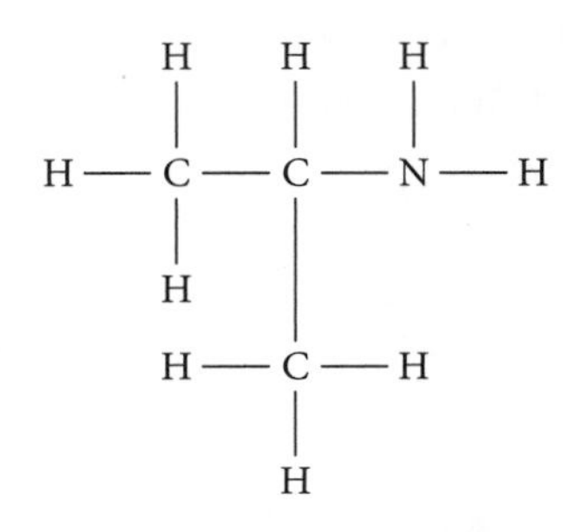

FIGURE 3.12 What is the name of the amine with this structural formula?

Step	Answer
1 Count the number of carbon atoms in the parent chain.	3
2 Write the stem name for this number of carbon atoms.	3 is prop-
3 Check the parent chain for single, double or triple bonds between carbon atoms.	Only carbon single bonds
4 Write the name ending for this type of carbon bonding.	Carbon single bonds: -ane
5 Check for any substituents.	$-NH_2$ group attached to carbon 2 (same number in either direction)
6 Adjust the name to indicate the substituent and position.	Adjust -ane to -an-2-amine
7 Put the parts of the name together.	Compound name is propan-2-amine

The compound represented in Figure 3.13 has a side chain at the nitrogen atom. What is its molecular formula?

FIGURE 3.13 What is the name of the amine with this structural formula?

Step	Answer
1 Count the number of carbon atoms in the parent chain.	4
2 Write the stem name for this number of carbon atoms.	4 is but-
3 Check the parent chain for single, double or triple bonds between carbon atoms.	Only carbon single bonds
4 Write the name ending for this type of carbon bonding.	Carbon single bonds: -ane
5 Check for any substituents.	–NH attached to carbon 1 (numbering from right gives smaller number) $-CH_3$ group attached to nitrogen atom
6 Adjust the name ending to indicate the substituent and position.	Adjust -ane to -1-amine Use a prefix to indicate $-CH_3$ group: -methyl Methyl group is not attached to parent chain, so element symbol rather than a number is used: *N*-methyl
7 Put the parts of the name together.	Compound name is *N*-methylbutan-1-amine

Amides

The easiest way to think of amides is as derivatives of a carboxylic acid, where the –OH group has been replaced by an $-NH_2$ group. Like the amines, amides can be primary, secondary or tertiary and may include a number and/or an *N* to indicate the position of substituents.

An example is shown in Figure 3.14. What is its molecular formula?

FIGURE 3.14 What is the name of the amide with this structural formula?

Step	Answer
1 Count the number of carbon atoms in the parent chain.	2
2 Write the stem name for this number of carbon atoms.	2 is eth-
3 Check the parent chain for single, double or triple bonds between carbon atoms.	Only carbon single bonds
4 Write the name ending for this type of carbon bonding.	Carbon single bonds: -ane
5 Check for any substituents.	An O and an $-NH_2$ group attached to terminal carbon atom
6 Adjust the name ending to indicate the substituent and position.	Adjust -ane to -amide
7 Put the parts of the name together.	Compound name is ethanamide

Note

For more complex organic compounds, multiple name endings are added to the name stem, in alphabetical order, with their appropriate numbers.

Multiples of the same substituent are indicated by prefixing with 'di' (2), 'tri' (3), 'tetra' (4), and so on.

Naming halogenated organic compounds

Halogenated organic compounds are organic compounds in which one or more hydrogens have been replaced by halogen atoms. We keep the same rules as used previously, and add the halogens alphabetically (including numbers and prefixes) to the front of the name.

If there is more than one possible name, choose the name that gives the lowest numbers to the halogens (when in alphabetical order).

9780170465281

An example is provided in Figure 3.15. What is its molecular formula?

```
      H    F    Cl        Br   H    H    H
      |    |    |         |    |    |    |
 H —  C —  C —  C  =  C — C —  C —  C —  C — H
      |    |          |   |    |    |    |
      H    F          Br  H    Br   H    H
```

FIGURE 3.15 What is the name of the halogenated organic compound with this structural formula?

Step	Answer
1 Count the number of carbon atoms in the parent chain.	8
2 Write the stem name for this number of carbon atoms.	8 is oct-
3 Check the parent chain for single, double or triple bonds between carbon atoms.	One carbon bond is C=C.
4 Write the name ending for this type of carbon bonding.	Carbon double bond: -ene
5 Determine the position of the carbon double bond.	Between carbons 3 and 4 (numbering from left) Between carbons 5 and 6 (numbering from right)
6 Check for any substituents.	Several halogens are present (2 fluorine atoms, 3 bromine atoms, 1 chlorine atom)
7 Adjust the name to indicate the substituent and position.	Numbering halogen positions from left (-3-ene): 2,2,3,4,5,6 Numbering halogen positions from right (-5-ene): 3,4,5,6,7,7 Choose the set with smaller numbers: -3-ene Use prefixes and numbers to indicate halogen substituents: • 3 bromine atoms (at carbons 4–6): 4,5,6-tribromo- • 1 chlorine atom (at carbon 3): 3-chloro- • 2 fluorine atoms (at carbon 2): 2,2-difluoro-
8 Put the parts of the name together.	With halogens in alphabetical order, compound name is 4,5,6-tribromo-3-chloro-2,2-difluoro-oct-3-ene

Functional group naming hierarchy

You are unlikely to be asked to name compounds with several functional groups but be aware that there is a hierarchy. A simplified hierarchy is presented in Table 3.4.

TABLE 3.4 The naming hierarchy for organic compounds

Priority	Organic compound	Functional group
1	Carboxylic acid	–COOH
2	Amide	$–CONH_2$
3	Aldehyde	C=O (at terminal carbon)
4	Ketone	C=O (not at terminal carbon)
5	Alcohol	–OH
6	Amine	$–NH_2$
Equal priority is given to carbon double bonds, carbon triple bonds and halogens, so alphabetical rules apply (e.g. -ene before -yne, chloro- before fluoro-).		

3.1.2 Isomerism

Numbers (or **locants**) (and sometimes element symbols) are needed when there is ambiguity about where an atom or substituent is located. A chlorine atom that is a substituent of a methane molecule can only be attached to one carbon. A chlorine atom that is a substituent of an ethane molecule could be attached to either of two carbons, but by convention it is always assigned the smallest number, so the compound would be called 1-chloroethane, which we simplify to chloroethane.

If you were to draw chloroethane, you could draw different structures, but they would all have the chlorine atom on a terminal carbon, so there is no ambiguity. However, what if you had to draw dichloroethane? There are now two chlorine atoms, so placement is important. You could place both chlorine atoms on the same carbon, or one on each. These two options are not the same, no matter how we rotate the molecules – we must distinguish between 1,1-dichloroethane and 1,2-dichloroethane. These are shown in Figures 3.16 and 3.17.

```
     H    Cl
     |    |
H —  C —  C — Cl
     |    |
     H    H
```

FIGURE 3.16 The structural formula of 1,1-dichloroethane

```
     Cl   H
     |    |
H —  C —  C — Cl
     |    |
     H    H
```

FIGURE 3.17 The structural formula of 1,2-dichloroethane

These two compounds have the same molecular formula ($C_2H_4Cl_2$), and we refer to them as **isomers**. Here, we will consider an isomer as one of a series of compounds that have the same molecular formula but different structural formulae. The three types of structural isomers are chain isomers, positional isomers and functional group isomers.

Chain isomers

Chain isomers are isomers in which the structural differences are in the carbon chain. They may have different chain lengths and include one or more side branches. For example, the molecular formula for both butane and methylpropane is C_4H_{10}, but the compounds are structurally different. Butane (Figure 3.18) has four carbons in its longest chain, while methylpropane (Figure 3.19) has only three.

```
    H   H   H   H
    |   |   |   |
H — C — C — C — C — H
    |   |   |   |
    H   H   H   H
```

FIGURE 3.18 The structural formula of butane

```
    H   H   H
    |   |   |
H — C — C — C — H
    |   |   |
    H   |   H
        |
    H — C — H
        |
        H
```

FIGURE 3.19 The structural formula of methylpropane

Positional isomers

Positional isomers are isomers in which structural differences are located at a particular functional group. This could be as simple as the location of a double bond in an alkene. An example is but-1-ene (Figure 3.20) and but-2-ene (Figure 3.21). Both of these compounds have the molecular formula C_4H_8 but the location of the double bond is different.

> **Note**
> C–C, C=C and C≡C are functional groups, but they are often not described as such unless they are substituents of a parent chain.

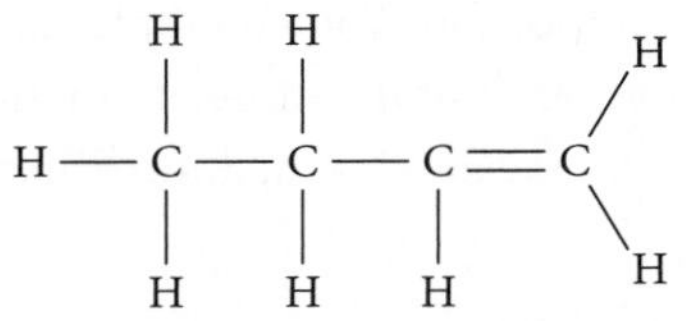

FIGURE 3.20 The structural formula of but-1-ene

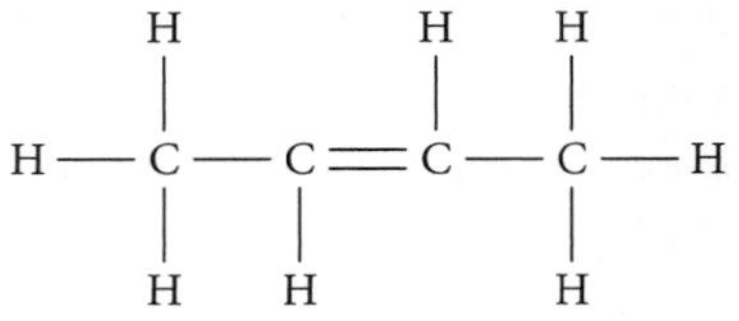

FIGURE 3.21 The structural formula of but-2-ene

Functional group isomers

Functional group isomers are isomers that contain functional groups with structural differences. An example is cyclopropane (Figure 3.22) and propene (Figure 3.23). Both have the molecular formula C_3H_6; however, cyclopropane has a ring structure (hence its name), and propene contains a double bond. These compounds do not belong to the same homologous series, so we classify them as functional group isomers.

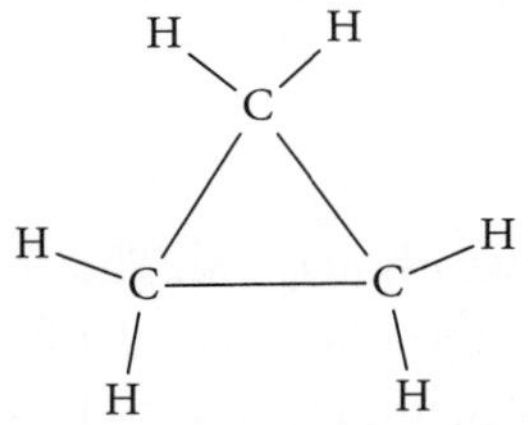

FIGURE 3.22 The structural formula of cyclopropane

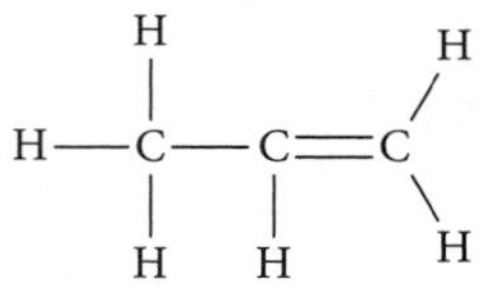

FIGURE 3.23 The structural formula of propene

3.2 Hydrocarbons

A **hydrocarbon** is a compound containing only hydrogen and carbon. We have already discussed naming the three groups of hydrocarbons: alkanes, alkenes and alkynes.

We have previously referred to organic compounds as belonging to a particular homologous series. Homologous series are groups of compounds characterised by:

- a common general formula
- a common functional group (if present)
- similar structures and chemical properties
- small and steady changes in their physical properties (as molar mass increases).

In this section we will take a closer look at the first eight members of the straight-chain hydrocarbons and make some general conclusions about their structures and properties.

3.2.1 Formulae for hydrocarbons

Hydrocarbons containing only single carbon bonds are **saturated hydrocarbons**. Their general formula is:

$$C_nH_{2n+2} \quad \text{or} \quad R'(CH_2)_nR''$$

where R′ and R″ represent H or an alkyl group (sometimes R^1, R^2 are used).

Using the symbol R helps us focus on the functional group and make some generalisations.

There are different ways to simplify the structural formula for hydrocarbons. Consider the five ways to represent hept-2-ene, as shown in Table 3.5.

TABLE 3.5 Representations of hept-2-ene

Molecular formula	C_7H_{14}
Condensed structural formula	$CH_3(CH_2)_5CHCH_2$
Structural formula	H_3C CH_2 CH_2 CH_2 CH CH CH_3
Skeletal formula	H_3C CH_3
Skeletal formula without terminal groups	

For Table 3.5, we assume a carbon at each vertex and/or end, and sufficient hydrogens to make four bonds around each carbon. For skeletal formulae without terminal groups, atoms other than carbon and hydrogen (i.e. functional groups) are indicated by their element symbols. Skeletal formulae do not usually appear in HSC exams.

There are many ways to construct models of organic compounds, both in written form and in 3D. A common way to model organic compounds is using an organic model kit. Figure 3.24 shows a molecular model of ethane made using a kit. This model shows the constituent atoms of ethane (carbon in black and hydrogen in white) and their orientation (we will look at shape in a later section). It also indicates the molecular formula.

Photo courtesy of Jessica Harrison

FIGURE 3.24 A molecular model of ethane

Note

The atoms in the model are connected by rigid plastic, and the connections are the same length. The atoms are shown as solid balls, although we know atoms are mostly empty space. If these models are inaccurate, why do we use them?

Organic model kits help us to explain our observations and/or the behaviour of substances in different states or in the presence of other substances. We can use them provided we understand their limitations.

9780170465281

Alkanes

Being saturated hydrocarbons, alkanes have the general formula C_nH_{2n+2}. Single bonds between carbon atoms are extremely stable so saturated hydrocarbons are relatively stable. However, in many combustion reactions they are important fuels; the alkanes are the principal compounds in natural gas and petroleum. They all burn in sufficient oxygen to produce carbon dioxide and water. Table 3.6 shows the first eight straight-chain alkanes and examples of their uses.

TABLE 3.6 The first eight straight-chain alkanes

<table>
<tr><th>Name and example of use</th><th>Molecular formula</th><th>Structural formula</th></tr>
<tr><td>Methane
A major component of natural gas, supplied for gas cooking and heating</td><td>CH_4</td><td>H
|
H—C—H
|
H</td></tr>
<tr><td>Ethane
A major component of natural gas</td><td>C_2H_6</td><td>H H
| |
H—C—C—H
| |
H H</td></tr>
<tr><td>Propane
A major component of liquefied petroleum gas (LPG), used as bottled gas</td><td>C_3H_8</td><td>H H H
| | |
H—C—C—C—H
| | |
H H H</td></tr>
<tr><td>Butane
The liquid fuel in cigarette lighters</td><td>C_4H_{10}</td><td>H H H H
| | | |
H—C—C—C—C—H
| | | |
H H H H</td></tr>
<tr><td>Pentane
A component of crude oil, often used to manufacture polystyrene foam</td><td>C_5H_{12}</td><td>H H H H H
| | | | |
H—C—C—C—C—C—H
| | | | |
H H H H H</td></tr>
<tr><td>Hexane
A component of crude oil, often used as a solvent</td><td>C_6H_{14}</td><td>H H H H H H
| | | | | |
H—C—C—C—C—C—C—H
| | | | | |
H H H H H H</td></tr>
<tr><td>Heptane
A component of crude oil, found in petrol</td><td>C_7H_{16}</td><td>H H H H H H H
| | | | | | |
H—C—C—C—C—C—C—C—H
| | | | | | |
H H H H H H H</td></tr>
<tr><td>Octane
A component of crude oil, found in petrol</td><td>C_8H_{18}</td><td>H H H H H H H H
| | | | | | | |
H—C—C—C—C—C—C—C—C—H
| | | | | | | |
H H H H H H H H</td></tr>
</table>

Alkenes

Alkenes contain at least one carbon double bond, and any other carbon bonds in their parent chain are single. Their general formula is:

$$C_nH_{2n} \quad \text{or} \quad R'CH{=}CHR''$$

where R′ and R″ represent H or an alkyl group.

Figure 3.25 shows a molecular model of ethene.

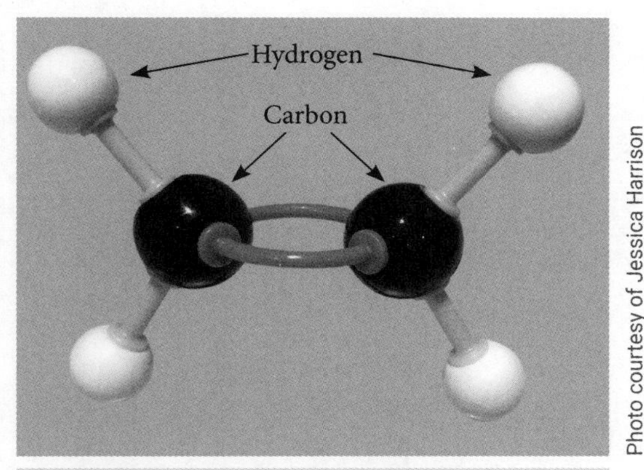

Photo courtesy of Jessica Harrison

FIGURE 3.25 A molecular model of ethene

For ethene, there is only one position for the carbon double bond, so we don't need to use a number to identify it.

For pentene (C_5H_{10}) the double carbon bond may or may not be at the terminal carbon. Using a number (e.g. pent-1-ene) indicates the position of the double bond.

When the parent chain of an alkene contains more than one carbon double bond, we can use numbers and prefixes, as we did earlier in the chapter, to indicate positions and how many there are. Consider oct-1,6-diene, represented in Figure 3.26.

```
    H    H         H    H    H          H
    |    |         |    |    |         /
H — C —  C ═ C  —  C  — C  — C  — C ═ C
    |        |     |    |    |    |    \
    H        H     H    H    H    H     H
```

FIGURE 3.26 The structural formula of oct-1,6-diene

The parent chain of oct-1,6-diene contains eight carbons; hence, oct-. It has carbon double bonds, so it is an -ene. There are two double bonds, so it is a diene. The name 'octdiene' would not indicate where the double bonds are located, so we need numbers:

- counting from the left: 2 and 7
- counting from the right: 1 and 6.

Using the smallest combination of numbers, the correct name for this compound is oct-1,6-diene.

The first eight straight-chain alkenes are shown in Table 3.7.

TABLE 3.7 The first eight straight-chain alkenes

Name	Molecular formula	Structural formula
Ethene	C_2H_4	H H C=C H H
Propene	C_3H_6	H H H C=C–C–H H H
But-1-ene	C_4H_8	H H H H C=C–C–C–H H H H
Pent-1-ene	C_5H_{10}	H H H H H C=C–C–C–C–H H H H H
Hex-1-ene	C_6H_{12}	H H H H H H C=C–C–C–C–C–H H H H H H
Hept-1-ene	C_7H_{14}	H H H H H H H C=C–C–C–C–C–C–H H H H H H H
Oct-1-ene	C_8H_{16}	H H H H H H H H C=C–C–C–C–C–C–C–H H H H H H H H

Alkynes

Alkynes contain at least one carbon triple bond. Their general formula is:

$$C_nH_{2n-2} \quad \text{or} \quad R'C{\equiv}CR''$$

where R′ and R″ represent H or an alkyl group.

Figure 3.27 shows a molecular model of ethyne.

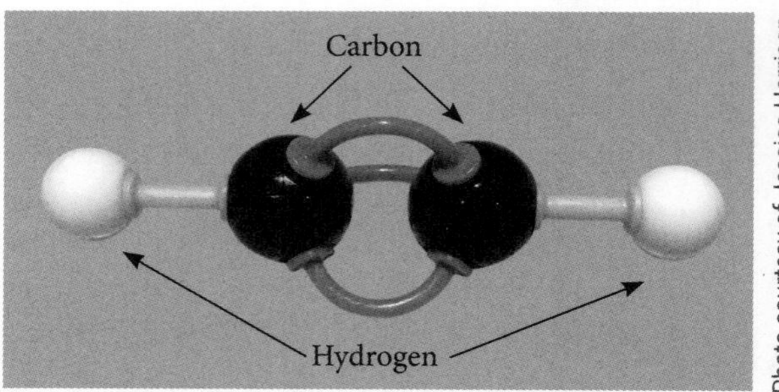

Photo courtesy of Jessica Harrison

FIGURE 3.27 A molecular model of ethyne

Both alkenes and alkynes are **unsaturated hydrocarbons** – the carbons at the double and triple bonds have fewer hydrogens attached than are the carbons in single bonds do. They can take part of addition reactions, during which they combine with other atoms at double or triple bonds.

The first eight straight-chain alkynes that have a triple bond at the second carbon (where possible) are shown in Table 3.8.

TABLE 3.8 The first eight straight-chain alkynes

<table>
<tr><th>Name</th><th>Molecular formula</th><th>Structural formula</th></tr>
<tr><td>Ethyne</td><td>C_2H_2</td><td>H—C≡C—H</td></tr>
<tr><td>Propyne</td><td>C_3H_4</td><td>H
|
H—C—C≡C—H
|
H</td></tr>
<tr><td>But-2-yne</td><td>C_4H_6</td><td>H H
| |
H—C—C≡C—C—H
| |
H H</td></tr>
<tr><td>Pent-2-yne</td><td>C_5H_8</td><td>H H H
| | |
H—C—C≡C—C—C—H
| | |
H H H</td></tr>
<tr><td>Hex-2-yne</td><td>C_6H_{10}</td><td>H H H H
| | | |
H—C—C≡C—C—C—C—H
| | | |
H H H H</td></tr>
<tr><td>Hept-2-yne</td><td>C_7H_{12}</td><td>H H H H H
| | | | |
H—C—C≡C—C—C—C—C—H
| | | | |
H H H H H</td></tr>
<tr><td>Oct-2-yne</td><td>C_8H_{14}</td><td>H H H H H H
| | | | | |
H—C—C≡C—C—C—C—C—C—H
| | | | | |
H H H H H H</td></tr>
</table>

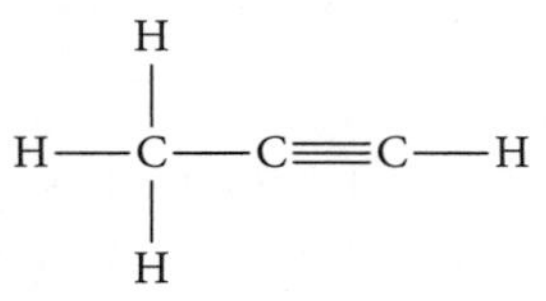

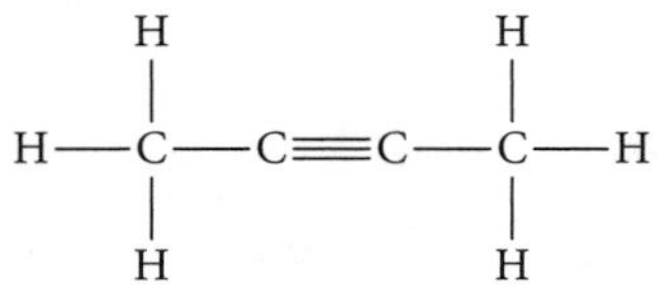

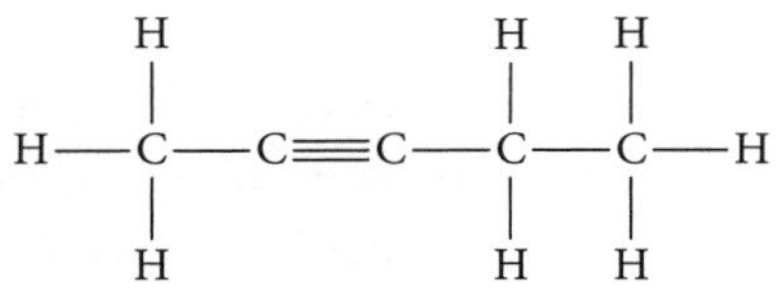

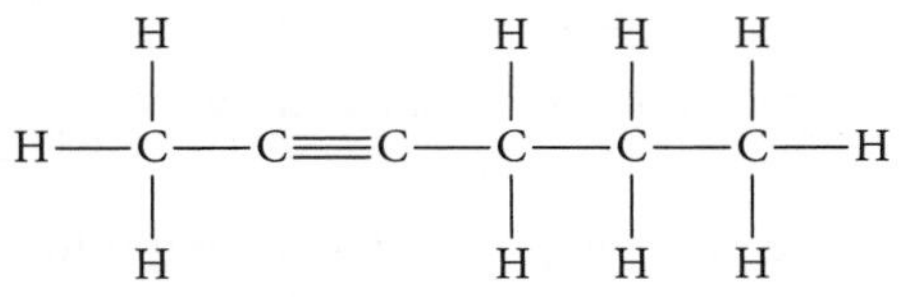

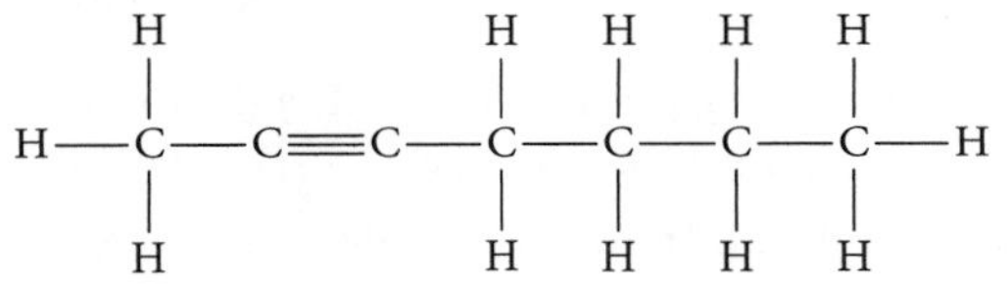

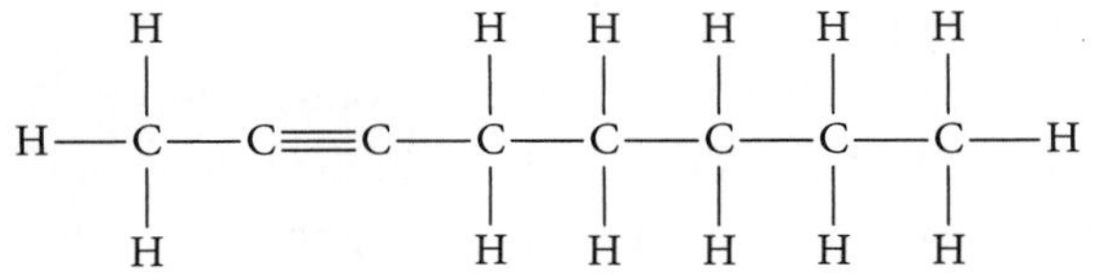

Hydrocarbons can exist as long chains or rings, with or without side chains. Compounds with side chains are named according to the number of carbons attached to the carbon in the parent chain (i.e. not counting the attached carbon in the parent chain). We use prefixes in the same way as we did earlier in this chapter (methyl, ethyl etc.) and list them in alphabetical order if more than one is present. The 'cyclo' prefix is used to indicate cyclic hydrocarbons; for example, methylcyclobutane (Figure 3.28).

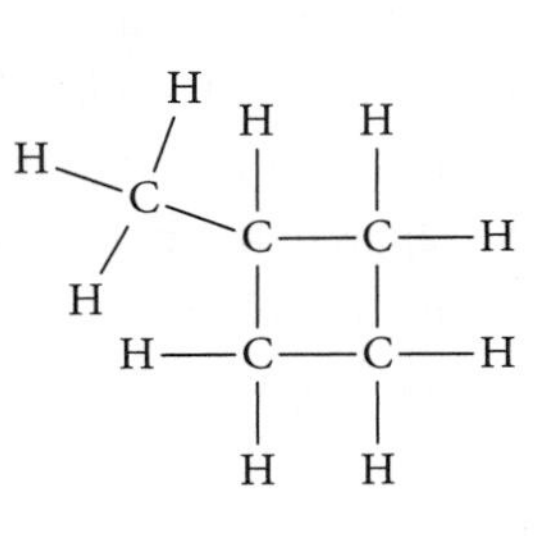

FIGURE 3.28 The structural formula of methylcyclobutane

Cyclic forms will not conform to the general formula for their homologous series. For example, methylcyclobutane contains only carbon and hydrogen atoms, so it is a hydrocarbon. It has only single bonds, so it is an alkane. It contains five carbons, four of which are in the parent chain, which would make it butane if the parent chain were straight. However, the four carbon atoms are part of a ring, and there is one side carbon group, a methyl group. This changes the molecular formula to C_5H_{10} and the name to methylcyclobutane. We do not have to number the methyl group because any orientation would have the number 1 for the carbon to which it was attached.

3.2.2 Properties of hydrocarbons

Both chemical and physical properties have significant effects on the behaviour of hydrocarbons.

Chemical properties

Carbon and hydrogen atoms have similar electronegativities, so hydrocarbons are non-polar. Chemical properties relate to the presence of functional groups, including double and triple bonds. Hence, alkenes and alkynes are chemically more reactive than alkanes. Hydrocarbon bonding and reactions will be covered later in this chapter.

Physical properties

Non-polar bonds in non-polar molecules affect the physical properties of a compound. As solids, non-polar molecules occur in covalent molecular networks and are held together by weak dispersion forces. For longer chains, there are more dispersion forces; hence, larger molecules have higher melting and boiling points. This is why small alkanes such as methane, ethane and propane exist as gases under standard laboratory conditions; pentane, octane and several others are liquids under the same conditions; and molecules with longer chains, such as paraffin waxes (20 or more carbon atoms), are solids at 25°C.

The non-polar nature of hydrocarbons makes them insoluble in polar substances such as water. They will dissolve in non-polar solvents.

Hydrocarbons are poor conductors because there are neither free electrons nor ions to respond to an electric field.

One way to compare the physical properties of compounds within a homologous series is to list the values of the compounds for a particular physical property and then use knowledge of bonding to explain the pattern.

Table 3.9 shows the boiling point of the first eight straight-chain alkanes. We could plot these points on a graph, but it is fairly easy to see an upward trend. When there are more carbons in the molecule (and hence molecular weight), the boiling point is higher.

At a compound's boiling point, the compound changes state from a liquid to a gas. The molecules in the compound separate completely from one another. Boiling the compound overcomes the forces of attraction between the molecules, and the heat energy is transformed into kinetic energy. The larger the mass of each molecule and the greater the number of dispersion forces between the molecules, the more energy is needed to separate them; hence, a trend of increasing boiling points for the first eight alkanes. A similar trend is observed for the alkenes and alkynes.

TABLE 3.9 The boiling points for the first eight straight-chain alkanes

Alkane	Boiling point (°C)
Methane	−164
Ethane	−89
Propane	−42
Butane	0
Pentane	36
Hexane	69
Heptane	98
Octane	126

3.2.3 Shapes of hydrocarbons

Different types of bonds influence the molecular shape (geometry) of hydrocarbons. This can affect the forces within molecules, and the chemical or physical properties of compounds.

We can explain trends in physical properties by comparing molecular geometries of hydrocarbons. Let's look at the molecular models of an alkane, an alkene and an alkyne (Table 3.10).

TABLE 3.10 Molecular models and geometry for ethane, ethene and ethyne

Hydrocarbon	Molecular model	Boiling point (°C)	Geometry
Ethane		−89	The four atoms around each carbon atom are arranged in the shape of a tetrahedron. Equal spacing as a result of the tetrahedral arrangement minimises the forces of repulsion between the electron pairs and confers greater stability on the molecule. The presence of non-polar carbon–carbon and carbon–hydrogen bonds ensures that only dispersion forces exist between the molecules. This results in a very low boiling point; hence, ethane is a gas at room temperature.
Ethene	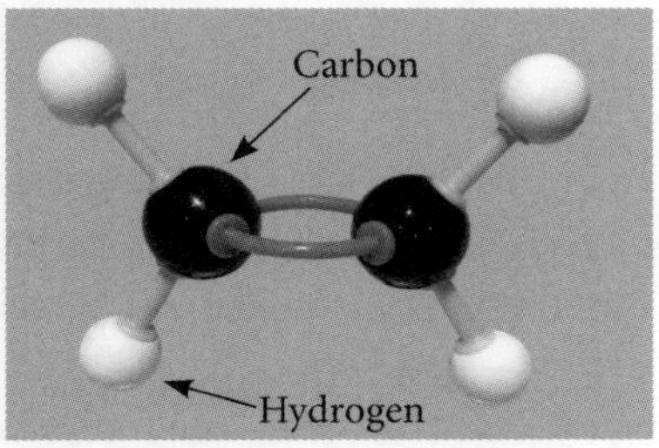	−104	The double bond between the two carbons has resulted in the distribution of the three atoms (one carbon and two hydrogens) in a trigonal planar arrangement around a central carbon. This planar arrangement enables closer packing, but fewer hydrogen atoms means less dispersion forces; hence, ethene has a slightly lower boiling point than ethane.
Ethyne	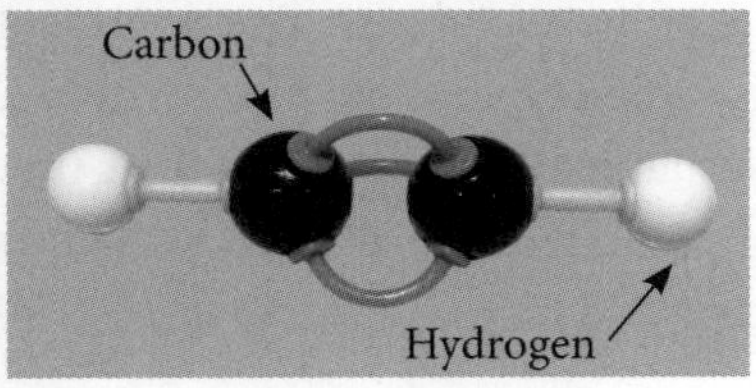 Photos courtesy of Jessica Harrison	−84	The triple bond between the carbon atoms pulls all four atoms of ethyne into a line. Linear molecules can be tightly packed and held firm by strong intermolecular forces, which can increase their melting or boiling points, despite a small decrease in the number of dispersion forces. Ethyne has a slightly higher boiling point than ethene.

There are various conventions for drawing organic molecules. For example, methane (CH_4) is often drawn as shown in Figure 3.29.

This is simple to draw, but it does not show the 3D arrangement of atoms in the molecule. We can also draw methane as is shown in Figure 3.30. The dashed line indicates a bond pointing back into the page and the solid wedge indicates a bond pointing out of the page.

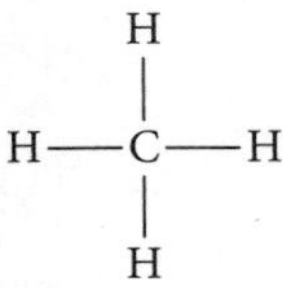

FIGURE 3.29 The 2D structural formula of methane

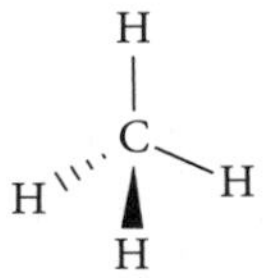

FIGURE 3.30 The 3D structural formula of methane

Hybridisation

You are unlikely to need anything more than a basic understanding of hybrid orbitals for your HSC course. Here are a few main points.

- Hybridisation is based on the concept of subshells. From Year 11, we know carbon has four outer-shell electrons, and a subshell configuration of $1s^22s^22p^2$.
- There is an energy difference between the s and p subshells, yet when carbon forms four bonds with four hydrogen atoms, all four bonds are of equal strength.
- The atomic orbitals combine to form molecular orbitals. We can use the concept of molecular orbitals to explain the geometry of tetrahedral, trigonal planar and linear structures.
- If one 2s electron jumps to a 2p orbital, the excited carbon atom can hybridise this state to form four identical hybrid orbitals, equivalent in energy, size and shape.
- Each electron occupies a lone region (1 in s, 1 in p_x, 1 in p_y and 1 in p_z). The result is four equal-strength bonds known as sp^3 orbitals. The formation of these hybrid orbitals slightly lowers the energy, conferring some stability. The orbitals are arranged so that they overlap, which accounts for the tetrahedral arrangement of hydrogen atoms around a central carbon atom.
- Excited carbon atoms may also stabilise through the formation of two or three hybrid orbitals. When these types of hybrids are formed, the geometry about the central atom changes and the additional p orbitals sit at right angles to the sp hybrids. They are weaker bonds and usually appear as the double or triple bonds in alkenes and alkynes.

3.2.4 Bonding and hydrocarbons

The chemical properties of a compound or a series of compounds are related to **intramolecular bonding** – the bonds within molecules. Alkanes are characterised by single carbon–carbon and carbon–hydrogen bonds. The paired electrons are part of a tetrahedral arrangement, which contributes to the stability of hydrocarbons.

The physical properties of alkanes are related to **intermolecular forces** – the forces between molecules. These forces help determine the size and shape of hydrocarbons. The dominant intermolecular forces are dispersion forces. Table 3.11 allows us to analyse trends in boiling point within a homologous series – and also between the alkanes and molecules of similar length – based on their functional groups.

Table 3.11 indicates a clear trend of increasing boiling point with molar mass of alkanes. It is even easier to see this trend if we plot the data on a graph (Figure 3.31).

TABLE 3.11 The boiling points of some alkanes and alkenes

Number of carbon atoms	Boiling point (°C)	
	Alkanes	Alkenes
1	−164	
2	−89	−104
3	−42	−48
4	−0.5	−6
5	36	30
6	69	64
7	98	94
8	125	121
9	151	146
10	174	172

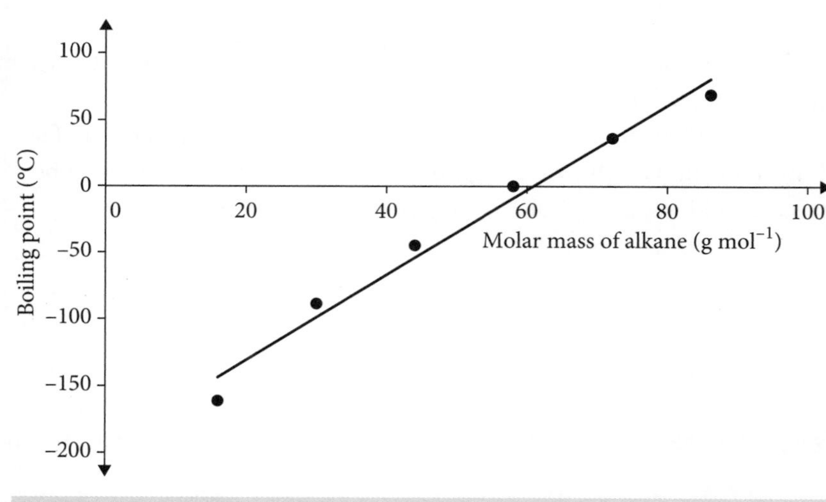

FIGURE 3.31 The trend in boiling point with molar mass for alkanes

The trend line is almost perfectly linear. It suggests a strong relationship between boiling point and molar mass (which for hydrocarbons is related to the number of carbon atoms) for alkanes. The only difference between each member of the alkanes is a $-CH_2-$ group. This group increases the size of each molecule and increases the strength of the dispersion forces between the molecules. These forces must be overcome for the molecules in a liquid to move independently as gas molecules.

The shape of a molecule can also affect the physical properties of hydrocarbons. Consider pentane, methylbutane and dimethylpropane, which all have the same molecular formula and hence the same molar mass (72.146 g mol^{-1}). We would call them chain isomers because they only differ in the length of the parent chain. However, the boiling points are as shown in Table 3.12.

TABLE 3.12 Comparing boiling points with chain branching in alkanes

Compound	Structural formula	Molecular mass (g mol^{-1})	Boiling point (°C)
Pentane	H H H H H H—C—C—C—C—C—H H H H H H	72.146	36
Methylbutane (isopentane)	H H—C—H H H H H—C—C—C—C—H H H H H	72.146	27.8
Dimethylpropane (neopentane)	H H—C—H H H H—C—C—C—H H H H—C—H H	72.146	10

This is a clear trend, but it cannot be explained by a change in molar mass because all three isomers have the same mass. Considering forces and molecular geometry, we can explain that, if a compound is more branched, the dispersion forces holding the molecules together are slightly weaker. The more linear compound, pentane, has the strongest intermolecular forces and the highest boiling point.

When analysing the trends in observed or recorded properties within a homologous series and/or between different functional groups (as we will cover later), it is important to:

1 look carefully at the data

2 identify a trend in the data

3 determine which factors might help you explain the observed trend (e.g. mass, degree of branching, number and type of intermolecular forces).

3.2.5 Safety and hydrocarbons

Hydrocarbons are often used as fuels, so they need to be carefully stored, transported and handled. The alkanes from methane to octane are gases or volatile liquids under standard laboratory conditions. They readily vaporise to form a gas mixture above a liquid hydrocarbon mixture such as petrol. When a vehicle's petrol tank is being filled, you can notice this vapour as a haze around the petrol nozzle. The vapour is very combustible – the gases in the mixture have low **flashpoints**. The warnings at petrol stations about smoking and using mobile phones near fuels are to minimise ignition sources for these volatile gases.

Weak dispersion forces are the reason that smaller alkanes have lower boiling points and higher **volatility**. Their low flashpoints mean they can be readily ignited, even at lower temperatures during winter. In general, the lower the molecular weight, the greater the rate of evaporation and the higher the volatility. Boiling point can be a good indicator of volatility, which increases with increasing temperature.

In general, the higher the boiling point of a hydrocarbon, the higher its flashpoint. For example, solid waxes have higher boiling points and flashpoints than liquid hydrocarbons. Stronger dispersion forces make them less volatile. Liquids such as octane have weaker dispersion forces between the molecules and are more volatile, with lower flashpoints.

We need to consider several safety measures when storing and transferring hydrocarbons.

- Store gaseous hydrocarbons in sealed, air-tight cylinders and check them regularly for leaks.
- Store liquid fuels in clearly labelled metal containers. The containers should have small openings with close-fitting lids, such as those used to store lawnmower fuel.
- Store fuels in cool, well-ventilated spaces.
- Transfer liquid fuels (e.g. filling up a petrol lawnmower) outside in well-ventilated areas to prevent any build-up of combustible gases.
- Check the RiskAssess or HazChem codes, risks and safety measures associated with the use of a particular substance.
- Place fire extinguishers near fuel storage areas and check them regularly.
- Substitute one fuel for a safer one where possible.

3.2.6 Implications of hydrocarbon use

Hydrocarbons commonly occur in high concentrations in **fossil fuels**. Two of the simplest hydrocarbons, methane and ethane, are the primary components of natural gas, but hydrocarbons can also be extracted from crude oil. Crude petroleum (crude oil) is a mixture of straight, branched and cyclic hydrocarbons and can be refined to separate the individual hydrocarbons and produce the chemicals to make polymers. These hydrocarbons and their products have major economic significance. Petroleum products are a critical global commodity and contribute significantly to political instability in countries where petroleum is a major export.

Fractional distillation is the process used to separate a mixture of substances with different boiling points. It is used to refine crude oil, which is heated in a column like the one shown in Figure 3.32. The lighter hydrocarbons – methane, ethane, propane, butane and pentane – boil first (some are already gases at room temperature), and some of these are liquefied to separate them. The hydrocarbons with higher boiling points are then separated and processed into products such as petrochemicals (e.g. plastics) and fuels.

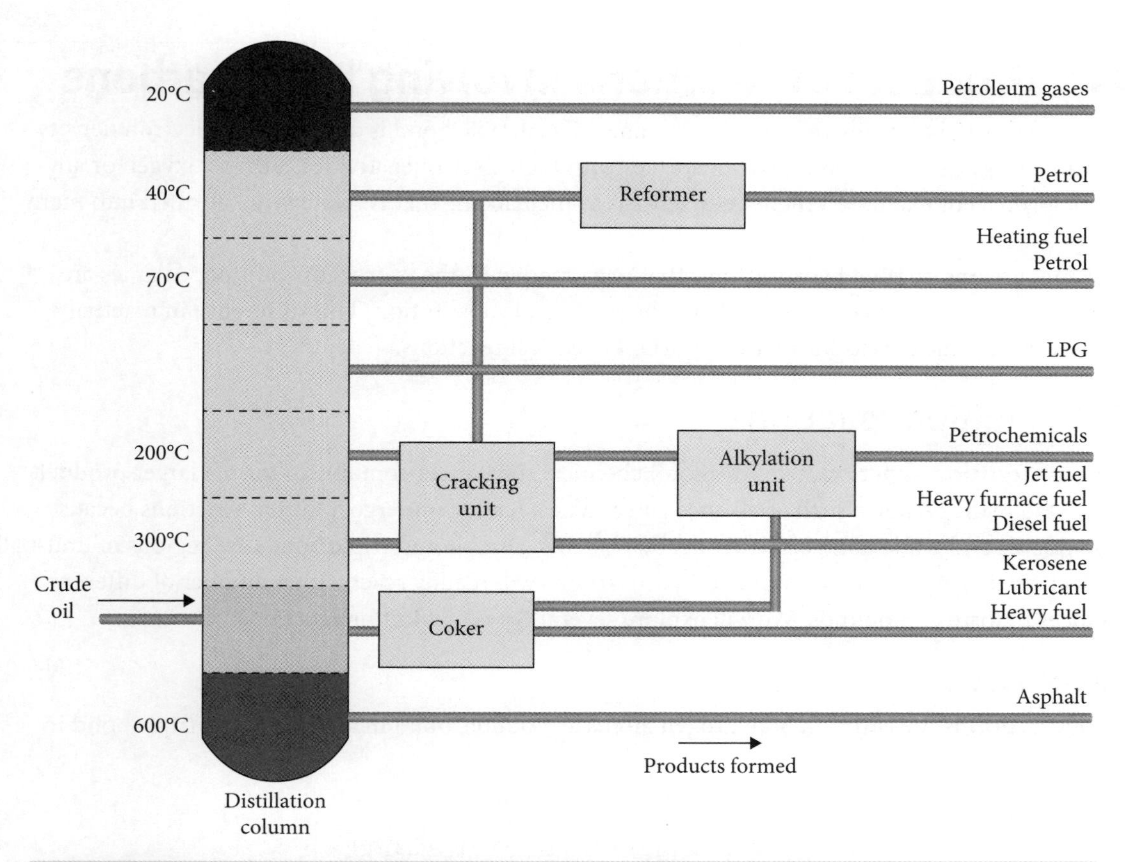

FIGURE 3.32 Fractional distillation of oil produces a range of hydrocarbon products.

Over many decades, we have become heavily reliant on products of the petrochemical industry, such as petrol and plastic, and on the employment that this industry has provided. The environmental costs of this reliance, such as climate change and plastics pollution, are significant and global. As awareness of these problems increased, research began into energy sources whose use might have fewer negative impacts on the environment, and new industries were established. For example, research into solar energy and wind energy led to the development of solar panels and wind turbines.

The use of any energy source will have environmental, economic and sociocultural drawbacks. Examples are the affordability of installing solar panels, the problems with the disposal of electric car batteries and community opposition to wind farms in certain areas.

When constructing a response relating to the implications of the use of hydrocarbons, make sure you distinguish between:

- *environmental implications*: extraction of fossil fuels (e.g. mining, fracking), combustion of fossil fuels (e.g. increased carbon particulates – carbon monoxide or carbon dioxide depending on oxygen availability during combustion), as well as other implications (e.g. sulfur in coal deposits and its link to acid rain) and the potential for waste (e.g. oil spills)
- *economic implications*: global distribution of fossil fuel deposits and their extraction, refinement and export, and the political implications of supply and demand
- *sociocultural implications*: countries that have thrived because of the fossil fuel industry, governments who have mandated changes to electricity generation, climate targets and how these can affect individuals and communities.

3.3 Products of reactions involving hydrocarbons

Alkenes are more chemically reactive than alkanes. The double bond is a site of high electron density and so alkenes can readily combine with species with high electronegativities, such as oxygen or any of the halogens. For example, ethene (ethylene) is a small alkene that is readily transformed into many useful products.

Alkenes are unsaturated hydrocarbons that react readily by the process of addition. Alkanes are saturated hydrocarbons that react slowly by the process of substitution. This difference in reactivity is used as the chemical basis for many important chemical reactions.

3.3.1 Addition reactions

During an **addition reaction**, two (or more) chemical substances combine to form a larger product. Unsaturated hydrocarbons, such as alkenes and alkynes, readily undergo addition reactions because carbon double and triple bonds are chemically unstable. Double and triple bonds are regions of unusually high electron density and so unsaturated hydrocarbons will readily react with a number of different substances to form compounds. We will explore several types of addition reactions.

Hydrogenation

Hydrogenation is the addition of hydrogen atoms at a double bond in an alkene or a triple bond in an alkyne.

For example, ethene can be hydrogenated to produce ethane:

$$CH_2{=}CH_2(g) + H_2(g) \rightarrow CH_3{-}CH_3(g)$$

A metal catalyst, such as one made of nickel or platinum, is often used to increase the rate of the reaction. Both reactants attach to the surface of the catalyst, breaking the H–H bond of H_2 and the C=C bond of ethene. Hydrogen atoms then react with ethene at the site where the carbon double bond was broken – they add 'across' it, one at each carbon.

Halogenation

Halogenation is the addition of halogen atoms (F, Cl, Br, I) across a carbon double bond in an alkene, or across a carbon triple bond in an alkyne, often to produce a dihaloalkane.

For example, ethene can be halogenated with chlorine to produce 1,2-dichloroethane:

$$CH_2{=}CH_2(g) + Cl_2(g) \rightarrow CH_2Cl{-}CH_2Cl(g)$$

This reaction doesn't require a catalyst.

Hydrohalogenation

Hydrohalogenation is the addition of one hydrogen atom and one halogen atom (F, Cl, Br, I) across a double bond in an alkene, or across a carbon triple bond in an alkyne, to produce a haloalkane.

For example, ethene can be hydrohalogenated with hydrogen chloride to produce chloroethane:

$$CH_2{=}CH_2(g) + HCl(g) \rightarrow CH_3{-}CH_2Cl(g)$$

Markovnikov's rule

When an alkene or alkyne is hydrogenated or halogenated, one atom bonds to each of the carbon atoms in the original double bond. How does this change for a hydrohalogen?

We can work this out using **Markovnikov's rule**, which says that hydrogen atoms preferentially add across a carbon double bond to the carbon with more hydrogen atoms.

Contrast the addition reactions of hydrogen fluoride with pent-1-ene and with pent-2-ene.

In pent-2-ene, the double bond is between the second and third carbon atom. Each of these carbon atoms has one hydrogen atom attached; hence, it is equally likely that the fluorine atom will attach to the second carbon (to produce 2-fluoropentane) as it will be to attach to the third carbon (to produce 3-fluoropentane). We would expect a roughly 1 : 1 ratio of these isomeric products:

$$CH_3CH{=}CHCH_2CH_3 + HF \rightarrow \underset{\text{2-fluoropentane}}{CH_3CHF(CH_2)_2CH_3} + \underset{\text{3-fluoropentane}}{CH_3CH_2CHFCH_2CH_3}$$

In pent-1-ene, the first carbon atom is attached to two hydrogen atoms and the second carbon atom is attached to only one. Markovnikov's rule says that the hydrogen bonds to the terminal carbon and the fluorine bonds to the second carbon, producing 2-fluoropentane. Although it is possible to produce 1-fluoropentane as well, this will be a substantially minor product:

$$CH_2{=}CH(CH_2)_2CH_3 + HF \rightarrow \underset{\substack{\text{2-fluoropentane} \\ \text{(major product)}}}{CH_3CHF(CH_2)_2CH_3} + \underset{\substack{\text{1-fluoropentane} \\ \text{(minor product)}}}{CFH_2(CH_2)_3CH_3}$$

Hydration reactions

During a **hydration reaction** of an alkene, a water molecule (or equivalent) adds across a carbon double bond, usually to produce an alcohol. Ethene can be hydrated to produce ethanol when heated with dilute sulfuric acid (H_2SO_4) catalyst:

$$CH_2{=}CH_2(g) + H_2O(l) \rightarrow CH_3{-}CH_2OH(l)$$

Markovnikov's rule can be used to determine how many different products may form and whether they are major or minor.

3.3.2 Substitution reactions

Under certain reaction conditions, one atom (or group of atoms such as a hydroxyl group, –OH) may be able to swap with another atom in an organic molecule in a **substitution reaction**. This results in more than one product. We will focus on substitution reactions of alkanes.

The low reactivity of alkanes means they do not readily react with halogens unless they are irradiated by ultraviolet light. When they do, they undergo substitution reactions. One atom in a halogen molecule substitutes for (replaces) a hydrogen atom to form a haloalkane. A second product will form from the combination of the substituted hydrogen with the remaining halogen atom.

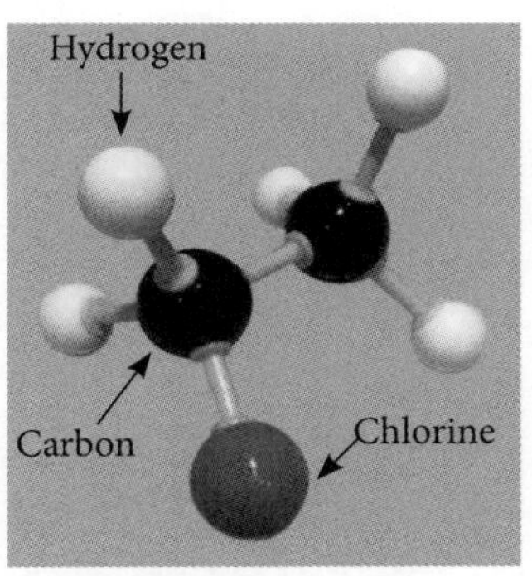

FIGURE 3.33 A molecular model of chloroethane

One example is chloroethane (Figure 3.33), which is produced from the reaction of ethane and chlorine in the presence of ultraviolet light:

$$CH_3{—}CH_3(g) + Cl_2(g) \xrightarrow{\text{UV light}} CH_3{—}CH_2Cl(g) + HCl(g)$$

In the presence of excess halogen, the substitution product (in this case chloroethane) may undergo additional substitution reactions, increasing the number of halogen atoms on the original hydrocarbon:

$$CH_3{—}CH_2Cl + Cl_2 \rightarrow CH_2Cl{—}CH_2Cl + HCl$$

3.3.3 Physical properties of organic compounds

Before we look at the properties of organic acids and bases that have the same functional group, it is worth revisiting their structures (Table 3.13).

TABLE 3.13 Comparing several important homologous series of organic compounds

Homologous series	General formula	Example	Structural formula	Description
Alcohol	$C_nH_{2n+2}O$	Pentan-1-ol	H H H H H H—C—C—C—C—C—O—H H H H H H	Carbon compound with hydroxyl functional group (–OH)
Aldehyde	$C_nH_{2n}O$	Hexanal	H H H H H O H—C—C—C—C—C—C H H H H H H	Carbon compound with carbonyl functional group (C=O) at end of parent chain
Ketone	$C_nH_{2n}O$	Heptan-3-one	H H H H O H—C—C—C—C—C H H H H H C H H C H H H	Carbon compound with carbonyl functional group (C=O) not at end of parent chain
Carboxylic acid	$C_nH_{2n+1}COOH$	Butanoic acid	H H H O H—C—C—C—C H H H O—H	Carbon compound with carboxyl functional group (–COOH)
Amine	$C_nH_{2n+1}NH_2$	Propan-1-amine	H H H H H—C—C—C—N H H H H	One or more atoms of hydrogen in ammonia molecule (NH_3) substituted with a carbon-containing group
Amide	$C_nH_{2n-1}ONH_2$	Octanamide	H H H H H H H O H H—C—C—C—C—C—C—C—C—N H H H H H H H H	Hydroxyl part (–OH) of –COOH in a carboxylic acid substituted with amino functional group (–NH_2 or –NR_2)

Alcohols

We have previously covered the alcohols so we will not cover them again in detail.

The high melting points and boiling points of alcohols relative to hydrocarbons of similar length can be explained by the hydrogen bonding between the polar –OH regions of adjacent molecules, as we discussed earlier. Hydrogen bonding between adjacent alcohol molecules is shown in Figure 3.34.

The hydroxyl group contributes to the aqueous solubility of short-chain alcohols and their relatively high melting and boiling points, and to their ability to act as solvents for polar and non-polar substances (the latter as a result of their hydrocarbon tails). As a general rule, organic compounds with the hydroxyl groups at the end of a chain (especially a long chain) have a lower solubility than those with a hydroxyl group at a non-terminal carbon atom or where there is some chain branching.

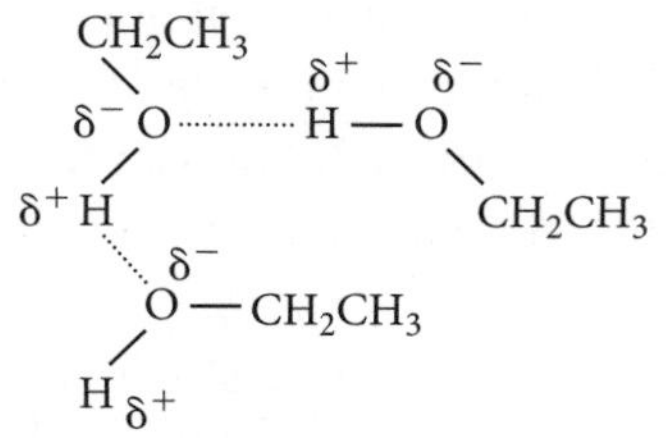

FIGURE 3.34 Physical properties of alcohols are related to the relatively strong hydrogen bonds between molecules.

Aldehydes and ketones

As we have discussed, both aldehydes and ketones contain a carbonyl group (C=O). The only difference between the compounds is whether the C=O bond is at the end of the molecule or not.

The aldehydes have the carbonyl bond at the end, as shown for propanal in Figure 3.35.

We can see in Figure 3.36 that propanone is an isomer of propanal – it has the same molecular formula. However, the carbonyl group is not located at the end of the molecule.

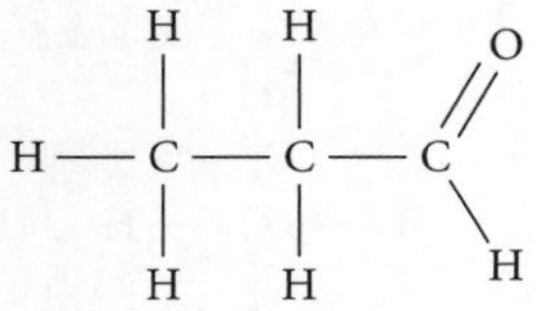

FIGURE 3.35 The structural formula of propanal

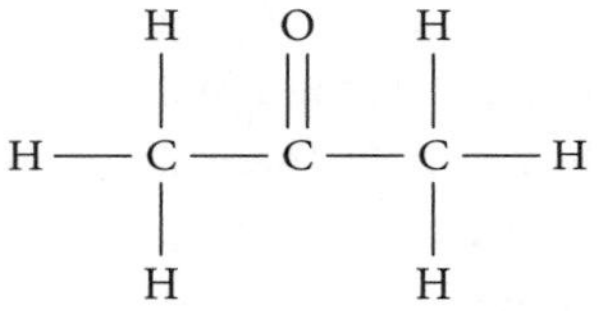

FIGURE 3.36 The structural formula of propanone

A carbonyl group of a ketone is more polar than the carbonyl group of an aldehyde, but only slightly. Hence a ketone has only a slightly higher boiling point than the aldehyde of the same carbon chain length (e.g. 48°C for propanal and 56°C for propanone). This means that aldehydes are slightly more volatile than ketones.

The carbonyl bond is polar, but there is no O–H bond in aldehydes and ketones, so there can only be dipole–dipole interactions between adjacent molecules. This explains the lower melting and boiling points for the aldehyde and ketone in Table 3.14, p. 113).

Carboxylic acids

Carboxylic acids are characterised by the –COOH functional group. The carbon of the carboxyl group has three bonds to two oxygen atoms, so the carboxyl group has to be at a terminal carbon. Carboxylic acids have quite a distinct aroma, which tells us something about their volatility. This is particularly true of butanoic (butyric) acid, which is shown in Figure 3.37.

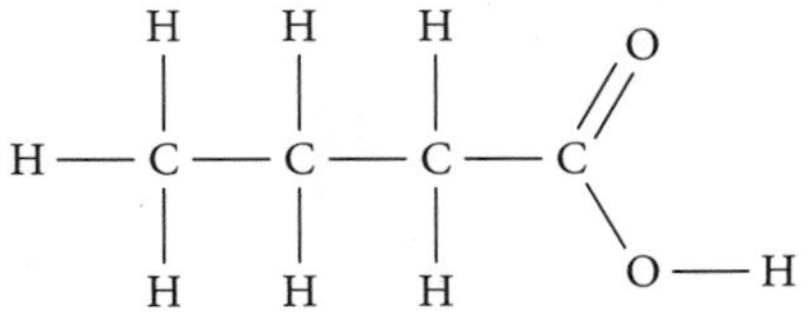

FIGURE 3.37 The structural formula of butanoic acid

9780170465281

Carboxylic acids have a non-polar region as well as C=O and C–O–H groups. Both of these groups have polar covalent bonds, and the C=O bond is particularly polar.

The oxygen atom in both the hydroxyl and carbonyl groups can be involved in hydrogen bonding with an adjacent molecule, which explains the higher melting and boiling points of carboxylic acids. It also accounts for the high solubility of the smaller carboxylic acids in water.

The ability of the –COOH group in carboxylic acids to be involved in two hydrogen bonds gives them even higher boiling points than those of similarly sized alcohols. Two hydrogen bonds can form between a pair of carboxylic acid molecules, as shown in Figure 3.38 for methanoic acid.

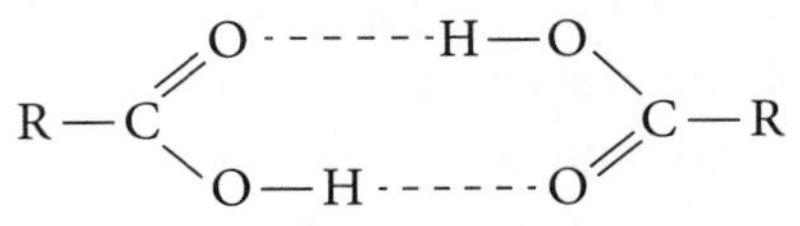

FIGURE 3.38 Hydrogen bonding between two molecules of a carboxylic acid (R is an alkyl group)

As previously mentioned, size is also important – larger molecules require additional energy for motion. As was the case for hydrocarbons, melting and boiling points are higher for longer carbon chains. This is because there are more electrons in each molecule and hence stronger dispersion forces between molecules.

Amines and amides

The presence of nitrogen in amines and amides affects both the chemical reactivity of the molecules and their physical properties. In an amine, the amine group (–NH_2) occupies the place where a hydrogen atom would be in an alkane of the same carbon chain length. So butane would become butan-1-amine (Figure 3.39) if the amine group were substituted for a hydrogen on a terminal carbon.

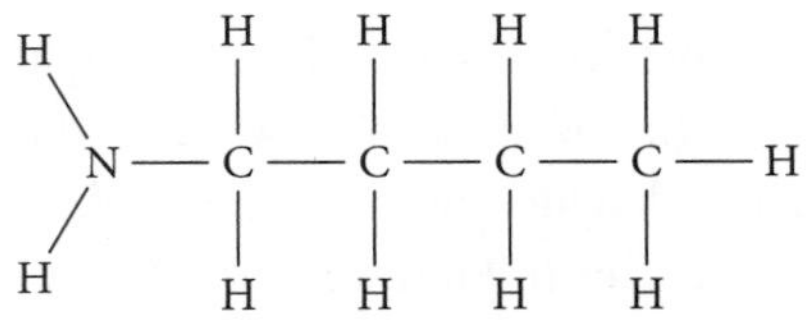

FIGURE 3.39 The structural formula of butan-1-amine

Just like the alcohols, there are primary, secondary and tertiary amines. Methylpropan-2-amine (Figure 3.40) is a tertiary amine.

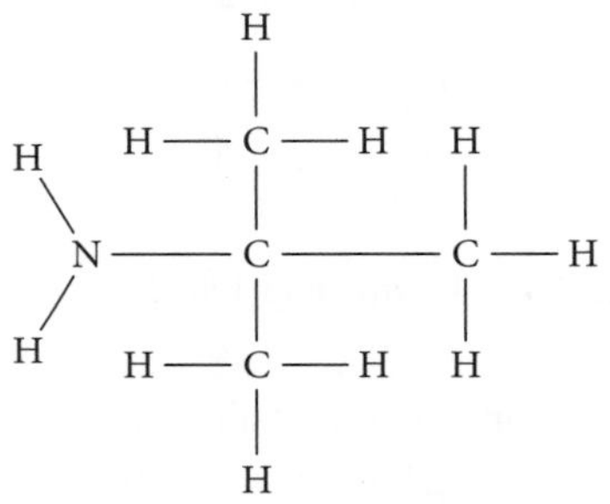

FIGURE 3.40 The structural formula of methylpropan-2-amine

Nitrogen atoms form polar covalent bonds with carbon atoms and strong polar bonds with hydrogen atoms. This means there can be hydrogen bonding between amine molecules. These hydrogen bonds are not as strong as those between alcohols, so an amine has lower melting and boiling points than an alcohol with the same carbon chain length.

The effect of hydrogen bonding is increased if a carbonyl group is also present. Pentanamide is one example, shown in Figure 3.41. This bonding involves a double bond in the carbonyl group and a single bond from carbon to nitrogen, so amide groups occur only on a terminal carbon.

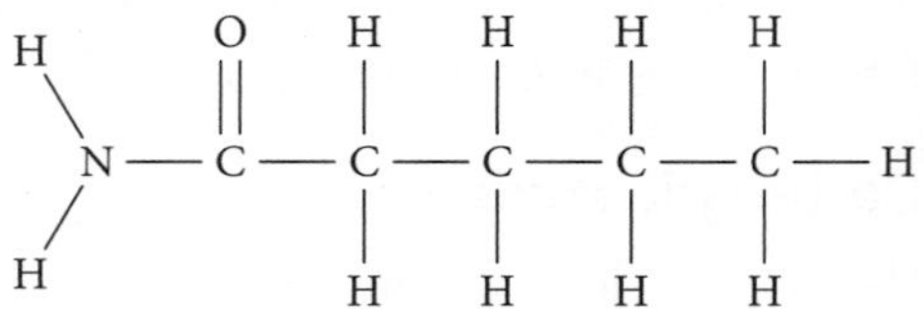

FIGURE 3.41 The structural formula of pentanamide

R groups

Figure 3.42 shows a common way to illustrate the characteristic structure of the members of a homologous series, in this case, carboxylic acids. The R group represents hydrogen or a carbon chain of unspecified length. For example, if the R group is hydrogen, the compound is methanoic acid (CHCOOH). A skeletal formula for carboxylic acids is shown in Figure 3.43.

Note
In HSC Chemistry exams, draw full (or graphic) structural formulae where possible.

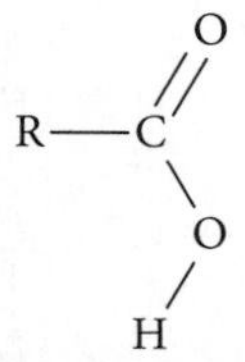

FIGURE 3.42 The structural formula of a carboxylic acid

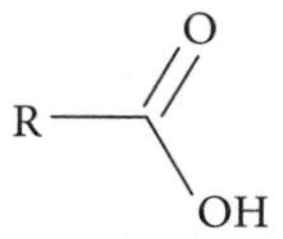

FIGURE 3.43 The skeletal formula of a carboxylic acid

3.3.4 Trends within and across homologous series

As we discuss organic acids and bases with different functional groups, Table 3.14 will help us compare their properties within and across homologous series. We can compare the properties considering only the functional groups because the molar masses of the compounds in the table are within 2 $g\,mol^{-1}$ of each other.

TABLE 3.14 Comparing properties of a range of organic acids and bases

Compound	Homologous series	Molar mass ($g\,mol^{-1}$)	Melting point (°C)	Boiling point (°C)
1-Butanol	Alcohol (primary)	74.1	−89.8	117.7
2-Butanol	Alcohol (secondary)	74.1	−114.7	99.5
2-Methyl-2-propanol	Alcohol (tertiary)	74.1	25.7	82.6
1-Butylamine	Amine	73.1	−49.1	77.9
Propanoic acid	Carboxylic acid	74.1	−20.7	140.8
Propanamide	Amide	73.1	77.5	213
Butanal	Aldehyde	72.1	−99	74.8
Butanone	Ketone	72.1	−86.6	79.6

Property trends within homologous series

Melting point

A solid melts when there is sufficient energy to overcome intermolecular forces between molecules. The stronger the bonds, the more energy will be required. For carboxylic acids, amides and tertiary alcohols, hydrogen bonding between molecules is very strong, which explains their higher melting points, as shown in Table 3.14. Amines and amides contain nitrogen atoms, which are highly electronegative and form polar bonds with carbon atoms. However, the C–N bond is not as strongly polar as the C–O bond, so hydrogen bonding for amines and amides is weaker than that for an alcohol with the same carbon chain length. Aldehydes and ketones have slightly lower melting points because there are no hydrogen bonds, only dipole–dipole interactions.

Boiling point

A liquid boils when there is sufficient energy to overcome intermolecular forces between molecules. The stronger the bonds, the more energy is required. For carboxylic acids and amides, hydrogen bonding between molecules is very strong, which explains their higher boiling points, as shown in Table 3.14.

In amines, the presence of the nitrogen atom provides opportunities for hydrogen bonding, and this is even more so for amides. Notice in Table 3.14 that propanamide has relatively high melting and boiling points. This is because the functional group contains three polar bonds: C=O, C–N and N–H. This makes additional opportunities for dipole–dipole interactions.

Aldehydes and ketones have slightly lower boiling points because there are no hydrogen bonds, only dipole–dipole interactions.

Density

Density relates to how tightly packed molecules are in the solid or liquid phase. Although stronger intermolecular forces (e.g. hydrogen bonds) hold molecules closer together than do weaker dispersion forces, side branches affect the ability of molecules to pack together, affecting density.

Comparing primary, secondary and tertiary alcohols of similar molar mass in Figures 3.44–3.46, we can see that the tertiary alcohol (Figure 3.46) has a side branch. Consequently, it has a lower density than the unbranched primary and secondary alcohols (Figures 3.44 and 3.45).

```
    H   H   H   H
    |   |   |   |
H — C — C — C — C — O — H
    |   |   |   |
    H   H   H   H
```

Density: 0.802 $g\,cm^{-3}$

FIGURE 3.44 The structural formula of butan-1-ol (primary alcohol)

```
            H
            |
    H   H   O   H
    |   |   |   |
H — C — C — C — C — H
    |   |   |   |
    H   H   H   H
```

Density: 0.806 $g\,cm^{-3}$

FIGURE 3.45 The structural formula of butan-2-ol (secondary alcohol)

```
        H
        |
    H   O   H
    |   |   |
H — C — C — C — H
    |   |   |
    H   C   H
        |
    H — C — H
        |
        H
```

Density: 0.741 $g\,cm^{-3}$

FIGURE 3.46 The structural formula of 2-methylpropan-2-ol (tertiary alcohol)

Property trends across homologous series

Solubility in water

A general solubility rule is that like dissolves like. Molecules with some polarity will interact with polar water molecules and will show some ability to mix. This is particularly so for acids, alcohols and amines, and less so for aldehydes and ketones. However, unless a functional group is present several times in a molecule, a hydrocarbon chain will not attract water molecules. This means the solubility of polar molecules in water is higher for molecules with shorter carbon chains.

Solubility in non-polar solvents

Molecules with some polarity will interact with other polar molecules and will show some ability to mix. Likewise, those with long non-polar regions can interact with other non-polar molecules. Polar solvents can dissolve acids and alcohols, but aldehydes and ketones are less soluble here. A molecule with a carbon–oxygen and oxygen–hydrogen bonds attracts more polar molecules than a molecule with a single carbonyl bond. In addition, a long hydrocarbon tail does not attract polar molecules but attracts non-polar molecules. This means the solubility of polar molecules in non-polar solvents is higher for molecules with longer carbon chains. Even if a molecule has a polar end, it will mix with a non-polar solvent if its carbon 'tail' is long enough.

3.4 Alcohols

Alcohols are an extremely important group of organic compounds. They are used as solvents, reactants in the production of esters and polyesters, as fuel sources and in alcoholic beverages.

You may recall that a functional group is either a type of bond, atom, or group of atoms that are found in all members of a homologous series, and that they help determine the chemical properties of each member of the series. For alcohols, the functional group is the hydroxyl (–OH) group.

3.4.1 Primary, secondary and tertiary alcohols

The general identifier for an alcohol is the presence of the hydroxyl (–OH) group. In its simplest forms, the hydroxyl group substitutes for one or more hydrogen atoms in an alkane to produce an alcohol. In the hydroxyl group, both the C–O and O–H bonds are polar. This affects the physical and chemical properties of alcohols.

Near the beginning of this chapter, we looked at the steps for naming alcohols. Before looking further at the properties of alcohols, we will further discuss the three main types of alcohols.

Primary alcohols

Primary alcohols are the most well known and the simplest of the alcohols to draw and explain. In a primary alcohol, the hydroxyl group is attached to a terminal carbon. This is shown in the model in Figure 3.47.

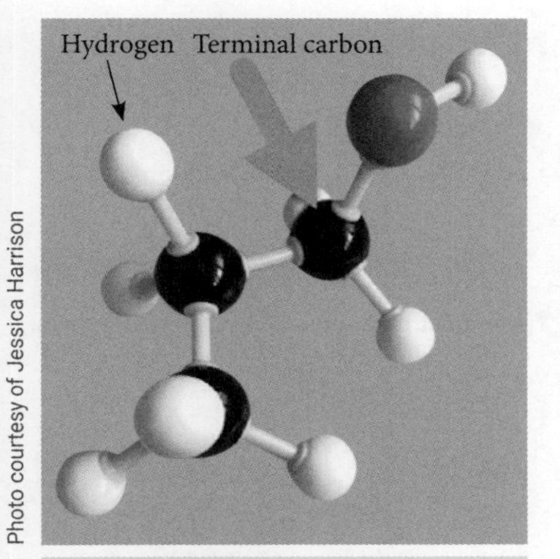

Photo courtesy of Jessica Harrison

FIGURE 3.47 A molecular model of propan-1-ol. The terminal carbon is identified by the arrow.

The terminal carbon is only attached to one other carbon atom; hence, this compound is a primary alcohol. There are three carbon atoms, and the hydroxyl group is attached to the terminal carbon, so this compound is propan-1-ol (or 1-propanol). The structural formula is shown in Figure 3.48.

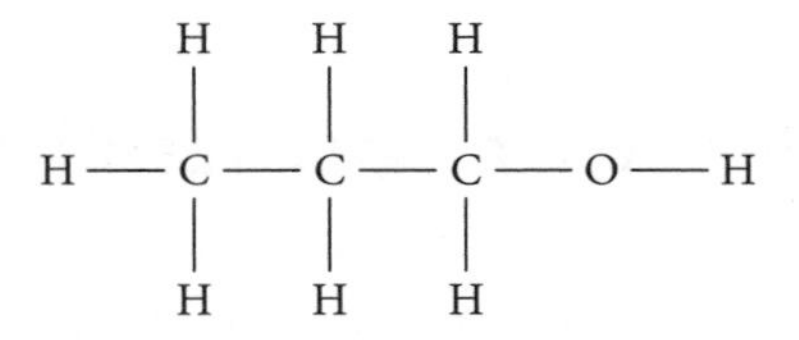

FIGURE 3.48 The structural formula of propan-1-ol

Secondary alcohols

A secondary alcohol has a hydroxyl group on a non-terminal carbon atom. This is shown in the model of propan-2-ol in Figure 3.49. The carbon atom is attached to two other carbon atoms; hence, the compound is a secondary alcohol. The hydroxyl group is attached to the second carbon (counting from either end), so this alcohol is propan-2-ol. Its structural formula is shown in Figure 3.50.

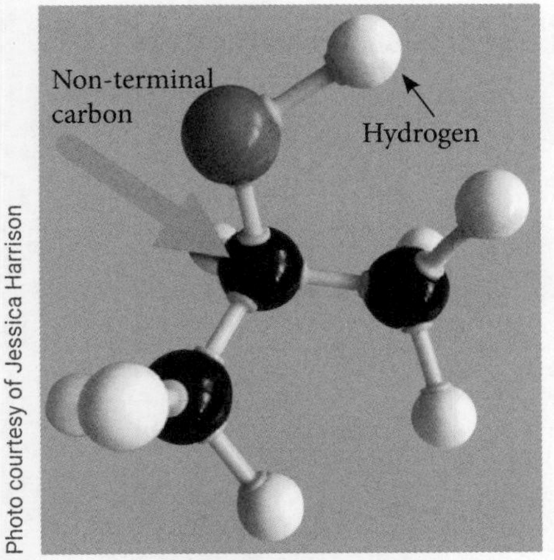

Photo courtesy of Jessica Harrison

FIGURE 3.49 A molecular model of propan-2-ol. The non-terminal carbon is identified by the arrow.

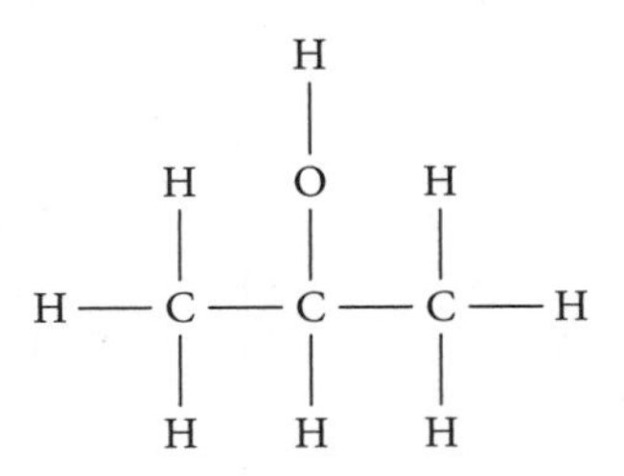

FIGURE 3.50 The structural formula of propan-2-ol

Tertiary alcohols

In a tertiary alcohol, the hydroxyl group is attached to a non-terminal carbon that also has a side chain attached to it.

The non-terminal carbon of the parent chain is shown in the model in Figure 3.51. This carbon is attached to three other carbon atoms; hence, the compound is a tertiary alcohol.

This alcohol has three carbon atoms in the parent chain, and the hydroxyl group is attached to the second carbon (counting from either end), so part of this compound's name is propan-2-ol. However, we need to account for the methyl group side chain. We do not need to number this methyl group – if it was located on either of the other two carbons, we would have an isomer of butanol (butan-2-ol). Therefore, the full name of this compound is methylpropan-2-ol (or methyl-2-propanol). Its structural formula is shown in Figure 3.52.

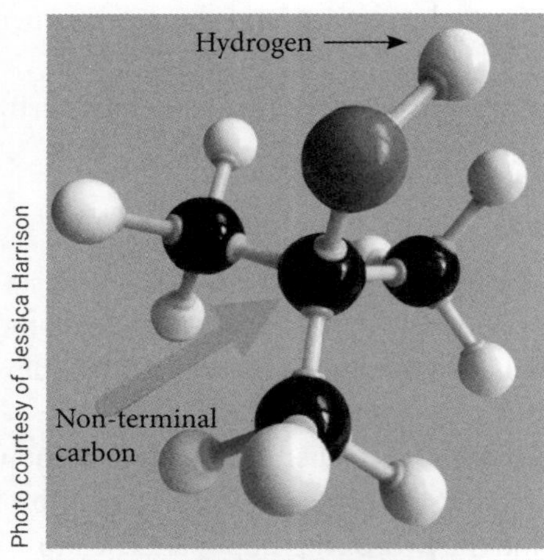

FIGURE 3.51 A molecular model of methylpropan-2-ol. The non-terminal carbon is identified by the arrow.

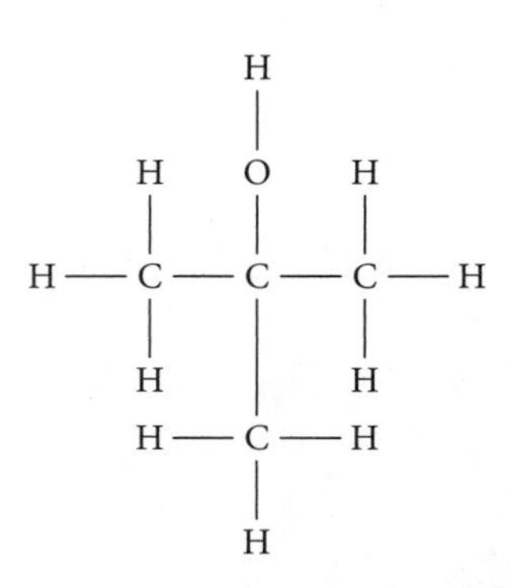

FIGURE 3.52 The structural formula of methylpropan-2-ol

3.4.2 Patterns in properties of alcohols

To see that the behaviour of substances with a common functional group is part of a pattern, we can look again at boiling points. Table 3.15 shows the boiling points for three homologous series.

The trend in boiling point for the alcohols is the same trend we noticed for the alkane and alkenes: the boiling points are higher for alcohols with longer carbon chains (greater molecular weights).

The impact of the hydroxyl group is more compelling when we glance across the rows of the table. For example, the two-carbon alkane (ethane) has a boiling point of −89°C; however, the two-carbon alcohol (ethanol) has a boiling point of 78°C. Oxygen atoms are larger than hydrogen atoms, but this is not sufficient to explain these marked differences.

When discussing hydrocarbons, we explained the boiling point trend in terms of the dispersion forces between molecules in a liquid, which must be overcome for the molecules to change state to become gases. To explain the pattern of boiling points for alcohols (as well as their properties such as solubility in water), we need to consider **intermolecular forces**.

TABLE 3.15 The boiling points for some alkanes, alkenes and alcohols

Number of carbon atoms	Boiling point (°C)		
	Alkanes	Alkenes	Alcohols
1	−164		65
2	−89	−104	78
3	−42	−48	97
4	−0.5	−6	118
5	36	30	138
6	69	64	157
7	98	94	178
8	125	121	195
9	151	146	214
10	174	172	229

3.4.3 Explaining the properties of alcohols

Ethanol is the least toxic of the alcohols (although it is poisonous in large amounts), which makes it suitable for use in industry and consumer products. In these products, ethanol is the second most important solvent after water. It is commonly used for this purpose in cosmetics (e.g. perfumes), food colourings and flavourings (e.g. vanilla essence), medicinal preparations (e.g. antiseptics and rubbing alcohol), as well as in some cleaning agents.

We can explain the properties of ethanol and other alcohols by looking at their intramolecular bonds and intermolecular forces.

The presence of the hydroxyl group changes some of the physical and chemical properties of the alcohols, primarily because of the polarity of the C–O and O–H bonds. This is particularly the case when they are interacting with other alcohols or water, which also have O–H groups. These strong interactions make primary alcohols highly water soluble.

Intramolecular bonds in alcohols

Intramolecular bonds in alcohols are covalent bonds. C–C and C–H are non-polar covalent bonds, and C–O and O–H bonds are polar covalent bonds. The polarity is due to differences in electronegativity between oxygen and carbon and between oxygen and hydrogen. If there is more than one hydroxyl group, this effect is increased, and the geometry of the molecule and the accessibility of the hydroxyl group have an effect. For example, a hydroxyl group's position at the terminal carbon of a primary alcohol makes it more likely to interact with other substances than a hydroxyl group of a tertiary alcohol.

Intermolecular forces in alcohols

Alcohols have a dual nature. One part of their structure contains only C–C or C–H covalent bonds. These are non-polar and hence can only interact with adjacent molecules by dispersion forces.

The other part of their structure is polar because of the polarity of the C–O and O–H covalent bonds. The relatively high electronegativity of oxygen allows alcohols to be part of dipole–dipole interactions as well as stronger hydrogen bonding with other molecules.

For example, one end of ethanol (CH_3CH_2OH) has non-polar C–H bonds, and the other end has a hydroxyl (–OH) group. The hydroxyl end of the ethanol molecule attracts polar and ionic substances. The ethyl (C_2H_5) group in ethanol is non-polar, so this end attracts non-polar substances. This enables ethanol to dissolve both polar and non-polar substances to some extent.

The δ^+ and δ^- annotations on Figure 3.53 indicate polarity in the molecule. The charged regions are not ionic charges, but they are permanent, and they attract oppositely charged regions. They can form dipole–dipole interactions with other molecules and hydrogen bonds, as shown by dotted lines in Figure 3.53. These intermolecular forces need to be overcome if the molecules are to be separate in the gaseous state.

Some important property trends in the alcohols are higher boiling point and lower solubility for longer carbon chain lengths. The higher boiling points for longer chain lengths are due to an increase in dispersion forces between molecules. The lower water solubility for longer carbon chain lengths (larger molecular weights) is due to the dominance of dispersion forces over hydrogen bonding. The small primary alcohols (e.g. methanol, ethanol and propan-1-ol) are readily soluble in water. However, octan-1-ol and longer chain primary alcohols are insoluble.

A secondary or tertiary alcohol with an identical molecular weight to a primary alcohol (i.e. a positional or chain isomer) will have a lower boiling point because hydrogen bonding between the molecules is weaker. This is usually a result of the relative accessibility of the C–O bond and O–H bond between molecules to form the hydrogen bonds. Bond accessibility at a terminal carbon of a primary alcohol is much better than it is at a carbon surrounded by three other carbons of a tertiary alcohol.

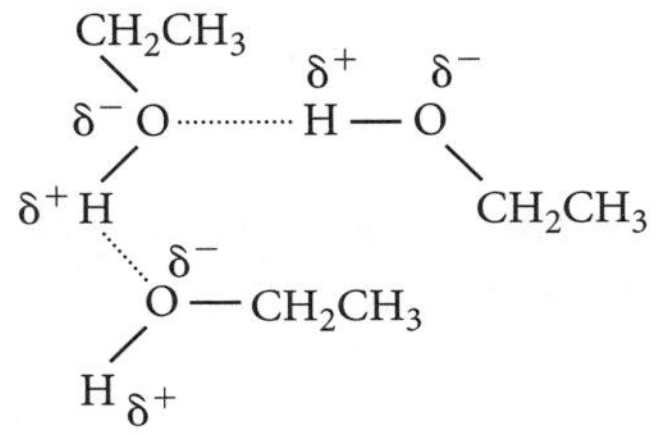

FIGURE 3.53 Hydrogen bonding in alcohols

9780170465281

3.4.4 Enthalpy of combustion of alcohols

The molar heat of combustion refers to the energy released when 1 mole of a compound undergoes complete combustion with oxygen at a constant pressure of exactly one atmosphere (100 kPa) and at 25°C (298 K), and the final products are carbon dioxide gas and liquid water.

Values for heats of combustion are positive numbers, and values for enthalpy changes of combustion reactions (ΔH) are negative numbers. This is because combustion reactions are always exothermic. Heats of combustion are expressed in kilojoules per mole (kJ mol^{-1}).

One accepted value for the molar heat of combustion of ethanol is 1360 kJ mol^{-1}. The reaction for the complete combustion of ethanol is:

$$C_2H_6O(l) + 3O_2(g) \rightarrow 2CO_2(g) + 3H_2O(l) \qquad \Delta H = -1360\,kJ\,mol^{-1}$$

To calculate the enthalpy of combustion of an alcohol, we can substitute experimental results into the following formulae to determine the enthalpy change:

$$\Delta H = -q \qquad \text{and} \qquad q = mC\Delta T$$

where:

ΔH = enthalpy change of the combustion reaction (J)
q = heat energy (J)
m = mass of water (kg)
C = specific heat capacity for water, as per the formulae sheet: $4.18 \times 10^3\,J\,K^{-1}\,kg^{-1}$
ΔT = change in temperature (K), which will be equal to Δ°C

The most important thing to remember is that the variables of 'mcat' ($mC\Delta T$) all relate to the same substance – water in this case. Do not use any value relating to the alcohol in your $mC\Delta T$ formula.

Let's say we heat 200 mL of water from 21°C to 56°C:

$$\begin{aligned} q &= mC\Delta T \\ &= 0.2\,kg \times 4.18 \times 10^3 \times (56 - 21) \\ &= 29\,260\,J \\ &= 29.26\,kJ \end{aligned}$$

Note
56 – 21 (°C) = 329 – 294 (K)

It is a positive value because the energy has been absorbed by the water produced in the combustion reaction, increasing the temperature of the system.

The next step is to determine the number of moles of the fuel combusted. Let's say the initial mass (burner and ethanol) is 190.16 g and the final mass is 186.84 g.

Change in mass:

$$190.16 - 186.84 = 3.32\,g \text{ of ethanol combusted}$$

Molar mass of ethanol:

$$2 \times 12.01 + 6 \times 1.008 + 1 \times 16 = 46.068\,g\,mol^{-1}$$

$$\begin{aligned} n &= \frac{m}{MM} \\ &= \frac{3.32}{46.068} \\ &= 0.072 \text{ mol of ethanol combusted} \end{aligned}$$

To calculate the molar heat of combustion, divide the heat energy (q) by the number of moles of fuel combusted:

$$\begin{aligned} \Delta H &= -\frac{q}{n} \\ &= -\frac{29.26}{0.072} \\ &= -406.0\,kJ\,mol^{-1} \end{aligned}$$

In these types of calculations, we are assuming that:

- *complete combustion occurs*: be aware that soot indicates incomplete combustion, which will give a calculated value less than the real value
- *all of the energy produced in the combustion reaction was used to heat the water*: in reality, energy can be 'lost' to the air or container, during the time between lighting the fuel and starting the heating, and the time between taking the heat source away and stopping the combustion reaction
- *measurements are accurate*: this includes the number of decimal places recorded for the mass of the spirit burner before and after the reaction, the temperature and the volume of water used. (Were measurements done with a beaker or more accurately with a measuring cylinder or graduated pipette?)

Due to heat loss to surroundings, the experimental value of the molar heat of combustion of ethanol will often be significantly lower than the accepted value. In our example, we calculated $-406\,\text{kJ mol}^{-1}$, compared to the theoretical value of $-1360\,\text{kJ mol}^{-1}$. The enthalpy change is now negative, indicating that the combustion reaction released energy into the surroundings, which we used to heat the water.

If we were to repeat the same experiment for the same or another alcohol, such as methanol, we may find the differences between the experimental value and accepted value are reasonably consistent. This difference could be used to calibrate the experimental results for ethanol, recognising some of the factors outside of our control.

To reliably compare the enthalpy of combustion for a range of alcohols, we would need to carry out this procedure several times, under controlled conditions.

3.4.5 Reactions involving alcohols

Alcohols are involved in several important types of reactions. Alcohols can undergo combustion, dehydration and substitution reactions with a hydrogen halide and oxidation reactions.

Combustion reactions of alcohols

In the previous section we looked at the enthalpy of combustion alcohols. The type of combustion depends on the amount of oxygen present.

For example, complete combustion of ethanol occurs in the presence of abundant oxygen to produce carbon dioxide and water:

$$C_2H_5OH(l) + 3O_2(g) \rightarrow 2CO_2(g) + 3H_2O(l) \qquad \Delta H = -1360\,\text{kJ mol}^{-1}$$

If there is insufficient oxygen (which often happens when using spirit burners), ethanol (or any fuel) will not combust completely, and carbon (and/or carbon monoxide) will be produced:

$$C_2H_5OH(l) + O_2(g) \rightarrow 2C(g) + 3H_2O(l) \qquad \Delta H < -1360\,\text{kJ mol}^{-1}$$

For longer carbon chain lengths, the tendency towards incomplete combustion increases. It is worth considering the trend in molar enthalpy of combustion for the alcohols and comparing the enthalpy change per gram of fuel, as shown in Table 3.16.

The trends are once again clear. For primary alcohols, the amount of energy released per mole of fuel and per gram of fuel is greater for greater values of molar mass.

TABLE 3.16 Comparing enthalpy of combustion for different alcohols

Alcohol	Molar enthalpy of combustion (kJ mol^{-1})	Molar mass (g mol^{-1})	Enthalpy of combustion (J) per gram
Methanol	−726	32.042	22.7
Ethanol	−1360	46.068	29.5
Propan-1-ol	−2021	60.094	33.6
Butan-1-ol	−2676	74.121	36.1
Pentan-1-ol	−3329	88.146	37.8
Hexan-1-ol	−3984	102.172	39.0
Heptan-1-ol	−4637	116.198	39.9
Octan-1-ol	−5294	130.224	40.7

Dehydration of alcohols

Ethanol can be heated with a concentrated sulfuric acid (H_2SO_4) or phosphoric acid (H_3PO_4) catalyst to give a dehydrated product (ethene) and water. Insufficient heat may lead to the formation of ethers (which are beyond the scope of this course).

The conditions required to dehydrate alcohols are different for the three structural types.

- Tertiary alcohols dehydrate quickly at lower temperatures.
- Secondary alcohols react more slowly, often with some heating.
- Primary alcohols only dehydrate in this way with very strong heating.

This chemical equation represents the **dehydration** of propan-2-ol to produce propene and water:

$$C_3H_7OH(l) \rightarrow C_3H_6(g) + H_2O(l)$$

We can also examine the structural change, as shown in Figure 3.54.

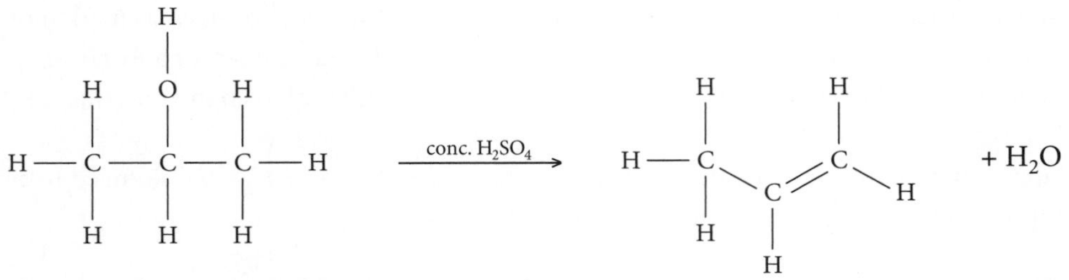

FIGURE 3.54 The dehydration of propan-2-ol to propene

Structural formulae make it easier to see that the –OH group at the central carbon atom combines with a hydrogen atom at a terminal carbon to form a water molecule. A double bond forms between the carbon atoms, producing propene. The location of the hydroxyl group in some secondary alcohols can allow two possible products. For example, the dehydration of butan-2-ol could produce water, as well as but-1-ene or but-2-ene.

Substitution reactions of alcohols

Alcohols can react with hydrogen halides to form an alkyl halide and water. This happens when a halogen substitutes for the hydroxyl group; hence, these reactions are called substitution reactions.

Consider the equation representing the substitution reaction of hydrogen bromide and propan-2-ol:

$$C_3H_7OH(l) + HBr(aq) \rightarrow C_3H_7Br(l) + H_2O(l)$$

Again, we can see this more easily by looking at the structural formula (Figure 3.55).

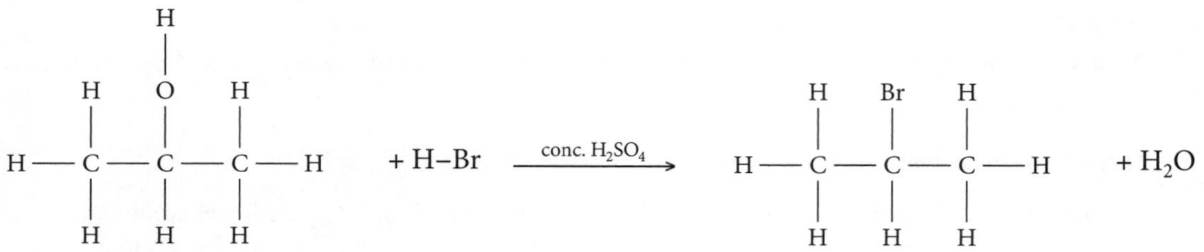

FIGURE 3.55 The substitution reaction of propan-2-ol to 2-bromopropane

Tertiary alcohols are the most reactive in substitution reactions, then secondary and primary alcohols. For the halogens, HI is the most reactive and HF the least reactive.

Oxidation reactions of alcohols

Primary and secondary alcohols readily react by oxidising agents. Tertiary alcohols do not react. The usual reagent for the oxidation of alcohols is an acidified solution of potassium permanganate ($KMnO_4$) or potassium dichromate ($K_2Cr_2O_7$).

Primary alcohols are oxidised initially to aldehydes and then to carboxylic acids. Secondary alcohols are oxidised to ketones. Primary alcohols can be converted directly into carboxylic acids when strongly heated with acidified $KMnO_4$ solution. At low temperatures, or without excess $KMnO_4$, the aldehyde will be produced as an intermediate product. The simplified reactions are written as:

$$CH_3CH_2OH + 2[O] \text{ (as hot acidified } KMnO_4) \rightarrow CH_3COOH + H_2O$$

$$CH_3CH_2OH + [O] \text{ (as cold acidified } KMnO_4) \rightarrow CH_3CHO + H_2O$$

Because these are true oxidation reactions, we can write them as redox couples. We shall investigate this further in both this module and in Module 8. Methanol can further oxidise from methanoic acid to CO_2 and H_2O under strong oxidising conditions.

3.4.6 Production of alcohols

Alcohols can be synthesised in a number of different ways. Methanol is synthesised industrially by the oxidation of methane or in a synthesis reaction involving carbon monoxide and hydrogen in the presence of a catalyst. The reactions we will focus on here are substitution reactions, hydration reactions and **fermentation**.

Synthesis of alcohols by substitution reactions

The reaction of water and a haloalkane is a substitution reaction. The hydroxyl group replaces the halogen to form an alcohol. This is because the carbon–halogen bond is highly polar and unstable compared with a carbon–hydrogen bond. The carbon–halogen bond is more stable for halogens with smaller atomic masses; for example, a fluoroalkane will not undergo substitution to form an alcohol.

The chemical equation in Figure 3.56 represents the reaction between 2-bromobutane and water to produce butan-2-ol and hydrogen bromide.

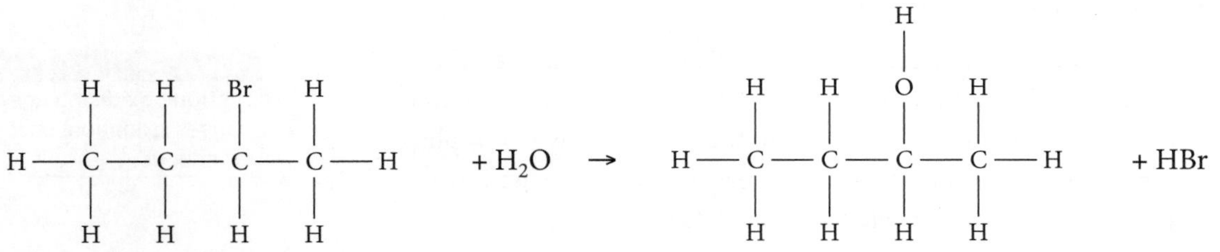

```
                                                      H
                                                      |
    H   H   Br  H                            H   H    O   H
    |   |   |   |                            |   |    |   |
H — C — C — C — C — H     + H2O    →     H — C — C —  C — C — H     + HBr
    |   |   |   |                            |   |    |   |
    H   H   H   H                            H   H    H   H
```

FIGURE 3.56 The substitution reaction of 2-bromobutane to produce butan-2-ol

As with alcohols, halogenated alkanes can be primary, secondary or tertiary. Tertiary haloalkanes are more likely than secondary and primary haloalkanes to react in substitution reactions.

You may have noticed that this reaction is the reverse of a similar reaction: the substitution reaction of hydrogen bromide and propan-2-ol (Figure 3.55). This is an example of the importance of reaction conditions to product formation. An acidic solution favours the substitution of the –OH group for the halogen, and a basic solution favours the substitution of the halogen for the hydroxyl group. This reaction is often carried out at higher temperatures and may involve reflux apparatus (a vertical condenser).

Synthesis of ethanol by hydration of an alkene

Ethanol, and many other alcohols, can be synthesised by hydration of an alkene. For example, ethene can be hydrated to produce ethanol when heated with a dilute sulfuric acid (H_2SO_4) catalyst:

$$C_2H_4(g) + H_2O(g) \xrightarrow{\text{dilute } H_2SO_4} C_2H_5OH(g)$$

The hydration of ethene is an example of an addition reaction. The double bond of ethene is electron rich, so it is liable to attack by any chemical species that is poor in electrons. The hydrogen ion (H^+) is the simplest example of such a species, and for this reason strong acids are used to hydrate ethene. A dilute acid solution (which has a higher proportion of water than a concentrated acid solution does) favours formation of the alcohol.

Synthesis of ethanol by fermentation

Fermentation is an anaerobic metabolic process that occurs in micro-organisms such as fungi and bacteria. Energy is produced when carbohydrates are broken down into simpler molecules.

Fermentation was one of the first chemical processes to have been utilised (at least 5500 years ago), but it wasn't until 1939 that the complex biochemistry of this deceptively simple reaction was worked out.

Fermentation is still a vital process in industries such as brewing, baking and pharmaceuticals. Ethanol is most commonly prepared by fermentation of glucose. This reaction is facilitated by yeast:

$$C_6H_{12}O_6(aq) \xrightarrow{\text{yeast}} 2C_2H_5OH(aq) + 2CO_2(g)$$

The fermentation of sugars into ethanol is promoted by:

- a catalyst (present in an organism such as yeast)
- absence of oxygen (to avoid oxidation of ethanol to ethanoic acid)
- warm conditions (about 30°C; exact temperature depends on the type of yeast used)
- slightly acidic conditions (pH of about 6.4).

Industrial fermentation chambers cannot usually produce solutions with concentrations greater than 15% alcohol v/v because the yeast would not survive. Distillation is used to produce alcoholic beverages with higher percentages of alcohol.

3.4.7 Oxidation of alcohols

To oxidise an alcohol, we need a strong oxidising agent, such as potassium permanganate or potassium dichromate. The oxidation and reduction reactions happen in acidified solution. In the following two examples, potassium is a spectator ion (it does not react). The two reduction half-equations (which are provided on the data sheet) are:

Note
A strong oxidising agent *causes* oxidation, so it is reduced.

$$8H^+(aq) + MnO_4^-(aq) + 5e^- \rightarrow 4H_2O(l) + Mn^{2+}(aq)$$

$$14H^+(aq) + Cr_2O_7^{2-}(aq) + 6e^- \rightarrow 7H_2O(l) + 2Cr^{3+}(aq)$$

Organic products and reactants are usually colourless and sometimes odourless. In the absence of a distinct odour, the only way we can tell the reaction has taken place is to note the colour change in the oxidising agent:

- Permanganate ions (purple) are reduced to manganese ions (colourless).
- Dichromate ions (orange) are reduced to chromium ions (green).

Oxidation reactions can be used to help identify whether an alcohol is primary, secondary or tertiary.

Oxidation of primary alcohols

Depending on the quantity of oxidising agent, primary alcohols may go through more than one oxidation step.

The first step is the formation of an aldehyde intermediate. Let's look at the reaction of ethanol with permanganate ion (as the oxidising agent) under acidified conditions to produce ethanal:

$$5C_2H_5OH(aq) + 6H^+(aq) + 2MnO_4^-(aq) \rightarrow 5C_2H_4O(aq) + 2Mn^{2+}(aq) + 8H_2O(l)$$

The second step is the formation of a carboxylic acid from the aldehyde intermediate in acid solution. Ethanal is oxidised by the permanganate ion:

$$5C_2H_4O(aq) + 6H^+(aq) + 2MnO_4^-(aq) \rightarrow 5CH_3COOH(aq) + 2Mn^{2+}(aq) + 3H_2O(l)$$

If the primary alcohol was methanol, this could further oxidise to carbon dioxide and water.

Oxidation of secondary alcohols

Secondary alcohols can be oxidised by the same oxidising agents to form a ketone. Let's look at propan-2-ol with the permanganate ion as the oxidising agent:

$$5C_3H_7OH(aq) + 6H^+(aq) + 2MnO_4^-(aq) \rightarrow 5C_3H_6O(aq) + 2Mn^{2+}(aq) + 8H_2O(l)$$

Oxidation of tertiary alcohols

Tertiary alcohols do not undergo oxidation. This is a very useful negative test to identify them.

3.4.8 Biofuels

Our reliance on fossil fuels (those fractions extracted from coal, crude oil and natural gas) has major economic and environmental impacts globally. The crude oil used to make petrol is a finite resource that is predicted to run out in coming decades, so alternatives are needed.

One idea to reduce this reliance is to use **biofuels**. Biofuels are fuels derived from animal and plant matter. They include ethanol and **biodiesel**.

One example is the production of bioethanol from cellulose, a natural polymer in the cell walls of plants. Enzymes are used to break down cellulose to glucose, followed by fermentation. Major sources of cellulose include:

- crops grown specifically for use as biofuels
- residue from agricultural activities
- forestry residue
- domestic green waste.

In Australia, ethanol is available in an E10 blend, which is 10% bioethanol and 90% petrol (Figure 3.57). The proportion of bioethanol in ethanol blends has not increased in Australia since E10 was introduced 30 years ago because car engines need significant modifications to be able to use higher percentages of ethanol. Higher percentages are used in some parts of the world, but there is not significant demand in Australia for suitable engines.

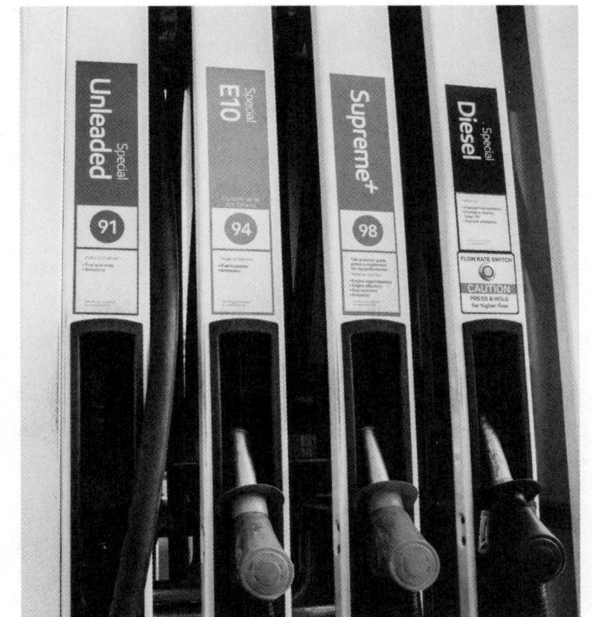

shutterstock.com/Daria Nipot

FIGURE 3.57 E10 is a fuel blend containing 10% ethanol.

The use of bioethanol in fuels has several advantages and disadvantages.

Advantages:

- Unlike the fossil fuel used to make petrol, bioethanol is a renewable resource.
- Ethanol burns more cleanly in air than petrol, producing less carbon (soot) and less carbon monoxide.
- Using bioethanol rather than petrol can reduce carbon dioxide emissions, provided that a renewable energy resource is used to produce crops required to obtain ethanol and to distil fermented ethanol.

Disadvantages:

- Ethanol has a lower heat of combustion (per mole, per unit of volume and per unit of mass) than petrol.
- Large amounts of land are needed to grow the crops required to produce bioethanol. This imposes environmental challenges such as soil erosion, deforestation, fertiliser run-off and salinity. Also, it uses land that may otherwise be used to produce food for people and animals.
- Major environmental problems could arise out of the disposal of fermentation wastes.
- Current engines in Australia would require modification to use high concentrations of bioethanol.

Table 3.17 compares the properties and costs of fossil fuels to those of a range of biofuels.

TABLE 3.17 Comparing fossil fuels and biofuels

	Fossil fuels	Biofuels
Chemical compound	Mostly carbon (coal) Natural gas (primarily CH_4) Petrol/diesel (primarily alkane chains of varying length)	C_2H_5OH (ethanol) Esters of varying chain length (biodiesel)
Source	Mining	Agricultural crops (sugar cane, wheat etc.) are a source of ethanol Fats and oils can be used to make biodiesel
Enthalpy of combustion (kJ g^{-1})	53.6 (natural gas) 9.8–27.9 (coal) 48 (petrol) 44.9–46.3 (crude oil) 42.6 (diesel)	37.2 (biodiesel) 29.6 (bioethanol) (Values depend on source of crops)
Emission of CO_2	High net release during both production and use (combustion)	Lower net release during combustion because recently removed from atmosphere through photosynthesis Some release associated with production
Vehicle modification required	None	Modification of vehicles to use higher percentages of ethanol or biodiesel
Running costs	No modifications required; fuel more expensive than biofuels, but similar to hydrogen	No modification required for vehicle use up to E10; cheaper than regular petrol No modification for up to 20% vehicle use for biodiesel; modification required for 100% use; cheaper than regular diesel

3.5 Reactions of organic acids and bases

We introduced the definition of acids and bases in Module 6 and discussed some of the common organic acids such as ethanoic acid.

Carboxylic acids have the ability to donate a proton. The highly polar O–H bond can ionise, fulfilling the definition of a Brønsted–Lowry acid, although organic acids are weak acids. Amines can likewise accept a proton by donating their electron pair (using the Lewis definition of bases) to a hydrogen atom, which means they are Brønsted–Lowry bases. A form of acid–base interaction occurs between an organic acid and an alcohol.

3.5.1 Production of esters

An ester is an organic compound formed in the reaction between an organic acid and an alcohol, referred to as **esterification**. Water is also produced in this reaction, making it look like a neutralisation reaction. However, this reaction needs both a catalyst (or dehydrating agent) and heat to proceed.

The reaction of alcohols with carboxylic acids requires a sulfuric acid catalyst to produce an ester:

$$\text{carboxylic acid} + \text{alcohol} \xrightarrow{\text{conc. } H_2SO_4} \text{ester (alkyl alkanoate)} + \text{water}$$

In basic solutions such as aqueous sodium hydroxide, esters may be hydrolysed to form alcohols and alkanoate ions. This is the basis of the **saponification** reaction we will look at later in this module. The carboxylic acids may be recovered from the reaction by acidification with mineral acids, such as HCl. This manipulation of the reactants or products suggests that this reaction does not go to completion, so the equation can be written as:

$$\text{carboxylic acid} + \text{alcohol} \rightleftharpoons \text{ester (alkyl alkanoate)} + \text{water}$$

Esterification is a condensation reaction because water is a product. Only a few drops of concentrated acid are needed to catalyse the reaction. Adding concentrated sulfuric acid in large amounts, say 5–10% of the reaction volume, has a significant effect on the position of equilibrium. Concentrated sulfuric acid is a dehydrating agent – it has a strong affinity for water. If a significant amount of sulfuric acid is present, it will remove water from the reaction, which shifts the equilibrium to the right to produce more water, and the desired product.

$$\text{alcohol} + \text{acid} \rightleftharpoons \text{ester} + \text{water}$$

Using very large amounts of sulfuric acid is wasteful and uneconomic and complicates the separation of ester from the reaction mixture.

Refluxing

Esterification requires heat for the reaction to reach equilibrium quickly (within an hour rather than after many days). If the reaction mixture was heated in an open system, volatile components, such as the reactant alcohol and the product ester, could escape. This problem is overcome by refluxing the reaction mixture using apparatus like that shown in Figure 3.58.

A condenser is placed vertically on top of the reaction vessel so that any volatile components pass into the condenser. The condenser can be cooled by air or water, and the cooling causes the volatile components to condense into a liquid and return to the reaction mixture. Refluxing also improves safety, especially in a school laboratory, because the volatile components (which are flammable) are further away from the heat source. Safety can also be increased by using a hot plate or heating mantle instead of a Bunsen burner.

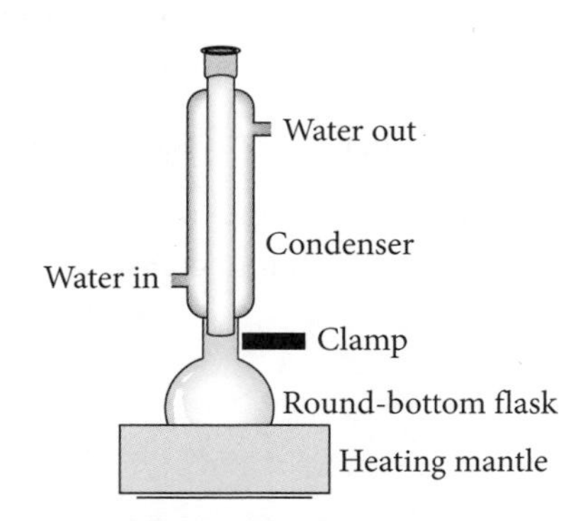

FIGURE 3.58 Reflux apparatus for esterification

Even under manipulated conditions, this reaction does not reach completion. The desired product, the ester, can be distilled to increase its purity.

The reactants used to produce esters are often very strong and unpleasant-smelling acids, such as butanoic acid. In contrast, the ester products are used in perfumes or as food additives, such as banana, pineapple or strawberry scents. Consider the following reaction and the structures of the reactants and products (except for water) in Figure 3.59:

ethanol + butanoic acid $\rightleftharpoons$ ethyl butanoate + water

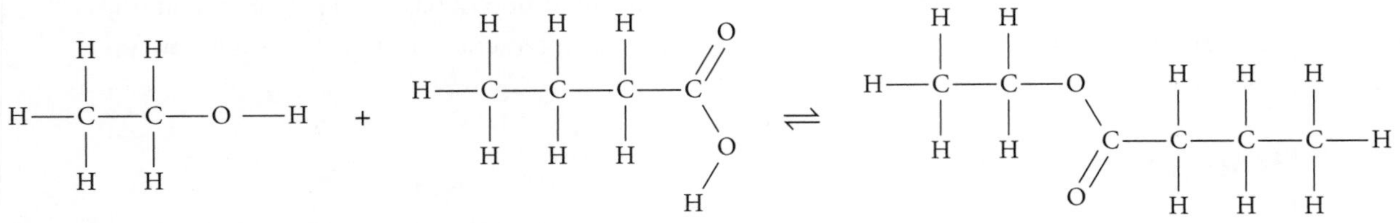

Ethanol + Butanoic acid $\rightleftharpoons$ Ethyl butanoate

FIGURE 3.59 The structural formula representation of an esterification reaction

> **Note**
> When naming esters, start with the alcohol and end with the anion of the acid. So propanoic acid and methanol would form the ester methyl propanoate. If you are unsure which part of the molecule is from the acid and which is from the alcohol, look for the O atom joining the two carbon atoms. Only one of these two carbons will have the carbonyl bond (C=O), and that was from the original acid.

3.5.2 Identifying organic acids and bases

The best way to analyse organic substances to distinguish between their acidic, neutral or basic nature is to use the Brønsted–Lowry or Lewis definitions of acids and bases. If a proton can be donated or a lone pair of electrons can be accepted, the substance is an acid. Carboxylic acids can donate a proton, according to the Brønsted–Lowry definition of acids. Ethanoic acid can donate a proton (H^+) to form the ethanoate ion:

$$CH_3COOH(l) + H_2O(l) \rightleftharpoons CH_3COO^-(aq) + H_3O^+(aq)$$

Likewise, a structure that can accept a proton is a base. The presence of the lone pair on the nitrogen atom of an amine indicates that these organic compounds can attract and accept a proton, and hence act as Brønsted–Lowry bases. Methylamine can accept a proton (H^+) to form the protonated methylamine ion (methyl ammonium):

$$CH_3NH_2(l) + H_2O(l) \rightleftharpoons CH_3NH_3^+(aq) + OH^-(aq)$$

We can generalise and suggest that most organic acids and bases are weak, but to determine strengths we need to look at the K_a and K_b or pK_a and pK_b values.

3.5.3 Soaps and detergents

Saponification is the production of soap by converting fats and oils to the salts of carboxylic acids and to alcohols such as glycerol. This happens in a basic solution, so it is described as alkaline hydrolysis. For example:

methyl octadecanoate + NaOH $\rightleftharpoons$ sodium octadecanoate (soap) + methanol

The salts of long-chain carboxylic acids are called soaps. They can form a lather in water, and they can be used to remove oils and grease, which don't otherwise mix with water.

Structure of soaps

A soap ion has a charged 'head' (the negatively charged carboxylate ion) and a non-polar region (tail) as shown in Figure 3.60. The head is **hydrophilic** ('water loving') and the tail is **hydrophobic** ('water fearing'). Together, the soap ions act as a bridge between polar or ionic substances and non-polar substances.

Note
The Na^+ ion associated with the carboxylate 'head' of a soap molecule dissociates in solution.

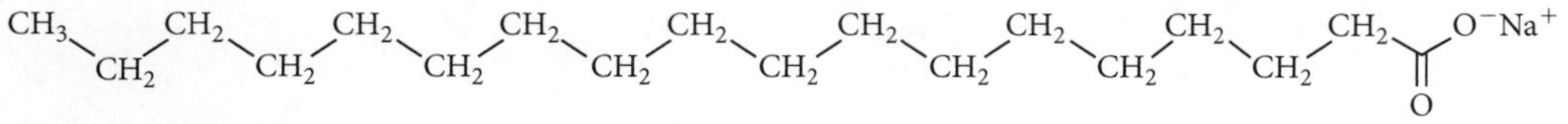

$$CH_3-CH_2-CH_2-CH_2-CH_2-CH_2-CH_2-CH_2-CH_2-CH_2-CH_2-CH_2-CH_2-CH_2-CH_2-CH_2-CH_2-C(=O)-O^-Na^+$$

FIGURE 3.60 The skeletal formula of a soap ion

Cleaning action of soaps

A **surfactant** (surface active agent) is a chemical substance that decreases the **surface tension** of water. It does this by breaking the hydrogen bonds between water molecules, allowing water to spread over a surface.

When water has a lower surface tension, the hydrophilic head of a soap ion can more readily dissolve and the hydrophobic tail can mix with oil, as shown in Figure 3.61. A group of soap ions can surround a particle of oil or grease, with their tails attached to the oil and their heads sticking out, in a structure known as a micelle. This physically separates the oil and grease particles, and the spheres of the negative heads repel one another. In this way, soap is able to move oil and grease away from fabrics, ceramics and skin.

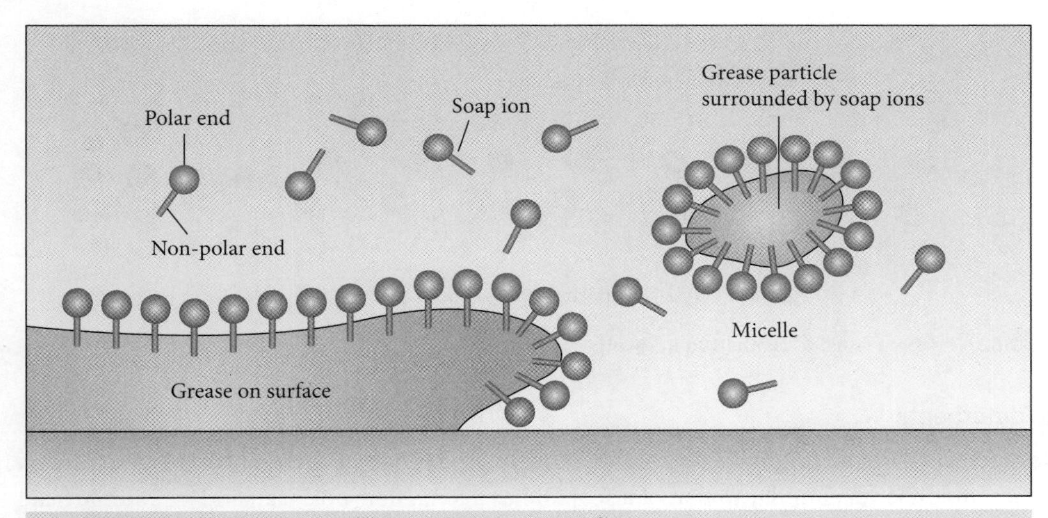

FIGURE 3.61 Formation of a micelle around a grease particle

Comparing soaps and detergents

Detergents differ from soaps in their structure, chemical composition and action in water. Both soaps and detergents have a hydrophobic tail and hydrophilic head; however, synthetic detergents have anionic, cationic or non-ionic (polar) heads (Figure 3.62).

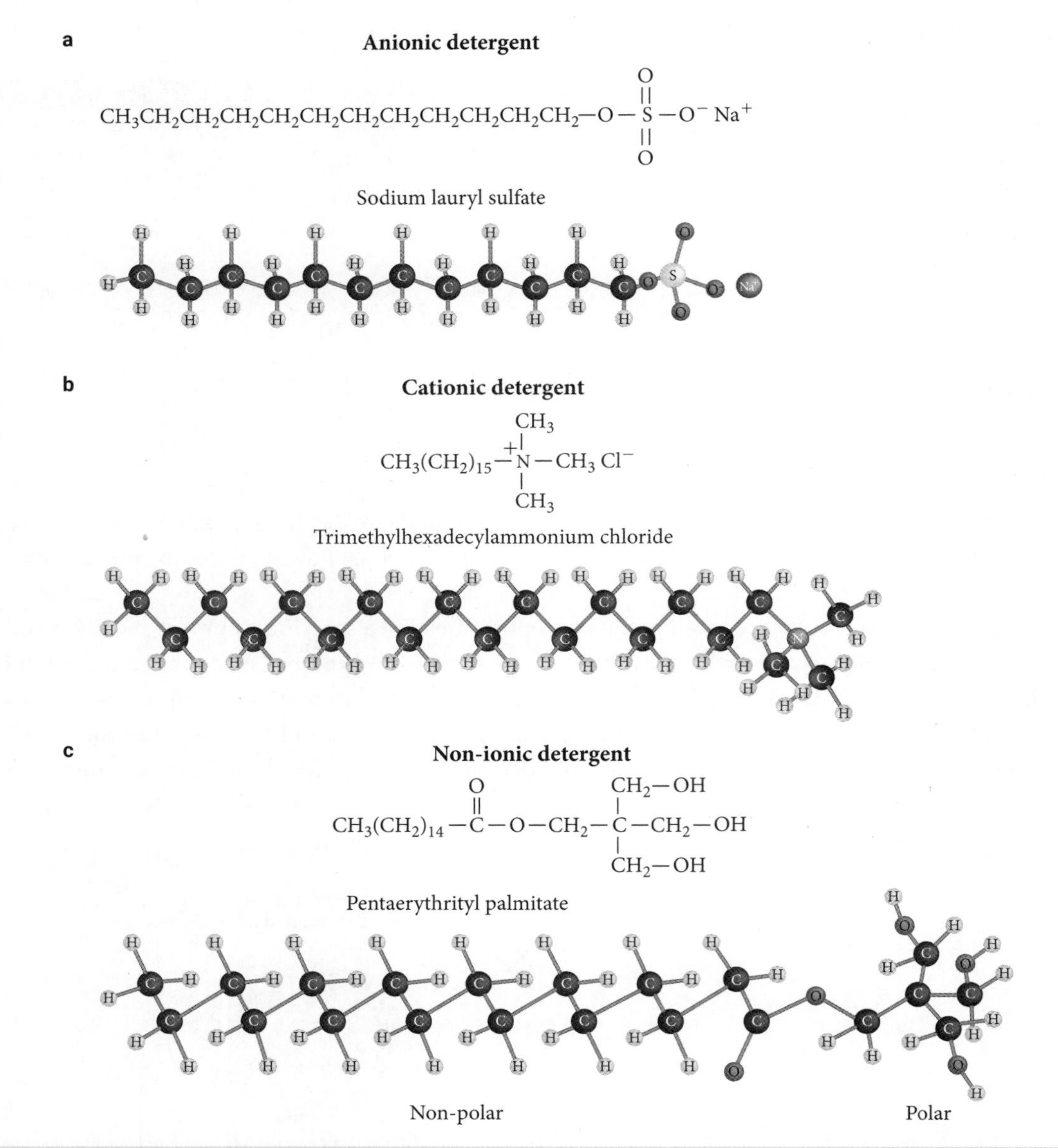

FIGURE 3.62 The three types of synthetic detergents

Anionic detergents

Anionic detergents are the original and still most widely used group of detergents. The structure of an ionic detergent ion is very similar to that of a soap. It has a long, non-polar tail and an anionic head (a sulfonate, C–O–SO_2–O–). These detergents work just like soaps but are slightly more effective. They are used in laundry detergents and dishwashing liquids. These also produce a lot of foam, which does not relate to the cleaning ability of the surfactant.

Cationic detergents

A cationic detergent ion has a region that is a derivative of ammonium (H replaced with alkyl groups, at least one, if not two of which are long chains.) The long chain is non-polar, and the nitrogen region is water soluble. These detergents are the preferred cleaners for plastics, and they are used in hair conditioners and fabric softeners because they reduce friction and static. They are a component of many disinfectants and antiseptics.

Non-ionic detergents

A non-ionic detergent has a tail that is similar to the other types of surfactants, but the head is different. These detergents contain a sequence of carbonyl groups and an alcohol at the end of the chain. This creates polarity within the head end and hence is water soluble. These detergents are molecules, not ions. They produce less foam and are used in cosmetics, paints, adhesives and pesticides.

3.5.4 Reaction pathways

We have now discussed a very large number of organic compounds and their reactions to produce other organic compounds. A flow chart is a convenient way of summarising this information and overviewing the key functional groups and organic series, and how they relate to one another. It can also show what (if any) reagents are needed and any conditions required for particular reaction pathways. A simplified example is shown in Figure 3.63.

You can see in the flow chart that we can use an addition reaction to produce a haloalkane from an alkene, but we need UV light for a similar reaction with an alkane. This is a substitution reaction rather than an addition reaction.

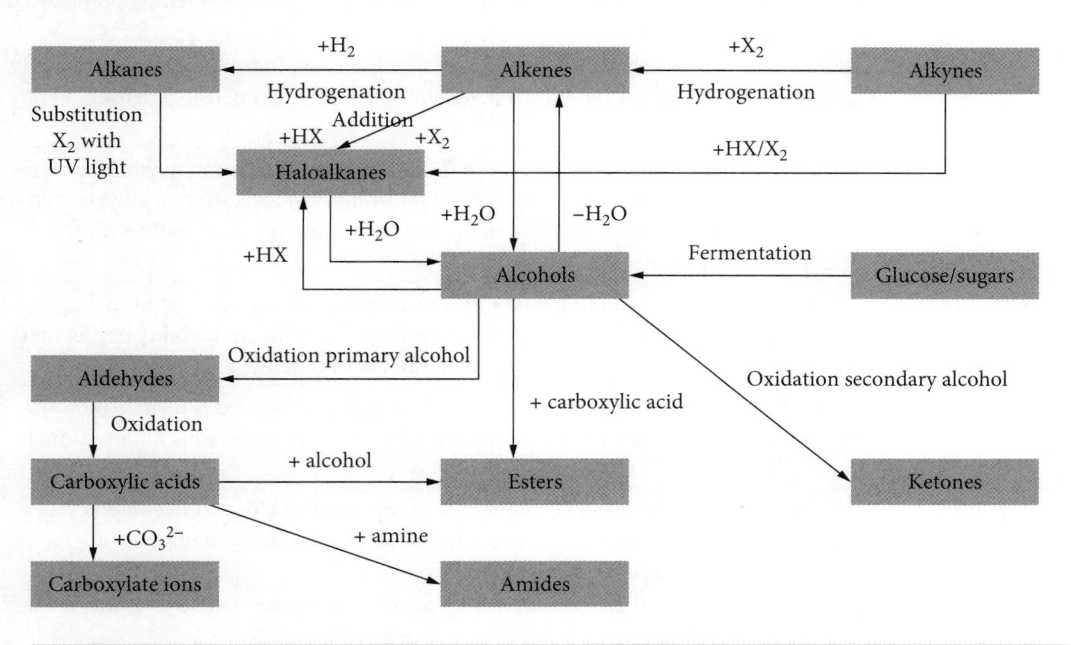

FIGURE 3.63 An overview of some reaction pathways for organic reactions

3.6 Polymers

Life would not exist without **polymers**. From the complex carbohydrates we eat for food to the proteins that regulate our internal reactions and protect our cells to the DNA of our nuclei, polymers are vital organic compounds.

The common name for synthetic polymers is plastic, although there are many types (Figure 3.64). They are very useful and convenient, but their manufacture and disposal are having serious consequences for the natural environment. Synthetic polymers are petroleum products, which means they are sourced from fossil fuels. Many types are not biodegradable and can last for years on land and in waterways and oceans. If they are combusted, they release toxic substances into the atmosphere.

Chemists and other scientists are researching and developing polymers derived from plant or animal matter; these biopolymers are more readily biodegradable than traditional plastics.

Polymers are often described in terms of their structure. The primary structure of a polymer comprises identical units, called **monomers**, linked together.

Science Photo Library/Alamy Stock Photo

FIGURE 3.64 Some household items made from polymers

3.6.1 Making polymers

Polymerisation is a chemical reaction in which monomer subunits are added together to produce a long chain called a polymer. There are two types of polymerisation processes: addition polymerisation and condensation polymerisation. The different processes produce polymers with a range of properties suited to a variety of uses (Table 3.18).

TABLE 3.18 Polymer properties and structures

Property	Aspect of polymer structure
Melting/softening point	Average molecular weight – the longer the chain, the higher the melting point and the harder the polymer Branching – unbranched polymer chains tend to intertwine and line up very closely, resulting in a highly crystalline arrangement. This increases density, melting point and hardness
Stability to heat and/or light	A higher proportion of carbon–carbon double or triple bonds can make the polymer more vulnerable to breakdown by UV light, through the formation of alkyl or peroxyl radicals. The addition of chemical stabilisers can reduce this consistent with the previous table cell
Chemical stability	Chemical stability relates to the type and distribution of different functional groups in the polymer chains. Certain additives can increase the biodegradability of polymers and hence reduce the time they remain in the environment
Mechanical strength	Cross-linking (the joining of two or more linear chains) increases the hardness of a polymer. Vulcanised rubber has sulfur bridges between the chains for additional strength
Flexibility/rigidity	Chain stiffening – addition of a large side group into a chain – reduces a polymer's flexibility. Additives such as chlorine (in polyvinyl chloride) or benzene (in polystyrene) gives a polymer more rigidity Branching – higher degrees of branching mean an irregular structure, which makes a polymer softer and more flexible

3.6.2 Products of addition polymerisation

Addition polymerisation primarily occurs between monomers of the type $CH_2{=}CHX$, where X is a hydrogen atom, halogen atom or benzene ring. For these monomers to add together in long chains, a carbon double bond in each monomer is broken. In addition polymerisation, all the atoms in the reaction are present in the desired polymer; there are no other products. The simplest monomer is ethene, shown in Figure 3.65.

FIGURE 3.65 The structural formula of ethene

The general reaction for the polymerisation of ethene is:

$$n(CH_2{=}CH_2) \rightarrow (-CH_2{-}CH_2-)_n$$

Notice that although the starting monomer is unsaturated, the polymer product is saturated.

Some commercially significant examples of addition polymers are:

- polyethene (polyethylene)
- polychloroethene (polyvinyl chloride)
- polyethenylbenzene (polystyrene)
- polycyanoethene (polyacrylonitrile)
- polytetrafluoroethene
- polypropene (polypropylene).

The IUPAC names of the polymers are shown first in the list, but they are more commonly referred to by the names in brackets.

Polyethylene

In the production of polyethylene from the monomer ethene, the double bond is broken and ethene units are added together to make long, straight chains or highly branched chains (Figure 3.66).

Polymers with few or no polar bonds are insoluble in water. The carbon bonding provides stability in the polymers. This is particularly true for polymers of ethene.

Two industrially important polymers are made from polyethylene: low-density polyethylene (LDPE) and high-density polyethylene (HDPE). Reaction conditions determine the form of polyethylene produced.

LDPE is highly branched and cannot be packed closely together; hence, its lower density. LDPE is used in materials such as cling wrap, milk bottles and squeeze bottles. It is soft and flexible, with a lower melting and boiling point than HDPE.

HDPE is often synthesised in the presence of a catalyst to reduce the amount of branching. Its long, straight chains increase its density, rigidity and melting point. It is used in kitchen utensils, toys, plastic grocery bags and some building materials.

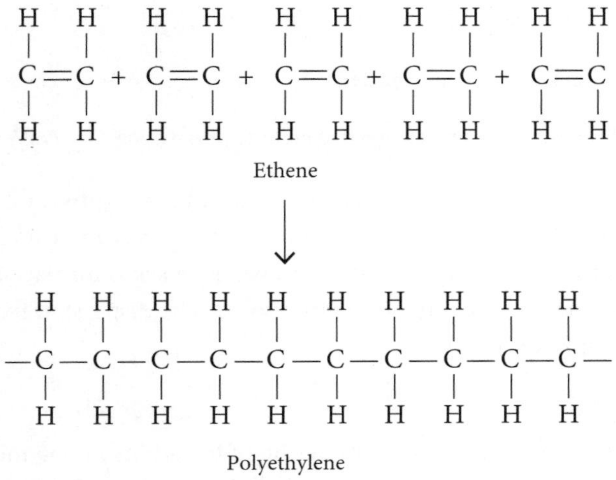

$$5\,CH_2{=}CH_2 \;(\text{Ethene}) \rightarrow -CH_2-CH_2-CH_2-CH_2-CH_2-CH_2-CH_2-CH_2-CH_2-CH_2- \;(\text{Polyethylene})$$

FIGURE 3.66 The polymerisation of ethene to produce polyethylene

Polyvinyl chloride

Chloroethene is the monomer used to produce polyvinyl chloride (PVC). It is formed in a substitution reaction, in which a hydrogen atom in ethene is replaced with a chlorine atom. Its structural formula is shown in Figure 3.67.

$$CH_2{=}CHCl$$

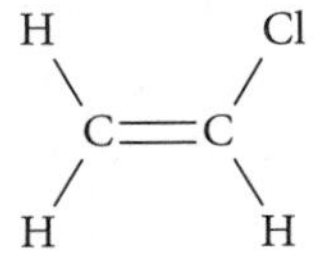

FIGURE 3.67 The structural formula of chloroethene

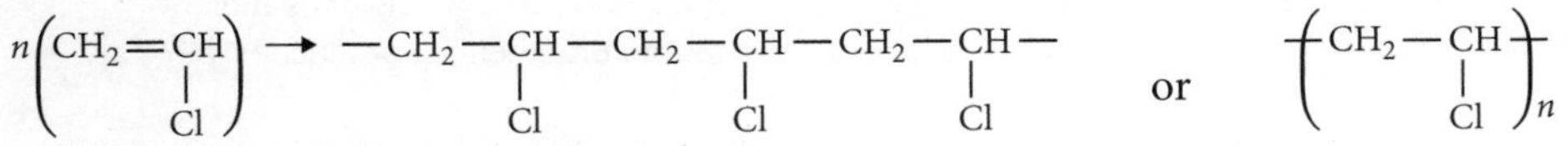

Figure 3.68 shows a simplified version of the production of PVC.

The C–Cl bond in PVC is unstable in UV light and when heated. Under these conditions, PVC may emit hydrogen chloride gas, which in the presence of water becomes the very strong acid hydrochloric acid. PVC is water resistant and flame resistant and is used in electrical insulation, appliance leads, wastewater pipes and garden hoses.

$$n\,(CH_2{=}CHCl) \rightarrow -CH_2-CHCl-CH_2-CHCl-CH_2-CHCl- \quad \text{or} \quad -\!\!\left(CH_2-CHCl\right)\!\!-_n$$

FIGURE 3.68 The polymerisation of chloroethene to produce polyvinyl chloride

9780170465281

Polystyrene

Styrene (ethenylbenzene) is the monomer used in the production of polystyrene. It has a benzene ring (six carbons in a ring, with alternating single and double bonds between the carbons) in place of one of the hydrogen atoms. It is a much larger monomer than ethene, as shown in Figure 3.69. The polymerisation process for polystyrene is shown in Figure 3.70.

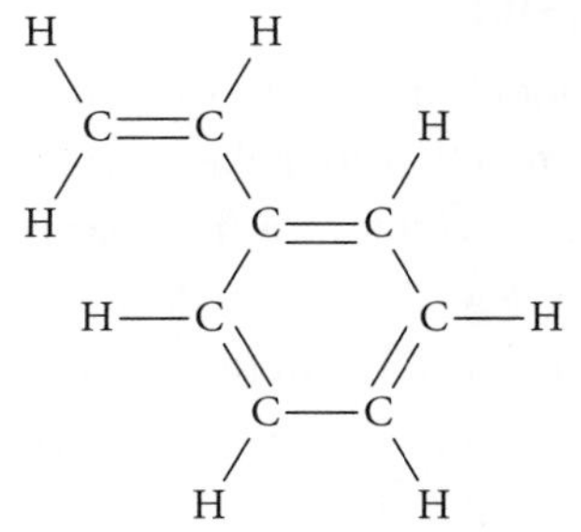

FIGURE 3.69 The structural formula of styrene

> **Note**
> The usual convention for depicting benzene rings is to show the six-carbon ring as a hexagon and the three double bonds as a circle, as shown in Figure 3.70. This makes it quicker to draw.

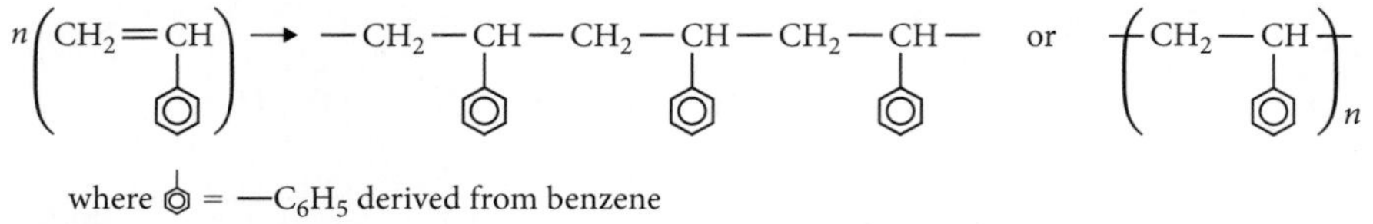

$$n\,(CH_2{=}CH(C_6H_5)) \rightarrow -CH_2-CH(C_6H_5)-CH_2-CH(C_6H_5)-CH_2-CH(C_6H_5)- \quad \text{or} \quad \left(-CH_2-CH(C_6H_5)-\right)_n$$

where $\text{⌬} = -C_6H_5$ derived from benzene

FIGURE 3.70 The polymerisation of styrene to produce polystyrene

Polystyrene is used in tool handles and screen protectors and as a lightweight alternative to glass. The large benzene ring means that the polymer has a low density. Gas can be added to polystyrene during production to manufacture the foam used in surfboards, bean bags and insulation. The long chain of only carbon and hydrogen atoms makes this polymer hydrophobic. The dispersion forces between the chains make polystyrene brittle, so it snaps easily.

Polytetrafluoroethylene

Tetrafluoroethene is the monomer used in the production of the addition polymer polytetrafluoroethylene (PTFE). The monomer has four fluorine atoms substituted for the four hydrogen atoms of ethene. It is not synthesised in this way, but we still use the derivative name. The structural formula is shown in Figure 3.71.

The polymerisation process is shown in Figure 3.72.

PTFE is commonly known by its brand name, Teflon, of which there are several types. Teflon is renowned for its low friction, non-stick properties and high resistance to chemical damage, and is a popular cookware coating. PTFE is used as an insulator for electrical wiring, in bearings and gears as well as for stain-resistant carpets and fabrics. It is also used in Gore-Tex textiles and as a membrane separating the half-cells for the electrolytic production of sodium hydroxide. The presence of C–F bonds provides polarity and strength because of the lack of dispersion forces.

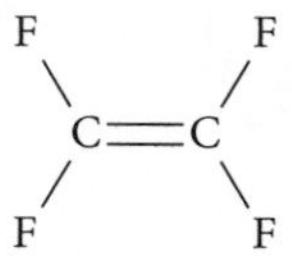

$$F_2C{=}CF_2$$

FIGURE 3.71 The structural formula of tetrafluoroethene

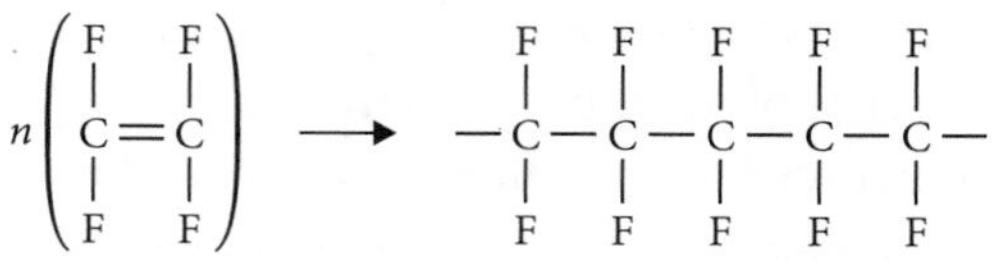

$$n\,(CF_2{=}CF_2) \rightarrow -CF_2-CF_2-CF_2-CF_2-CF_2-$$

FIGURE 3.72 The polymerisation of tetrafluoroethene to produce PTFE

3.6.3 Products of condensation polymerisation

Condensation polymerisation reactions happen between monomers with functional groups that can react with each other. During the reaction, a small molecule such as water is released as a **by-product**.

Condensation polymers are more common than addition polymers in biological systems and are particularly important in our diet. Starch is a condensation polymer that our bodies break down to glucose monomers. Conversely, the differently structured cellulose polymer is impossible for us to digest, even though it is made of the same monomer.

In the formation of cellulose, *n* glucose molecules combine to form the cellulose chain and (n–1) molecules of water. When two glucose monomers react, a water molecule is released and the monomers are linked through an oxygen atom. This combination process repeats to form a cellulose chain. As you can see in Figure 3.74, every second glucose molecule has the opposite orientation. The geometry of the glucose rings is such that the polymer is very linear.

You do not need to know the structure of cellulose or starch for the HSC course, but it is a good example of the condensation polymerisation process. The monomers are shown in Figure 3.73 and the polymer is shown in Figure 3.74.

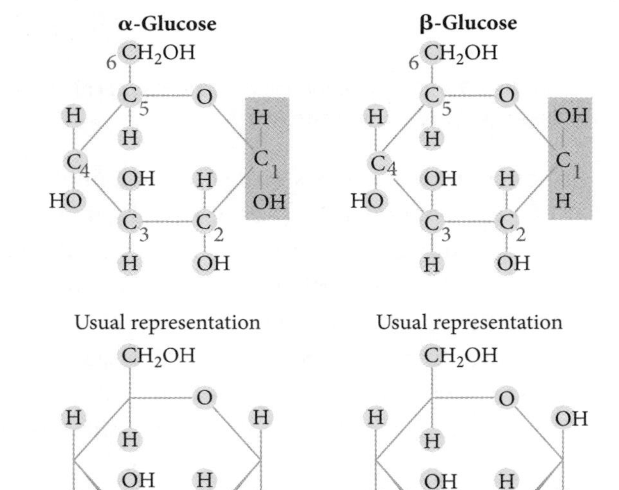

FIGURE 3.73 The structural formula for glucose monomers

Cellulose is a major component of biomass. It can be used as a source of hydrocarbons to produce petrochemicals and from which we may be able to synthesise biopolymers.

Cellulose

FIGURE 3.74 The polymerisation of glucose monomers produces cellulose.

Nylon

Nylon has different names depending on the starting monomers. One form of nylon is nylon-6,6. It is formed by the condensation polymerisation of 1,6-hexanediamine and adipoyl dichloride. These two monomers are shown in Figure 3.75. You should be able to recognise carbonyl groups (–C=O) at the ends of one compound and amine groups ($–NH_2$) at the ends of the other.

Adipoyl chloride ($COCl(CH_2)_4COCl$)

1,6-Hexanediamine ($NH_2(CH_2)_6NH_2$)

FIGURE 3.75 Two of the monomers that can polymerise to produce nylon

The bond formed during the polymerisation process is an amide bond, with HCl (rather than water) the small by-product that is produced.

Drawing a **dimer** is an effective way to show how these monomers join together (Figure 3.76). However, remember that polymerisation joins more than two molecules together; a condensation polymer is a long chain of dimers.

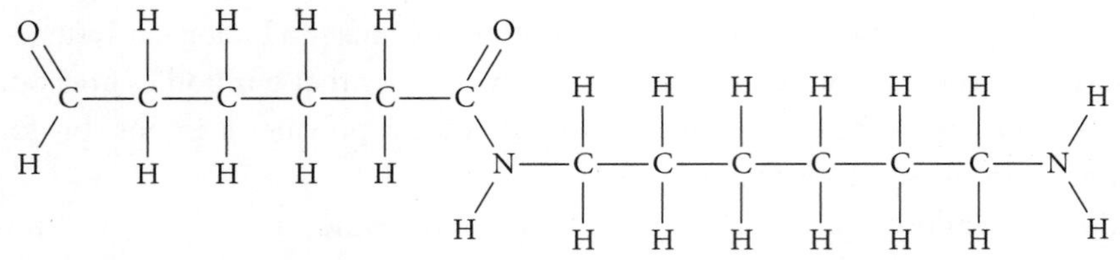

FIGURE 3.76 The dimer formed from an adipoyl chloride monomer and a 1,6-hexanediamine monomer, showing the amide bond in the centre

Another form of nylon is made from 6-aminohexanoic acid. This form of nylon is made from a single type of monomer rather than two alternating monomers. The bond formed during the polymerisation process is an amide bond, and a water molecule is released. The repeating amide bonds are responsible for some of nylon's properties; however, monomers with long carbon chains may decrease nylon's water absorbance and melting point.

Nylon was first produced in 1938 and its early uses were in toothbrushes and pantihose. Nylon's early claim to fame was to be 'as strong as steel, as fine as the spider's web'. Nylon is used commercially today in the fabrics industry, in the manufacture of cookware (e.g. spatulas, spoons) and for plastic fasteners and circuit board components.

Polyesters

We have already seen that esters are produced when an organic acid and an alcohol combine. Reacting a monomer containing a carboxyl group (–COOH) with a monomer containing a hydroxyl group (–OH) produces long chains of esters known as polyesters. One example of a reaction in polyester synthesis is shown in Figure 3.77.

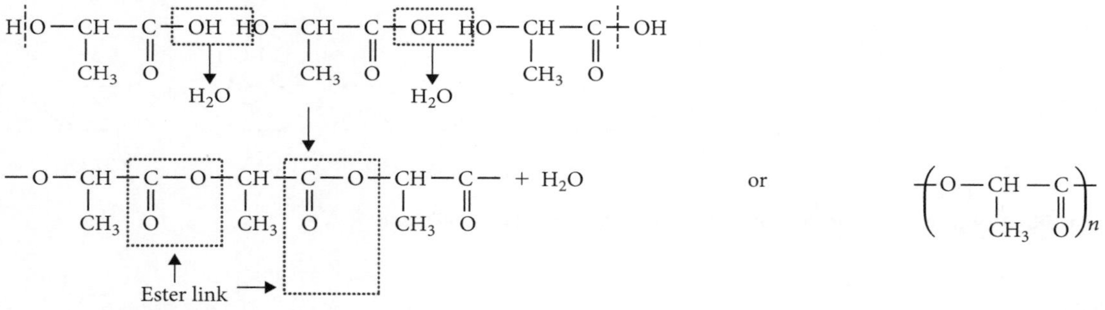

FIGURE 3.77 The polymerisation of a carboxylic acid group and an alcohol group to produce the ester bond characteristic of polyesters

Like nylon, polyethylene terephthalate (PET) has two monomers. For PET the monomers are terephthalic acid (benzene-1,4-dicarboxylic acid) and ethylene glycol (ethane-1,2-diol), shown in Figure 3.78. The ester bond forms when the acid and alcohol combine, releasing a water molecule.

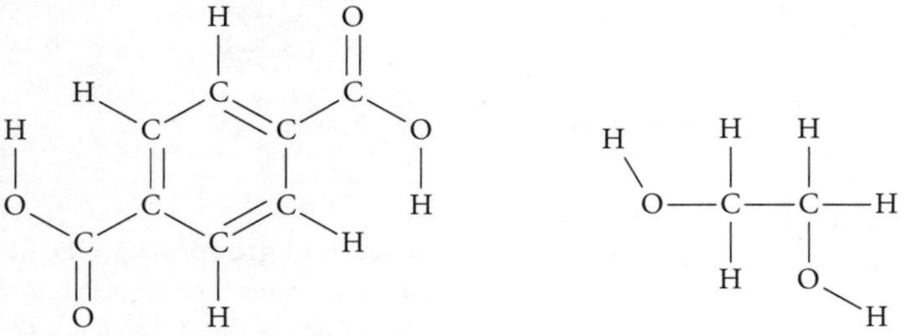

Benzene-1,4-dicarboxylic acid

Ethane-1,2-diol

FIGURE 3.78 Two of the monomer units used to produce a polyester (PET)

A dimer of these two monomers is represented in Figure 3.79.

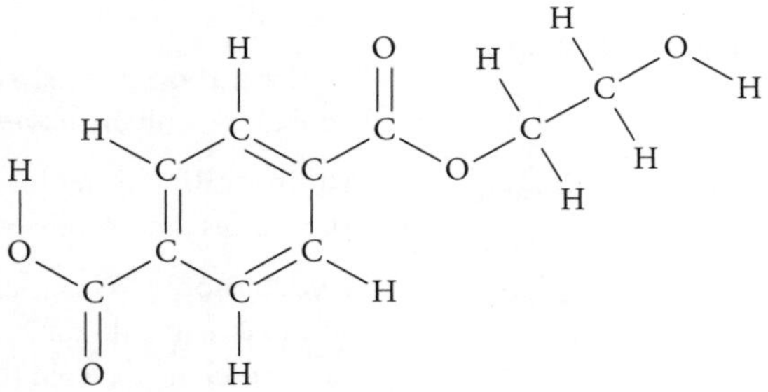

FIGURE 3.79 The dimer of benzene-1,4-dicarboxylic acid monomer and ethane-1,2-diol monomer. The ester bond joins the two monomers.

Polyesters have great strength, elasticity and electrical resistance and are used widely in the fabrics industry and for food and beverage storage. They are thermoplastic, which means they can be melted and reshaped. The most common type of polyester is PET, often used in soft drink bottles and t-shirt fabric. Another type of polyester is used in surgical theatres in the form of synthetic sutures (stitches).

Glossary

Addition reaction A type of reaction in which two or more reactants join chemically to form a single product

Alcohol An organic compound containing a hydroxyl (OH) group

Aldehyde An organic compound containing a C=O group on a terminal carbon

Alkane An organic compound containing only hydrogen and carbon and with only single bonds between the carbon atoms

Alkene An organic compound containing only hydrogen and carbon and with at least one double bond between two carbon atoms

Alkyne An organic compound containing only hydrogen and carbon and with at least one triple bond between two carbon atoms

Amide An organic compound containing an amine group (C–NH_2) on a carbon atom

Amine An organic compound containing an oxy group (C=O) and an amine group (NH_2) on the same carbon atom

Biodiesel A fuel sourced from a biological organism, which could replace diesel fuel

Biofuel A liquid fuel that has been produced from biological materials, e.g. plant and animal matter

By-product A secondary product in a chemical reaction that is not the desired product

Carboxylic acid An organic compound containing an oxy group (C=O) and a hydroxy group on the same terminal carbon

Dehydration The process of removing water molecules from a compound

Dimer Two monomer units joined together

Esterification The chemical reaction between an organic acid and an alcohol to produce an ester (an alkyl alkanoate)

Fermentation A chemical reaction often carried out by micro-organisms in the absence of oxygen where glucose is broken down into alcohol and carbon dioxide gas

Flash point A measure of the minimum temperature at which the vapour pressure of a hydrocarbon is sufficient to combust in air

Fossil fuel A fuel derived from fossilised plant or animal material

Fractional distillation The separation of multiple components of a mixture on the basis of their different boiling points

https://get.ga/aplus-hsc-chemistry-u34

A+ DIGITAL FLASHCARDS

Revise this topic's key terms and concepts by scanning the QR code or typing the URL into your browser.

Functional group A specific identifier (type of bond and/or presence of oxygen and/or nitrogen atoms) in an organic compound that confers specific chemical and/or physical properties

Homologous series A group of similar organic compounds united by their common functional group. The alkanes are a homologous series.

Hydration reaction A chemical reaction involving the addition of water

Hydrocarbon A group of organic compounds containing only hydrogen and carbon

Hydrophilic 'Water loving' – refers to chemical substances that readily bond to water molecules

Hydrophobic 'Water fearing'– refers to chemical substances that do not readily bond to water molecules, rather seeking to avoid them

Halogenation A chemical reaction involving the addition of a halogen (e.g. fluorine, chlorine, bromine or iodine)

Hydrogenation A chemical reaction involving the addition of hydrogen

Hydrohalogenation A chemical reaction involving the addition of a hydrogen and a halogen atom

Intermolecular force An interaction between molecules, e.g. dispersion force, dipole–dipole interaction or hydrogen bond

Intramolecular bond A bond within a molecule, e.g. a polar covalent or non-polar covalent bond

Isomers Two chemical compounds with the same molecular formula but different structural formulae, e.g. 1,1-dichloroethane and 1,2-dichloroethane

Ketone An organic compound containing a C=O group on a non-terminal carbon

Locant A number used to indicate the position of a functional group or substituent (i.e. a substituted halogen)

Markovnikov's rule A rule that states that the hydrogen atom from a hydrogen halide will bond with the carbon that has the greater number of hydrogens before the addition reaction. 'The rich get richer.'

Monomer A simple organic compound which forms the basis of a polymer when repeating units are joined together

Nomenclature A set of rules for naming chemical substances

Parent chain The longest continuous carbon chain of a hydrocarbon

Polymer A long chain of smaller units (monomers) linked together by chemical bonds

Polymerisation A chemical reaction between a large number of small monomer molecules to produce a large polymer molecule

Saponification The reaction between a base and an ester to produce soap (may also be applied to the manufacture of detergents)

Saturated hydrocarbon A compound containing carbon and hydrogen that only has single carbon–carbon bonds

Substituent An atom or group of atoms that has been substituted into an organic compound in place of a hydrogen atom

Substitution reaction A type of reaction in which two or more reactants join chemically to form a major organic product and an additional by-product

Surface tension The force of attraction between adjacent molecules at the surface of a liquid and the molecules below them that can slightly distort the surface of the liquid (e.g. the meniscus of water)

Surfactant A chemical substance that decreases the surface tension of water; also known as a surface acting agent

Terminal carbon A carbon atom at the end of a hydrocarbon parent chain

Unsaturated hydrocarbon A compound containing carbon and hydrogen that has at least one carbon double and/or triple bond

Volatility Refers to substances that vaporise at room temperature to produce a high vapour concentration above the solid or liquid phase (often in relation to fuels)

Exam practice

Multiple-choice questions

Solutions start on page 193.

Question 1

According to the rules of nomenclature for organic compounds, what is the name of the following compound?

```
      H   OH   H
      |   |    |
  H — C — C  — C — H
      |   |    |
      H   |    H
          |
      H — C — H
          |
          H
```

A Methylpropanol **B** Methylpropan-1-ol **C** Methylpropan-2-ol **D** Butan-2-ol

Question 2

Which of the following is a chain isomer of but-1-ene?

A But-2-ene **B** Cyclobutane **C** Methylpropene **D** Butan-1-ol

Question 3

Which of the following substances would you expect to have the highest boiling point?

A Pentane **B** Pentan-1-ol **C** Pentanal **D** Pentanoic acid

Question 4

The shape of the atoms around a carbon atom involved in the double bond of an alkene is

A linear.

B trigonal planar.

C trigonal pyramidal.

D tetrahedral.

Question 5

In a fermentation reaction, 18.0 g of glucose was completely broken down to form ethanol. The volume of carbon dioxide produced at SLC (100 kPa and 298 K) is closest to

A 2.27 L.

B 4.54 L.

C 4.95 L.

D 9.90 L.

Question 6

Propene was reacted with hydrogen chloride in a controlled experiment where the products of the reaction could be analysed. Which of the following statements is most correct for this reaction?

A The only product would be 2-chloropropane.

B There would be a significantly higher proportion of 1-chloropropane than 2-chloropropane.

C There would be equal amounts of 1-chloropropane and 2-chloropropane.

D There would be a significantly higher proportion of 2-chloropropane than 1-chloropropane.

Question 7

Which of the following substances would not change the colour of dichromate ions during an oxidation reaction?

A Ethanol

B Butan-2-ol

C 2-Methylbutan-2-ol

D Pentanal

Question 8

Ethanol and pentanoic acid were mixed under reflux and an organic substance was produced. The organic substance would have which of the following structures?

A

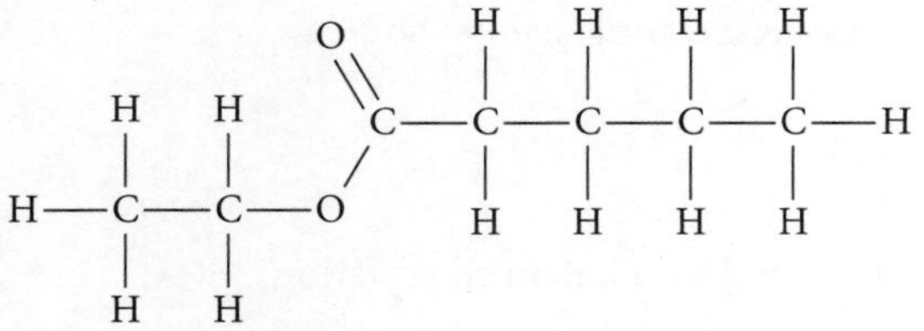

B

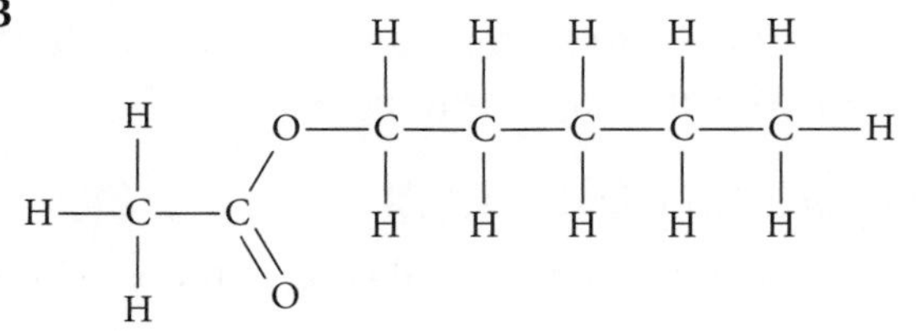

C

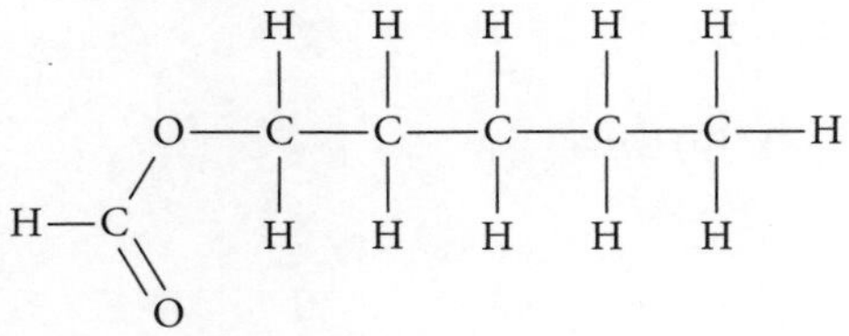

D

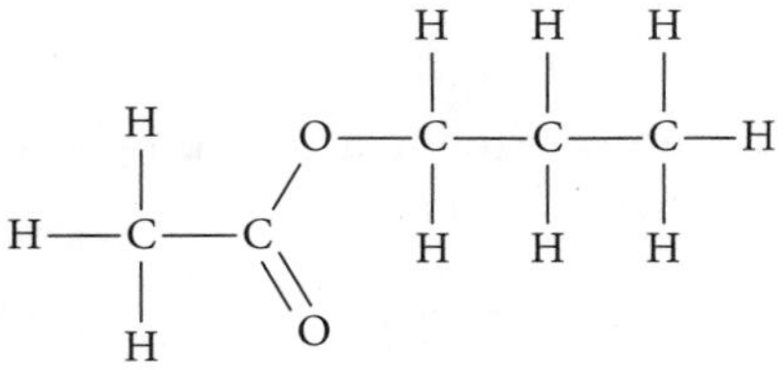

Question 9

Which of the following reactions represents the complete combustion of ethanol?

A $C_2H_6O(l) + O_2(g) \rightarrow 2C(s) + 3H_2O(l)$

B $C_2H_6O(l) + 3O_2(g) \rightarrow 2CO_2(g) + 3H_2O(l)$

C $2C_2H_6O(l) + 3O_2(g) \rightarrow 2C(s) + 2CO(g) + 6H_2O(l)$

D $C_2H_6O(l) + 2O_2(g) \rightarrow 2CO(g) + 3H_2O(l)$

Question 10

One version of nylon is made from 6-aminohexanoic acid. Which of the following structures corresponds to this name?

A

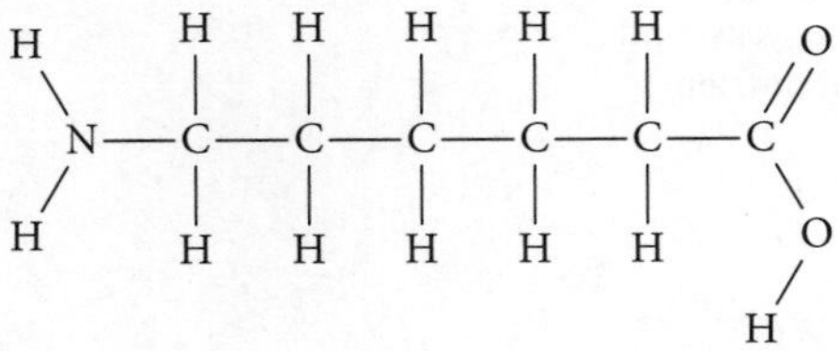

B

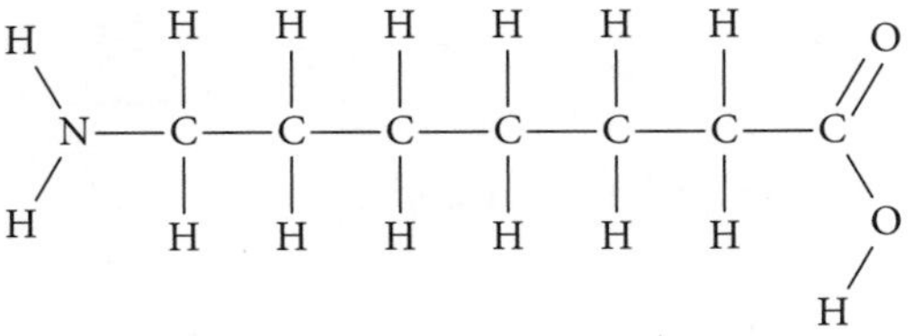

C

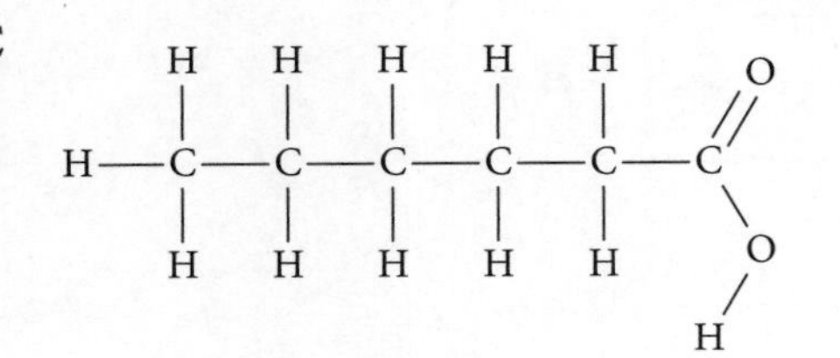

D

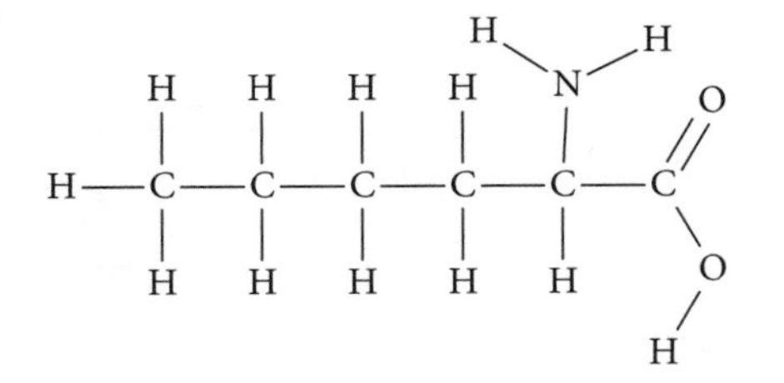

Short-answer questions

Solutions start on page 196.

Question 11 (4 marks)

Draw and name two isomers with the chemical formula C_4H_8O.

Question 12 (6 marks)

Draw a reaction pathway to show the reactants and conditions required to make the following chemical transitions: primary alcohol → alkene → alkane → haloalkane. You can use a specific starting reactant if you wish.

Question 13 (5 marks)

Discuss how the structure of soaps explains their effectiveness as cleaning agents. Use a labelled diagram in your response.

Question 14 (4 marks)

Discuss the importance of refluxing in the esterification reaction. Use a balanced equation in your answer.

Question 15 (6 marks)

Draw a table to compare the structure, properties and uses of polyethylene, polytetrafluoroethylene and polyester.

Question 16 (4 marks) ©NESA 2021 SII Q25

A student conducted an experiment in the school laboratory under standard laboratory conditions (25°C, 100 kPa) to determine the volume of carbon dioxide gas produced during the fermentation of glucose. The following apparatus was set up.

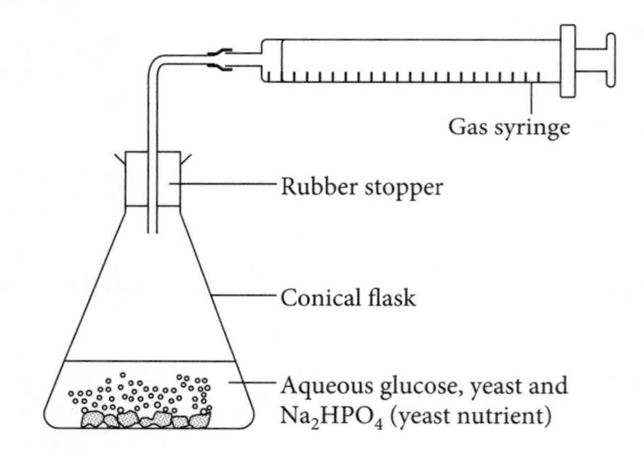

The following data were collected.

Day	Total volume of gas (mL)
1	489
2	677
3	899
4	1006
5	1006

Assume the total volume of gas produced was due to the production of carbon dioxide.

Calculate the mass of ethanol produced by the fermentation reaction. Include a relevant chemical equation in your answer.

CHAPTER 4
MODULE 8: APPLYING CHEMICAL IDEAS

Chapter 4
Module 8: Applying chemical ideas

Module summary

Chemistry is an inherently practical subject. It has a fundamental logic that can be applied in practical situations to help solve real problems.

The main challenge for students of chemistry is the jigsaw nature of the subject. As you begin to explore chemistry, you are given random pieces of the puzzle that often do not fit with your previous understandings. It is like doing a jigsaw puzzle where a blue piece could be part of the sea, the sky or a t-shirt. You feel as though you need more pieces to help you decide, but more pieces may only confuse you further.

Module 8 is about working out what you can do with your chemistry knowledge. In this module, you will apply many of the ideas and concepts you have learned in the previous modules, including tests to identify inorganic salts and organic functional groups. This module will also expand your current understanding, so you will find a few new things here too. You will extend your knowledge of identifying functional groups in organic chemistry into analysing various spectra, including mass, infrared and nuclear magnetic resonance spectroscopy. Real-world chemistry is used in many industries; it provides us with many of our goods and services, foods and textiles, and it is used to monitor the environment to help keep humans and ecosystems healthy.

The goals of this module are to investigate and evaluate the ways in which we can:

- identify and quantify the presence of inorganic ions in solutions
- chemically identify a range of organic substances containing specific functional groups
- draw conclusions from spectroscopic data
- see how chemistry is applied to the industrial production of a range of important chemicals to maximise yields and purity and minimise environmental impacts.

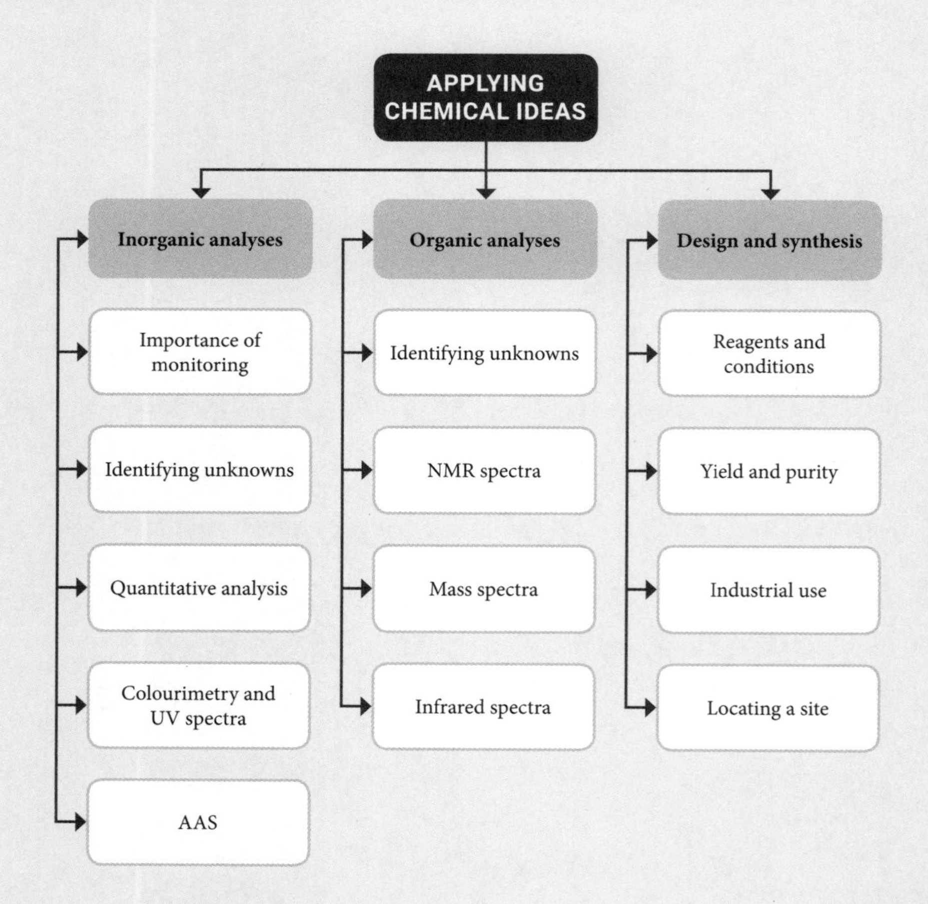

Outcomes

On completing this module, you should be able to:

- describe and evaluate chemical systems used to design and analyse chemical processes

Working Scientifically skills

In this module, you are required to demonstrate the following Working Scientifically skills:

- develop and evaluate questions and hypotheses for scientific investigation
- design and evaluate investigations in order to obtain primary and secondary data and information
- conduct investigations to collect valid and reliable primary and secondary data and information
- select and process appropriate qualitative and quantitative data and information using a range of appropriate media
- solve scientific problems using primary and secondary data, critical thinking skills and scientific processes
- communicate scientific understanding using suitable language and terminology for a specific audience or purpose

4.1 Analysis of inorganic substances

After the Module 7 topic of organic chemistry, it is logical to assume that 'inorganic' refers to anything that is not part of carbon's extended family. Here, we will consider an **inorganic substance** as an ionic salt dissolved in water. When ionic salts dissolve in water, they dissociate into their positively and negatively charged component ions: cations and anions. An example is copper sulfate, which exists as copper cations and sulfate anions in solution. A characteristic of copper ions in solution is their distinctive blue colour, but this is only the case at higher copper concentrations. What if we wish to identify the presence of copper ions in solution at lower concentrations?

If **qualitative analysis** such as observation of colour is not suitable – because of an ion that is colourless or at a very low concentration – we need other qualitative methods to determine which ions are present. To determine the concentrations of these ions, we use **quantitative analysis**, involving traditional methods or instrumentation.

4.1.1 Monitoring chemicals in the environment

Many reactions require some degree of monitoring; for example, we monitor reactions in the laboratory for safety reasons. We know that combustion reactions with enough oxygen produce carbon dioxide and that, at low oxygen concentrations, the combustion products of a **fuel** such as petrol will include carbon (soot). Either way, combustion reactions release **pollutants** into the atmosphere.

To monitor the waste products of these reactions and many others that affect air, water and land, we need to conduct **environmental monitoring**. The Environment Protection Authority (EPA) monitors and regulates environmental levels of many potential pollutants in Australia. It uses both qualitative and quantitative analytical techniques to characterise and measure chemicals being released into the environment. This helps the EPA to determine if groups and individuals are complying with environmental regulations.

9780170465281

Chemical pollution of the atmosphere

Excluding water vapour, Earth's atmosphere is composed primarily of nitrogen gas and oxygen gas. The rest of the gases (such as argon and carbon dioxide) make up less than 1% of the atmosphere. The composition of the atmosphere varies at different altitudes. For example, most of the ozone in the atmosphere is found in the upper atmosphere (stratosphere), where it filters out damaging UV radiation. In the lower atmosphere (troposphere), ozone can be poisonous to organisms and may contribute to the greenhouse effect.

The percentage of carbon dioxide in the atmosphere has been increasing over the last 150 years at an unprecedented rate, largely due to the combustion of fossil fuels, and is contributing to global warming.

There are many other pollutants in the troposphere, most of which are derived from the combustion of fuels in vehicles and in coal- or gas-fired power stations, and the smelting and purification of metals.

Polluted air is a significant cause of illness and death in many countries, so monitoring of gas pollutants in the atmosphere, especially in industrial regions, is very important.

Chemical pollution of land and water

Pollution of air, land and water are all related, because of cycling of chemicals in the environment. Carbon dioxide in the atmosphere can dissolve in water to produce weak carbonic acid, so unpolluted water is slightly acidic. Industrial plants that release oxides of nitrogen (grouped together as NO_x, where x could be 1 or 2) or sulfur (grouped together as SO_x, where x could be 2 or 3) into the atmosphere can decrease the pH of rain much further. This can have a devastating impact on waterways and their ecosystems.

Stormwater is often polluted with plastic litter, which flows into waterways and can harm or kill aquatic organisms. Microplastics from clothes are carried into waterways when washing machine water is discharged. These tiny chemical particles are easily swallowed by larger aquatic organisms or wash back up on to land.

The concentration of ions in water depends on the types and amounts of minerals in the rocks and soils that water flows through, the solubilities of those minerals (which depend on the acidity of the water) and the rate at which water enters a water body (higher rates can dilute the water as well as the rate of evaporation from the water body).

Human activities affect the concentration and type of ions in water. Some **heavy metals** such as mercury and lead are highly toxic, even in very small concentrations, and can cause serious health problems if people are exposed to them. Heavy metals are often used in mining activities. These activities have significant impacts on land, water and the atmosphere. There are many laws regulating the type and amount of chemicals that may be discharged during mining, and stipulating how mining sites must be remediated after they are no longer being used.

Other examples of land and water pollution include agricultural runoff from fertilised crops or from dairy farms, industrial waste discharged into waterways and leaching of chemicals in landfill into groundwater.

Anions such as nitrates and phosphates from agriculture are particularly problematic. Excess fertiliser washed from farms into waterways allows aquatic plants such as algae to grow abundantly. Eventually, these plants may use up all of the available nutrients that they require, and they die. As they decompose, plants use up all the dissolved oxygen during aerobic decay. After all the oxygen is gone, they decay anaerobically, producing chemicals that can kill other aquatic life. The decay causes sediments rich in nutrients such as nitrogen and phosphorus to accumulate at the bottom of water bodies such as lakes. This process is known as **eutrophication**. Nitrate and phosphate levels are monitored in waterways that are vulnerable to eutrophication.

4.1.2 Identifying cations in solutions

Some cations can be identified by their distinctive colours in solution, and their presence can be easy to confirm. However, some solutions are not coloured if cations are present at very low concentrations and others are not coloured at all. Some, even at low concentrations, can damage the health of humans and ecosystems, so in some situations we need to test for them.

Flame tests are an example of qualitative analysis. The colour of a flame can indicate that a particular ion is present, although a flame test does not determine the ion's concentration. A **precipitate** can indicate the presence of a particular ion, if the ion concentration is high enough for a precipitate to form.

Flame tests

Flame tests can be used to identify the presence of some cations in solution. Flame tests are most useful to distinguish between group 1 metals, as well as between barium and calcium in group 2. Flame tests do not provide any information about the concentration of an ion.

When a solution of cations is heated in a flame, some of the cations regain their valence electrons. The electrons have enough energy to move from their normal (ground) state to higher orbitals (excited states). An excited state is an unstable one, so the electrons eventually return to their ground state, emitting energy as they do so (Figure 4.1). They do this in one or several 'jumps'; each jump is from one energy level to another and emits energy of a particular wavelength.

Those energies in the visible part of the electromagnetic spectrum are seen as a colour that is a combination of all the individual visible wavelengths of energy emitted. Usually in flame tests we need atoms to produce wavelengths in the visible spectra. Ions will often emit in the ultraviolet (UV) region of the electromagnetic spectrum. Table 4.1 shows the flame test colours for various cations.

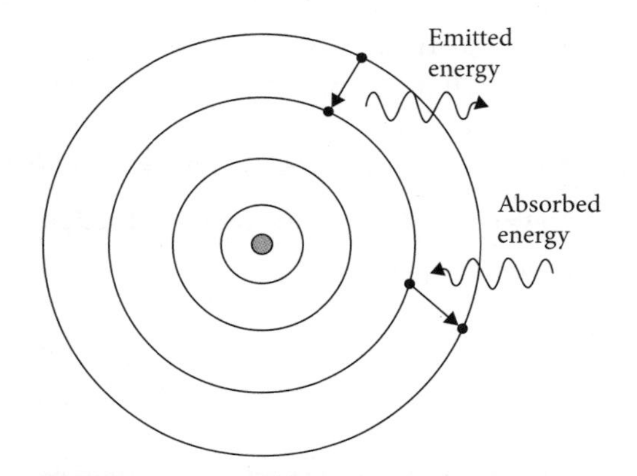

FIGURE 4.1 When an atom absorbs energy, it can jump from its ground state to an excited state. It emits energy when it jumps back to its ground state.

TABLE 4.1 Flame test colours for some cation solutions

Element name	Cation (in solution)	Colour in flame test
Sodium	Na^{+}	Yellow
Potassium	K^{+}	Lilac
Calcium	Ca^{2+}	Brick-red (orange-red)
Iron	Fe^{2+}	Gold when very hot; bright blue or green turning to orange-brown
Iron	Fe^{3+}	Orange-brown
Copper	Cu^{2+}	Blue (halides); green (others)
Strontium	Sr^{2+}	Red
Barium	Ba^{2+}	Pale green (apple green)
Lead (although not safe for flame testing)	Pb^{2+}	Light blue-grey

9780170465281

Precipitation reactions

Some metals do not produce a distinctive flame colour and it is dangerous to heat some others, such as lead, in an open flame. In such cases, or to confirm an ion detected in a flame test, we can use precipitation reactions.

> **Note**
> The HSC Chemistry data sheet includes a list of low solubility salts and their K_{sp} values.

Ionic salts are not equally soluble in water. From Module 5, we know that the solubility product is an expression of the equilibrium between a solid salt and its dissociated ions in solution. The solubility product for some of these – such as copper phosphate ($K_{sp} = 1.40 \times 10^{-37}$) – is so small that copper phosphate is effectively insoluble in water. It also means mixing a source of copper ions with a source of phosphate ions is very likely to form a precipitate because the concentration of ions in solution would have to be extremely low for no precipitate to form. It is more likely that the oppositely charged ions will form ionic bonds and precipitate out of the solution.

> **Note**
> The solubility mnemonics NAGSAG, LMS and CaStroBear were covered in Chapter 1.

Barium hydroxide has a K_{sp} of 2.55×10^{-4}. Although this value is still relatively low, compared to something like sodium chloride, a precipitate does not form if hydroxide or barium ion concentrations are sufficiently low. We might conclude, incorrectly, that there are no barium ions (or hydroxide ions) in the solution.

Recalling the solubility rules can help us to identify ions in solution. We can use precipitation reactions, sometimes in addition to flame tests, to identify a range of cations.

Table 4.2 shows some important precipitation reactions for the cations you need to be familiar with.

TABLE 4.2 Precipitation reactions used to identify cations

Element name	Cation (in solution)	Precipitate
Magnesium	Mg^{2+}	With OH^- forms a white precipitate No precipitate with SO_4^{2-}
Calcium	Ca^{2+}	With SO_4^{2-} and OH^- forms a white precipitate (if solution not too dilute, say 0.05 mol L^{-1}) With F^- forms a white precipitate
Iron	Fe^{2+}	With OH^- forms a green or white precipitate ($Fe(OH)_2$), which becomes brown if oxidised ($Fe(OH)_3$) Decolourises acidified dilute potassium permanganate solution
Iron	Fe^{3+}	With OH^- forms a brown precipitate With thiocyanate (SCN^-) forms a deep red solution
Copper	Cu^{2+}	With OH^- forms a blue precipitate, which dissolves in NH_3 to form a deep blue solution
Silver	Ag^+	AgOH decomposes rapidly in solution to form hydrated Ag_2O, a precipitate with a milk coffee colour With Cl^- forms a white precipitate that dissolves in NH_4OH solution
Barium	Ba^{2+}	With SO_4^{2-} forms a white precipitate No precipitate with OH^- or F^- (compare Ca^{2+})
Lead	Pb^{2+}	With OH^- forms a white precipitate that dissolves in excess OH^- With Cl^- forms a white precipitate (if solution not too dilute, say 0.05 mol L^{-1}) that does not dissolve in NH_4OH solution With I^- forms a yellow precipitate

Complexation reactions

A **complex** is a chemical combination of a metal atom or ion (usually a transition metal) bonded to one or more Lewis bases. This often occurs through the Lewis base donation of an electron pair to a metal atom or ion in a **complexation reaction**. The central atom is surrounded by ions or molecules and has a charge. Many complexes have coloured ions or form precipitates, enabling their ions to be identified in solution.

A **ligand** is an ion or a molecule bonded to a metal atom or an ion in a complex. The central metal in a complex is surrounded by two, four or six ligands. The most common ligands are CN^-, OH^- and NH_3.

The complex we have already encountered is iron(III) thiocyanate, $Fe(SCN)^{2+}$. This was formed in an equilibrium reaction involving ferric ions (Fe^{3+}) and the thiocyanate ion (SCN^-). The metal ion accepts an electron pair from the sulfur atom, forming a coordinate bond (Figure 4.2). This creates a complex with an overall charge of 2+.

Fe^{2+} — S—C≡N

FIGURE 4.2 The iron(III) thiocyanate complex

When silver chloride is added to aqueous ammonia, the nitrogen atom from each of two ammonia molecules donates an electron pair to a silver ion. This results in a complex with a charge of 1+.

> **Note**
> Note the square brackets used for $[Ag(NH_3)_2]^+$, showing that the overall complex has a charge.

The complex has the formula $[Ag(NH_3)_2]^+$ and the equation for the reaction is:

$$Ag^+(aq) + 2NH_3(aq) \rightleftharpoons [Ag(NH_3)_2]^+$$

Choosing appropriate tests for cations

We now have three methods of identifying the presence of cations: flame tests, precipitation reactions and complexation reactions. However, some tests are dangerous (e.g. identifying lead ions using a flame test) and some may be inconclusive (e.g. calcium and barium both form white precipitates with sulfate ions).

To choose the right test(s), we need to consider whether we want to identify the presence of a particular cation (qualitative analysis) or to measure the mass or concentration of a particular cation (quantitative analysis).

Flow charts are an excellent tool to use when developing a set of tests to identify cations. Figure 4.3 shows how a flow chart might look using qualitative methods to identify a cation in solution.

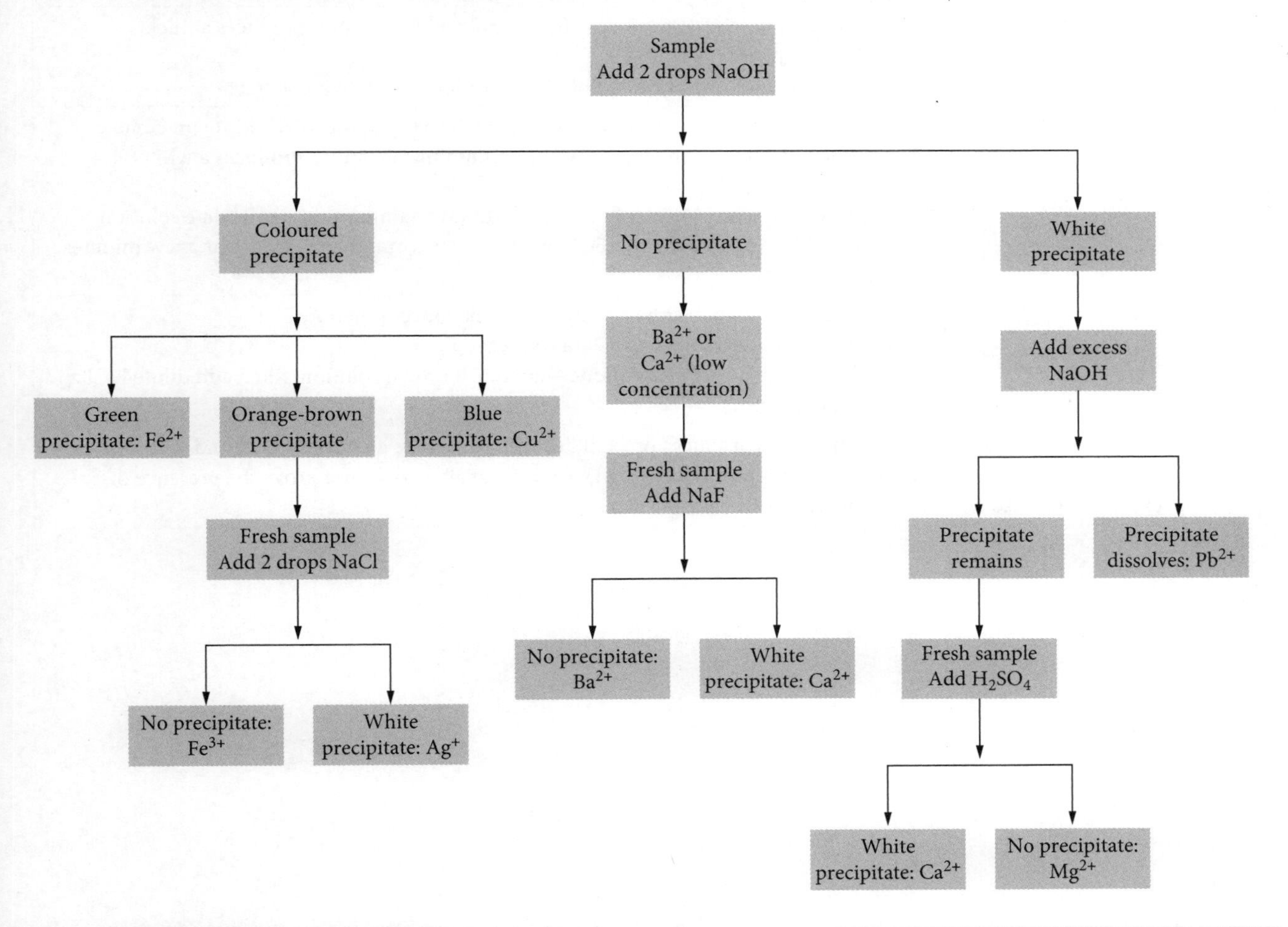

FIGURE 4.3 A flow chart for identifying cations in solution

4.1.3 Identifying anions in solutions

Precipitation or complexation reactions may both be appropriate to identify unknown anions in solution. We cannot use flame tests to identify anions.

To use a precipitation reaction to identify an unknown anion in solution, the solubility rules relating to NAGSAG, LMS and CaStroBear must again be employed. In combination with an acid test for carbonates, solubility tests enable us to identify a range of anions.

Table 4.3 shows some common tests for the anions you need to be familiar with. The flow chart in Figure 4.4 summarises the steps for identifying an unknown anion in a solution.

TABLE 4.3 Precipitation reactions used to identify anions in solution, and the acid test for carbonates

Anion	Test
Carbonate (CO_3^{2-})	1 Solution has a pH of 8–11 (pH paper suffices) 2 Addition of a dilute strong acid (e.g. HNO_3 but no Cl^- or SO_4^{2-}) produces bubbles of colourless gas (CO_2) that turns limewater milky
Hydroxide (OH^-)	1 pH > 7, changes red litmus to blue 2 Addition of NH_4^+ followed by gentle heating will produce ammonia gas
Chloride (Cl^-)	1 Addition of $AgNO_3$ to an acidified sample produces a white precipitate,* which dissolves in dilute ammonia solution (NH_4OH); solution darkens in sunlight
Bromide (Br^-)	1 Addition of $AgNO_3$ to an acidified sample produces a pale cream precipitate that dissolves in concentrated ammonia solution; solution darkens slowly in sunlight
Iodide (I^-)	1 Addition of $AgNO_3$ to an acidified sample produces a pale yellow precipitate that does not dissolve in ammonia solution (NH_4OH); solution not affected by sunlight 2 Addition of $Pb(NO_3)_2$ produces a yellow precipitate
Sulfate (SO_4^{2-})	1 Addition of $Ba(NO_3)_2$ to an acidified sample of the solution produces a thick, white precipitate 2 Acidification and addition of $Pb(NO_3)_2$ produces a white precipitate
Phosphate (PO_4^{3-})	1 Addition of ammonia solution followed by $Ba(NO_3)_2$ produces a white precipitate 2 Addition of Mg^{2+} in an ammonia/ammonium nitrate buffer produces a white precipitate, $Mg(NH_4)PO_4$ 3 Acidification with HNO_3 followed by addition of ammonium molybdate solution ($(NH_4)_2MoO_4$) produces a yellow precipitate; warming the mixture for a few minutes may be necessary
Ethanoate (CH_3COO^-)	1 Does not precipitate with any cations except concentrated Ag^+ 2 An aqueous solution may smell like vinegar 3 Addition of neutral $FeCl_3$ produces a reddish brown solution; filter, add dilute HCl and colour disappears

*In non-acidic solutions, silver nitrate also produces precipitates with carbonate and phosphate (and with sulfate at all pH values if sulfate concentration is moderately high), so this test alone does not prove the presence of chloride; it is also necessary to prove the absence of sulfate.

9780170465281

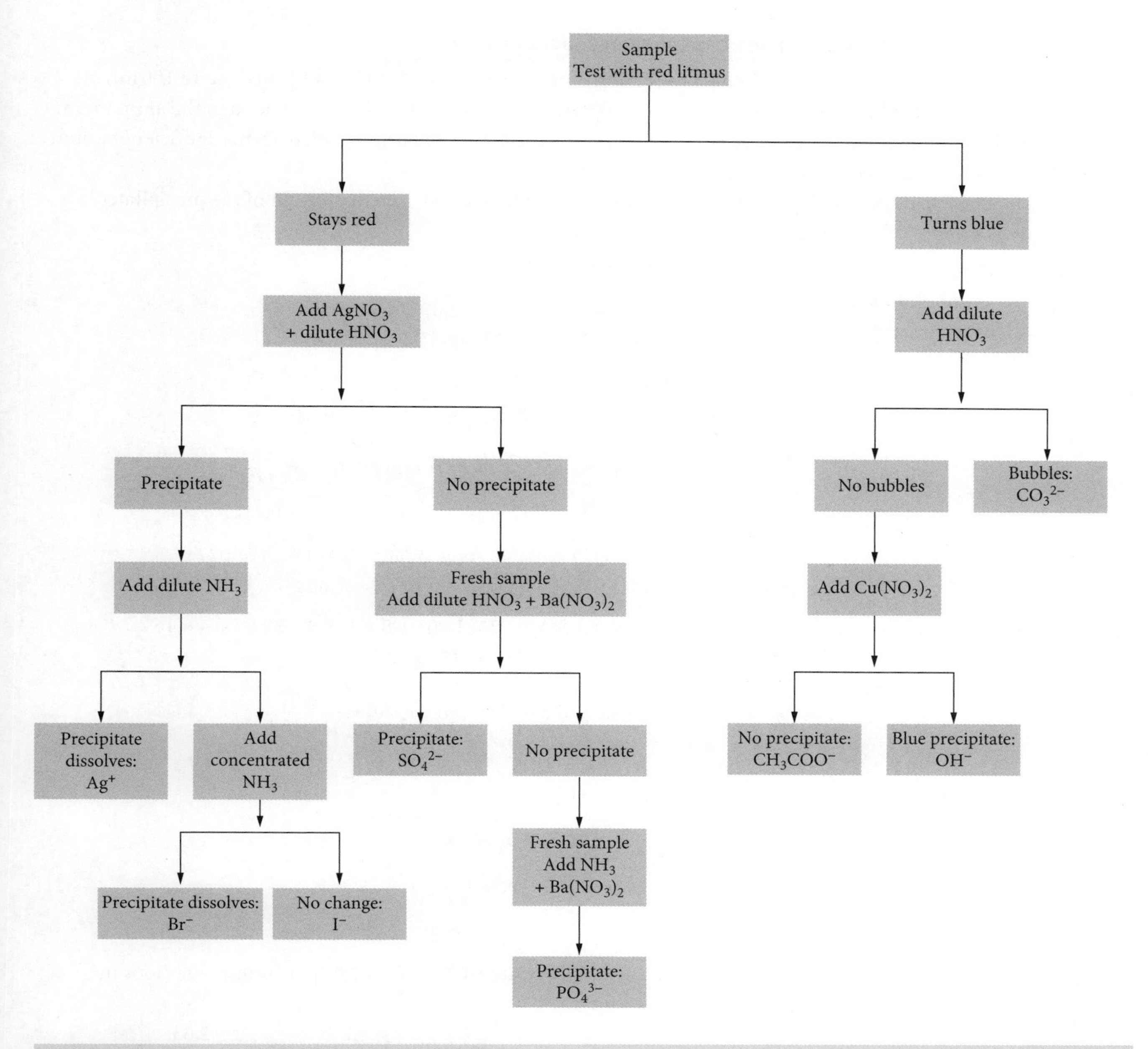

FIGURE 4.4 A flow chart for identifying anions in solution

4.1.4 Quantitative analysis of ions

Gravimetric analysis

Gravimetric analysis is a precise analytical quantitative technique used to determine the proportion, by mass, of a particular chemical substance in a compound or mixture. It is often used in conjunction with precipitation reactions. A precipitate can be dried and weighed, and then this mass used to calculate the number of moles present. Using the mole ratios for a chemical reaction, we can work backwards to find the initial mass or proportion by mass (percentage) in the system we are investigating.

One of the most common fertilisers used to increase the nitrogen content in soils is sulfate of ammonia (ammonium sulfate). To determine the amount of ammonium ions in a bag of fertiliser would be very difficult in the laboratory because ammonium ions are very soluble. We can use gravimetric analysis to determine the proportion of sulfate ions in a sample of fertiliser by precipitating the sulfate out as barium sulfate. From the mass of the isolated precipitate, we can calculate the percentage composition of the fertiliser in terms of both sulfate and ammonium ions.

Example: Purity of a sample of ammonium sulfate fertiliser

A 12.00 g sample of fertiliser known to contain ammonium sulfate ($(NH_4)_2SO_4$) is added to 100 mL of water and stirred until it completely dissolves. Barium nitrate solution is slowly added to the ammonium sulfate solution with a magnetic stirrer, and a white precipitate of barium sulfate forms. Sufficient barium nitrate is added to precipitate all of the sulfate ions.

The barium sulfate precipitate is finely filtered and then dried. The dry weight of the precipitate is 17.0022 g.

1 Calculate the mass of ammonium ions in the original fertiliser sample:

$m(BaSO_4) = 17.0022\text{ g}$

$$MM(BaSO_4) = 137.3 + 32.07 + 4 \times 16 = 233.37\text{ g mol}^{-1}$$

$$MM(SO_4^{2-}) = 32.07 + 4 \times 16 = 96.07\text{ g mol}^{-1}$$

$$\text{Proportion of } SO_4^{2-} \text{ in } BaSO_4 \text{ precipitate} = \frac{96.07}{233.37} = 0.4117$$

$$\therefore m(SO_4^{2-}) \text{ in } BaSO_4 \text{ precipitate} = 0.4117 \times 17.0022 = 7.00\text{ g}$$

2 Calculate the mass of ammonium sulfate in the original fertiliser sample:

$$MM((NH_4)_2SO_4) = 2 \times 14.01 + 8 \times 1.008 + 32.07 + 4 \times 16 = 132.154\text{ g mol}^{-1}$$

$$\text{Proportion of } SO_4^{2-} \text{ in } (NH_4)_2SO_4 = \frac{96.07}{132.154} = 0.7270$$

Total $m(NH_4)_2SO_4$:

$$0.7270 \times m = 7.00$$

So $$m = \frac{7.00}{0.7270} = 9.63\text{ g}$$

This means there are 9.63 – 7.00 = 2.63 g of ammonium ions in the fertiliser.

$$\text{Proportion of } NH_4^+ \text{ in original fertiliser sample} = \frac{2.63}{12.00} = 0.219 = 21.9\%$$

3 Calculate the purity of the fertiliser sample (compare the calculated mass of ammonium sulfate to that in the original sample):

$$\text{Purity} = \frac{9.63}{12.00} \times 100 = 80.3\% \text{ (3 significant figures)}$$

Note

For these calculations, we have assumed that the original sample did not contain any ions other than the sulfate. We have also assumed careful laboratory practices: no loss of precipitate during filtration or drying, no excess water after drying, no loss of the sample in the dissociation or transfer processes and sufficient barium nitrate to precipitate all of the sulfate ions.

9780170465281

Precipitation titrations

Identifying ion concentration using gravimetric analysis can go wrong if excess **reagent** is added. The mass of the dried precipitate will be inaccurate. It can be difficult to isolate a particular ion in the precipitate or to collect all of the ions in solution by precipitation. An alternative quantitative method to help resolve these challenges is a combination of precipitation and **titration**.

As with acid–base titrations, during precipitation titrations we record the volume of solution needed to reach an end point in order to calculate the concentration of particular ions in a solution. However, during precipitation titrations it is not possible to know when the last of the precipitate has formed, so we need an extra reaction. Usually, when conducting a precipitation titration, we select a reagent that will precipitate the desired ion and form a coloured complex with another ion when the target ion has been removed from the solution.

Example: Concentration of Cl^- in water

To identify the chloride ion concentration in a water sample, we can add silver nitrate. This precipitates silver chloride, which can be dried and then weighed:

$$Ag^+(aq) + Cl^-(aq) \rightarrow AgCl(s)$$

The problem with this technique is that silver chloride is a white precipitate and we do not know when all the chloride ions have precipitated out of solution.

If we add potassium chromate to our sample (in the same way we add an indicator for acid–base titrations) then, after the silver ions have precipitated all the chloride ions, they will start to form a precipitate with the chromate ions:

$$Ag^+(aq) + CrO_4^{2-}(aq) \rightarrow Ag_2CrO_4(s)$$

Silver chromate has a brick-red colour so we can easily determine when this second precipitate starts to form. However, there may be an overlap between the end point for the silver chloride precipitation and the starting point for the silver chromate precipitation. A calibration step can be done using a different salt that will not interact with the silver ions, allowing the precipitation of silver chromate to occur directly. When this volume has been tested reliably, it can be used to calibrate the actual precipitation titration result.

For example, if three repetitions of the calibration step have an average titre of 2.3 mL and the average titre calculated for the silver chloride precipitation is 27.8 mL, we subtract 2.3 from 27.8 to give a final average titre of 25.5 mL. This is the value we then use in our concentration calculations.

For any of these techniques to work, we need to ensure that the target ion is less soluble than the indicator ion. If silver chromate were less soluble than silver chloride, it would precipitate first, rendering our procedure invalid. This means there are several different methods for precipitation titrations.

4.1.5 Instrumental quantitative analysis of inorganic compounds

Spectroscopy is the study of the interactions between electromagnetic radiation and matter. We know from flame test investigations that vaporised elements absorb light of specific frequencies. When electrons absorb energy, the higher electron state can take the form of an increase in rotational or vibrational energy in and around bonds, or it can be electronic excitation. By studying changes in energy states, scientists are able to learn more about the physical and chemical properties of different molecules.

From our previous work with flame tests, we know that visible or UV light gives electrons sufficient energy to move from their ground state to excited states. The radiation needs to have a particular amount of energy to cause a particular transition.

It is important here to distinguish between an:

- **absorption spectrum**: shows wavelengths of electromagnetic radiation *absorbed* by an atom as electrons move from their ground state to excited states
- **emission spectrum**: shows wavelengths of electromagnetic radiation *released* by an atom as the electrons move from excited states back to their ground state.

In some spectroscopic techniques, we analyse emission spectra and in others we analyse absorption spectra. These techniques are sensitive to concentrations much smaller than we could detect through precipitation reactions.

The key in these types of instrumental analysis is to find wavelengths that are characteristic of the element(s) of interest.

Atomic absorption spectroscopy

Atomic absorption spectroscopy is used particularly for detecting the presence and concentrations of metal ions in solutions. A sample solution is vaporised and atomised, absorbing light of certain wavelengths, and the output (also called a spectrum) is analysed.

As shown in Figure 4.5, atomic absorption spectroscopy involves the following steps.

1. A hollow cathode lamp, with the cathode made of the same metal as the one being tested, emits light of certain frequencies.
2. The light emitted by the lamp is passed through the sample to be tested, which has been vaporised in a flame.
3. The light passes into a **monochromator**, which isolates the wavelength of light required for the sample.
4. The intensity of light of that wavelength that passes through the flame is measured and displayed.

By comparing the intensity of the specific wavelength with the intensity of the same wavelength of light produced from a control sample containing none of the metal ions being tested, we can measure the degree of absorption, known as the absorbance. We can prepare a series of solutions of different known concentrations and measure their absorbances. These values can be plotted on a set of axes and the relationship between absorbance and concentration represented by a straight line of best fit. From this line, we can calculate the concentration of ions in a solution by interpolating its measured absorbance.

Note

A calibration graph may only be linear over a small range of values. If a sample's absorbance value lies outside the range of a calibration curve, you may need to prepare solutions with a wider range of known concentrations.

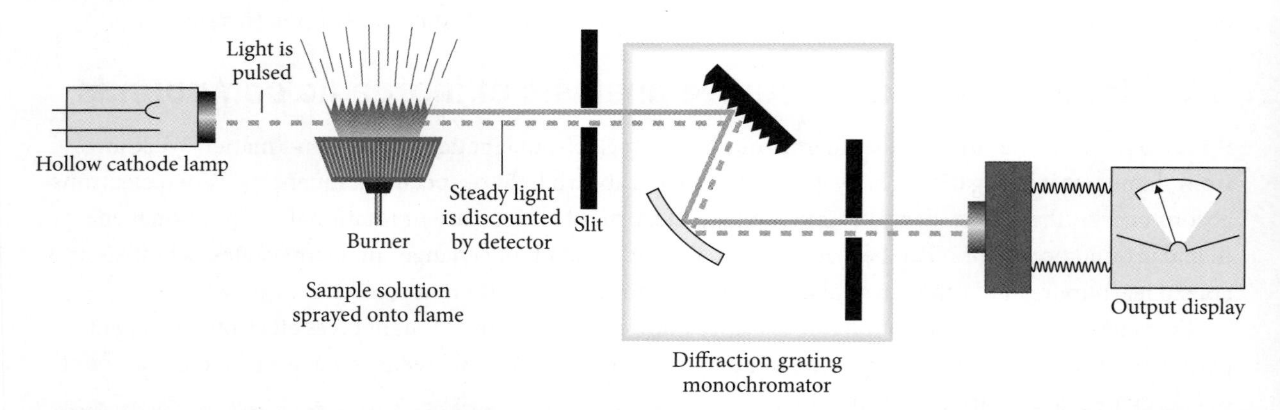

FIGURE 4.5 The main components of an atomic absorption spectrometer

Colourimetry

In Chapter 1, we investigated the determination of equilibrium constants using **colourimetry**. Colourimeters comprise a light source, a filter that transmits a particular wavelength of visible light, a sample cell and a light detector. The key components are shown in Figure 4.6. Colourimetry works well for measuring quantities of ions in solutions at very low concentrations; for example, in parts per million (ppm).

Colourimeters measure absorbance by comparing the light intensity before and after the light passes through the sample. The absorbance of a reference cell containing pure water is used as a control. A sample with a value for absorbance contains a high number of ions in the coloured solution that absorbed that particular wavelength of light (i.e. a high concentration).

To accurately determine the concentration of ions in a solution using colourimetry, we need to construct another calibration curve. If we carry out this procedure using copper ions of varying concentration (which produce a blue solution) and record their absorbances using a colourimeter, we can construct a calibration curve similar to the one in Figure 4.7.

If the sample had an absorbance of 1, we could interpolate from our calibration curve that the concentration of copper ions would be about 4.5 ppm.

Colourimeters often have several wavelength settings, and it may be useful to try several of these to produce the most valid calibration curve. Sometimes the wavelength with the highest absorbance may be absorbed by other elements in the mixture, so be careful to avoid interference from other substances in the sample.

The relationship between wavelength, absorbance and concentration is described by the Beer–Lambert law, which was discussed in section 1.3.4. The relevant equations are provided on the formulae sheet.

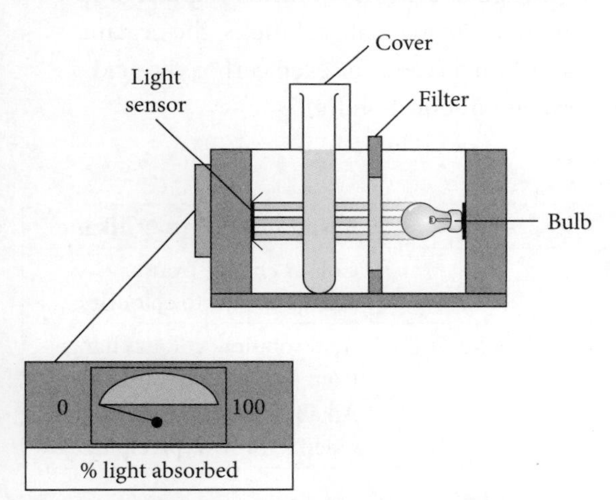

FIGURE 4.6 The main components of a colourimeter

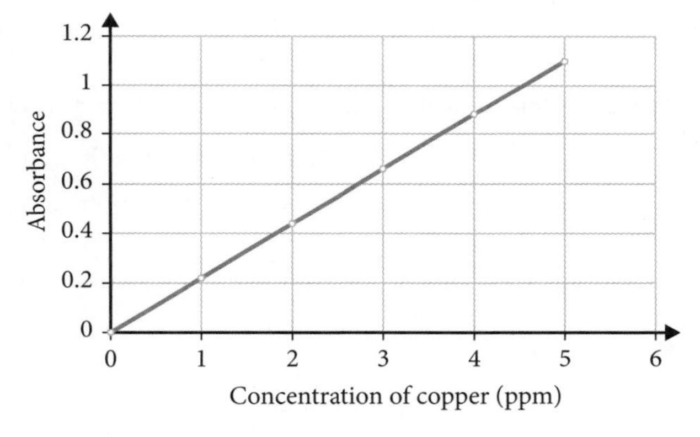

FIGURE 4.7 A calibration curve for copper ions in solution based on colourimetric data

Ultraviolet–visible spectroscopy

Absorption and emission bands are characterised by their wavelengths of maximum absorption and the intensity of absorption compared to a baseline or control. Flame tests can only be effective for elements with electrons that transition within the visible part of the electromagnetic spectrum. For some other elements, the UV part of the electromagnetic spectrum can be used.

UV–visible spectroscopy can determine ion concentrations in solutions with wavelengths in the ultraviolet and visible ranges of the electromagnetic spectrum. Although colourimetry can be used for solutions in the visible range too, it can only do so for a single wavelength at a time.

UV–visible spectroscopy can determine concentration of ions of inorganic substances such as transition metal complexes, but it is very widely used for carbon compounds, so we shall look at this spectroscopic technique in the next section.

4.2 Analysis of organic substances

In Module 7, we looked at the structure, properties and uses of hydrocarbons, alcohols, acids, esters and polymers. Here, we will look at ways in which we can distinguish chemically between similar groups of organic compounds. We will review some previously studied reactions and find out how spectroscopic analyses are employed to analyse organic compounds. Several technologies and techniques can help identify the important functional groups in organic compounds and the components of powders and environmental pollutants.

4.2.1 Identifying unsaturated hydrocarbons

The first functional group we will discuss is the C=C double bond in unsaturated hydrocarbons.

The high electron density of a double bond makes these compounds chemically less stable – they readily undergo an addition reaction across a double bond. Recall from Module 7 that the simplest way to identify the presence of a double bond is the bromine water test. If the addition reaction has a detectable colour change, an alkene is present (Table 4.4).

> **Note**
> For the bromine water test, we have considered non-polar bromine molecules migrating from the aqueous layer to the organic layer before adding across the double bond. In reality, bromine molecules will probably interact with water too, to form HBr and HOBr. Other products of addition reactions between an alkene and bromine water, namely a bromoalkane and a bromoalkanol, are possible.

An oxidation reaction can also be used to test for unsaturated hydrocarbons (Table 4.4). The chemistry associated with these tests is a little more complex. In alkaline solutions, the oxidation number of manganese changes from +7 (permanganate, MnO_4^-) to +6 (manganate, MnO_4^{2-}, which is green) and then to +4 (a manganese dioxide precipitate, MnO_2). In acidified permanganate solutions, the organic product, which is a diol initially, can be further oxidised to a carbonyl (C=O) or even carboxylic acid group, giving reaction products that were separated at the reactant double bond(s).

TABLE 4.4 Testing for the presence of unsaturated hydrocarbons

Test	Reaction (simplified)	Result indicating an alkane
Bromine water test	–C=C– + $Br_2(aq)$ → Br–C–C–Br	Colour change from orange-brown to colourless
Acidified permanganate test (use of strong oxidant)	–C=C– + $MnO_4^-(aq)$ → HO–C–C–OH + $MnO_2(s)$	Acid solution: colour change from purple to colourless Alkaline solution: purple to green to brown (precipitate)

4.2.2 Identifying hydroxyl groups

In Module 7, we looked at oxidation reactions involving alcohols. Many of the organic compounds we have looked at are colourless liquids, so we cannot use an identification technique that relies on colour. Provided that no water is present, sodium metal can be used to identify the presence of hydroxyl groups (Table 4.5).

If we identify the presence of the hydroxyl group, we can use an oxidation reaction to determine its position on the carbon chain (Table 4.5).

- Primary alcohols are oxidised to aldehydes. These can be further oxidised to carboxylic acids or even carbon dioxide and water, particularly in the case of methanol.
- Secondary alcohols are oxidised to ketones.
- Tertiary alcohols do not react and will not change the colour of permanganate solution.

TABLE 4.5 Testing for the presence of hydroxyl groups and alcohol type (primary, secondary or tertiary)

Test	Reaction (simplified)	Result indicating a hydroxyl group/position of group
Sodium metal test for presence of –OH	$C–OH + Na(s) \rightarrow C–O^-Na^+ + H_2(g)$	Release of hydrogen gas
Acidified permanganate test (use of strong oxidising agent) to identify position of –OH	$–C–OH + MnO_4^-(aq) \rightarrow –C{=}O–$	Colour change from purple to colourless (unless –OH group is part of a tertiary alcohol)

We could substitute the dichromate ion (orange) for the permanganate ion and it too would oxidise the alcohol and result in a green solution (due to the chromium ion). In cold solutions, the reactions proceed more slowly, and low oxidising agent concentrations may favour formation of the aldehyde from the primary alcohol. If we warm the solution with a high concentration of oxidising agent, then primary alcohols may continue to be oxidised beyond the aldehyde to the carboxylic acid.

4.2.3 Identifying carboxylic acid groups

An indicator test is an easy starting point to test for carboxylic acids. We also know that acids and carbonates produce carbon dioxide gas, so we can add a carbonate (Table 4.6) or test for the presence of carbon dioxide using the limewater test (Table 4.3, p. 148).

Another test that can be done is esterification. If we follow the process we used with an alcohol such as ethanol and notice a sweet smell from the ester, this indicates a carboxylic acid.

TABLE 4.6 Testing for the presence of carboxylic acid groups

Test	Reaction (simplified)	Result indicating a carboxylic acid group
Sodium bicarbonate test	$COOH + HCO_3^-(aq) \rightarrow COO^- + CO_2(g) + H_2O(l)$	Release of carbon dioxide gas
Esterification	$C_2H_5OH + CH_3COOH \xrightleftharpoons{\text{Conc. } H_2SO_4} CH_3—C(=O)—O—C_2H_5$ Ethanol Ethanoic acid → Ethyl ethanoate (ester)	Sweet smell

4.2.4 Instrumental quantitative analysis of organic molecules

The tests we have examined so far are primarily qualitative tests to help us determine the presence or absence of a particular functional group. Sometimes, we may want more information; for example, we may confirm the presence of an acid with a carbonate test, but how can we tell which acid it is?

Chemical instrumentation covers a wide range of wavelengths so that we can use electromagnetic radiation to analyse atoms, molecules and bonds. Table 4.7 summarises the various techniques of instrumental analysis, including three important techniques for studying organic compounds: **nuclear magnetic resonance (NMR)**, **mass** and **infrared (IR) spectroscopy**.

TABLE 4.7 Analytical techniques and their purpose

Electromagnetic radiation	Technique	Target	Effect	Description
Radio waves	Nuclear magnetic resonance (NMR) spectroscopy	Nuclei of atoms	Alters magnetic orientation if an odd total number of protons and neutrons in the nucleus	Focused on ^{1}H and ^{13}C Identifies number and position of atoms of a particular isotope
Microwaves	Microwave spectroscopy	Molecules	Rotation	Identifies bond lengths, angles and structures of molecules
Infrared waves	Infrared spectroscopy	Bonds between atoms	Vibrations (atomic flexing)	Identifies functional groups in organic molecules
UV–visible light waves	Atomic absorption spectroscopy (AAS) or colourimetry	Electrons	Transitions between ground states and excited states	Identifies specific cations and their concentrations
X-rays	X-ray crystallography	Atoms	Atoms scatter the rays in different ways	Produces 3D image of the organisation of atoms in a molecule
None (electric field and magnetic field)	Mass spectroscopy	Molecules	Molecules are broken into smaller fragments	Mass-to-charge ratios can be used to determine molecular mass and identify different parts of the molecule

Note

The techniques in Table 4.7 that are most relevant to the HSC Chemistry course are shaded.

Nuclear magnetic resonance spectroscopy

NMR spectroscopy is an important analytical technique for determining the structure of organic molecules. We can consider NMR spectroscopy as a kind of neighbourhood watch where we try to work out the types of environments housing the carbon or hydrogen atoms in a molecule and what sorts of neighbours they might have.

An atom's nucleus may have 'spin' if it has an odd mass number. Nuclei of the isotopes ^{1}H and ^{13}C do this. The HSC Chemistry course requires you to analyse the structure of simple organic molecules using carbon NMR (^{13}C NMR) spectroscopy and proton NMR (^{1}H NMR) spectroscopy.

It is useful to note these points when we analyse NMR spectra.

- Most NMR spectral outputs note 'TMS' (tetramethylsilane) at the zero mark. It is the standard reference point and does not provide any information about the sample.
- The numbers on the *x*-axis increase to the left, away from the TMS reference point, and are usually measured in ppm.
- The spectral peaks correspond to a particular carbon (or group of carbon atoms), or a hydrogen (or group of hydrogen atoms) within a molecule.
- Each peak represents a different environment.

To detect spin, we need to place the sample in a strong external magnetic field. In this way some of the nuclei will align with the direction of the magnetic field while others will oppose it. This allows us to measure the absorbed frequencies of radiation and make conclusions about the different environments associated with different nuclei. A simplified diagram is shown in Figure 4.8.

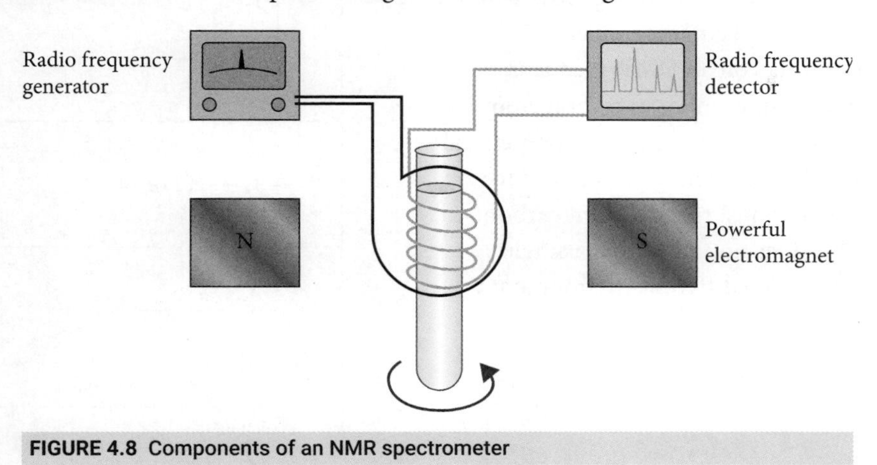

FIGURE 4.8 Components of an NMR spectrometer

Table 4.8 compares what is happening in the nucleus during NMR spectroscopy with what happens to electrons in UV–visible spectroscopy.

TABLE 4.8 A comparison of UV–visible spectroscopy and NMR spectroscopy

Electrons (UV–visible)	Nuclei (NMR)
Ground state is the low-energy state for the electron	Alignment to the magnetic field is the low-energy state for the nucleus
Electron takes in energy of a particular frequency (UV–visible region) to move to an excited state	Nucleus takes in energy of a particular wavelength (radio wave region) to spin around and reverse the direction of the magnetic field
Excited state for the electron occurs after it takes in energy	Unaligned state for the nucleus occurs after it takes in energy
The electron emits energy as it returns to its ground state, which is often detected as visible light	Nucleus emits energy as it returns to its lower aligned state; this energy will correspond to the radio wave frequency it absorbed

If this were the whole story, every ^{1}H nucleus should produce a consistent output. So should every ^{13}C nucleus, despite the fact that most organic molecules are dominated by ^{12}C atoms. Similar nuclei should take in the same amount of energy and release it as they return to their low-energy states. However, the interactions between the internal and external magnetic fields are complicated by shielding effects. Despite their tiny mass, moving electrons also induce magnetic fields and this happens both within atoms (unbonded electrons) and between atoms (bonded electrons). The variation in the number and type of atoms present in different compounds will influence the way the various ^{1}H or ^{13}C atoms respond to different frequencies of electromagnetic radiation. Hence, different environments will show slightly different patterns.

Interpreting NMR spectra

Your HSC Chemistry data sheet includes ^{13}C NMR chemical shift data, as shown in Figure 4.9. We need to use the data to interpret ^{13}C NMR spectra, without having a structural or molecular formula.

The number of peaks on an NMR spectrum represents the minimum number of carbon atoms (or hydrogens) present in a molecule. A symmetrical molecule may have four carbons and yet produce an NMR spectrum with just two different carbon environments. This does not mean that a single carbon atom exists in two different environments.

^{13}C NMR chemical shift data

Type of carbon	δ/ppm
—C—C—	5–40
R—C—Cl or Br	10–70
R—C(=O)—C—	20–50
R—C—N	25–60
—C—O— alcohols, esthers or esters	50–90
C=C	90–150
R—C≡N	110–125
(benzene ring)	110–160
R—C(=O)— esters or acids	160–185
R—C(=O)— aldehydes or ketones	190–220

FIGURE 4.9 ^{13}C NMR chemical shift data (from HSC Chemistry data sheet)

Analysing ^{1}H NMR spectra

Let's consider ethanol, whose structural formula is shown in Figure 4.10.

On a ^{1}H NMR spectrum at low resolution we see three different environments for hydrogen:

- *environment 1*: three hydrogen atoms of $-CH_3$
- *environment 2*: two hydrogen atoms of $-CH_2-$
- *environment 3*: single hydrogen atom bonded to the oxygen atom next to $-CH_2-$.

FIGURE 4.10 The structural formula of ethanol with hydrogen environments numbered

Three different environments for hydrogen appear as three peaks. Protons in the same environment have similar peaks, which add together to increase the height of the peak (or, more accurately, the area under the peak). This is referred to as their intensity. Their location can be seen by the relative shift to the left (downfield) from the zero mark. The values are a result of different amounts of shielding of the protons by the surrounding electrons.

High-resolution NMR spectroscopy shows an additional level of detail. Depending on the molecule, the peaks appear as clusters. If there is more than one hydrogen nucleus, the spin of a hydrogen nucleus can couple with that of its neighbouring protons in **spin–spin coupling**. Their spins can either align (magnify) or oppose (cancel) one another. The greater the number of neighbouring protons, the greater the range of possibilities of alignment or opposition.

We can use the $n+1$ (or neighbours + 1) rule to describe splitting. This splitting can affect the intensity as the area under the peak may now be represented by many peaks, so our simple rule that three times the number of hydrogens has a peak area three times that of a single hydrogen will now depend on how the signal is split by the neighbours.

For ethanol, splitting can be identified for the hydrogen atoms as follows:

- *environment 1*: each of the three hydrogen atoms of $-CH_3$ has two neighbouring hydrogens on the central carbon. Using the rule, 'neighbour (n) + 1' is 2 + 1 = 3. So these three hydrogens have a higher peak (because all three hydrogen atoms on the same carbon atom have the same neighbours). Hence we would have a triplet split.
- *environment 2*: $-CH_2-$ is bonded to an oxygen atom and to $-CH_3$. This means that each of the two hydrogen atoms has three hydrogen neighbours. So $n + 1$ is 3 + 1 = 4. This means the two central hydrogens are seen as four peaks (a quartet).
- *environment 3*: the hydrogen bonded to the oxygen does not interact with hydrogen atoms on a neighbouring atom and will always appear shifted further left and as a singlet: $n = 0$, and 0 + 1 = 1.

The features of the ^{1}H NMR spectra for ethanol are summarised in Table 4.9.

TABLE 4.9 Hydrogen environments for ethanol

Component of ethanol	Number of ^{1}H	Intensity	Location (ppm)	Number of neighbours (n)	$n+1$	Splitting
CH_3	3	3 × peak size	1.5	2	3	Triplet (1:2:1)
CH_2	2	2 × peak size	3.5	3	4	Quartet (1:3:3:1)
OH	1	1 × peak size	5.5	0	1	Singlet (1)

Note

The way in which the peak is split can be found using the 'neighbours + 1' value and then using Pascal's triangle (singlet = 1, doublet = 1 : 1, triplet = 1 : 2 : 1, quartet = 1 : 3 : 3 : 1, quintet = 1 : 4 : 6 : 4 : 1, and so on).

The ^{1}H NMR spectrum for ethanol is shown in Figure 4.11. You can read the environments from right to left. The large triplet peak of the protons in the $-CH_3$ group occurs at about 1.5 ppm, followed by the smaller quartet of the $-CH_2-$ protons at 3.5 ppm and finally the singlet of the hydrogen on the $-OH$ group at about 5.5 ppm.

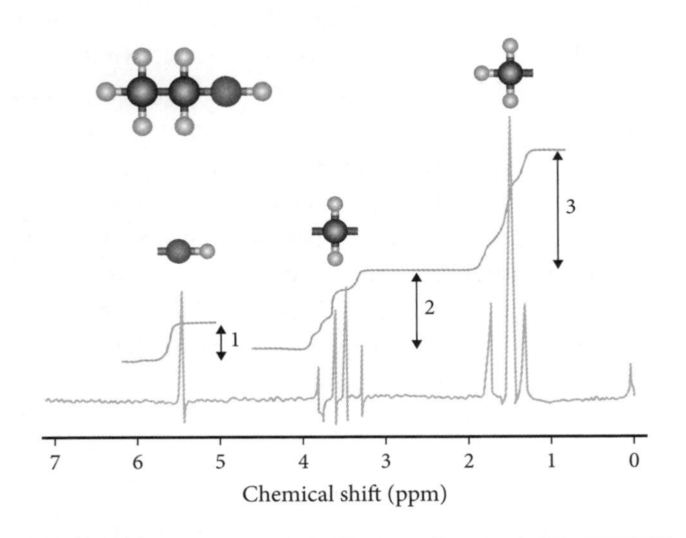

FIGURE 4.11 The high-resolution proton NMR spectrum for ethanol

Analysing ^{13}C NMR spectra

Now let's look at the ^{13}C NMR spectrum for ethanol. The ^{13}C isotope is much rarer than the ^{1}H isotope, so this does not always provide the same quality or quantity of data as ^{1}H. However, there are some important features to observe.

Consider the structure of ethanol with carbons numbered in Figure 4.12. We can see two different environments for carbon on an ^{13}C NMR spectrum:

- *environment 1*: a carbon atom is bonded to two hydrogen atoms, a second carbon atom and an –OH group on the first carbon atom
- *environment 2*: the second carbon atom is attached to the first carbon atom as well as to three hydrogen atoms.

Note

The carbon chain for ethanol is numbered so that C1 is the carbon closest to the –OH functional group.

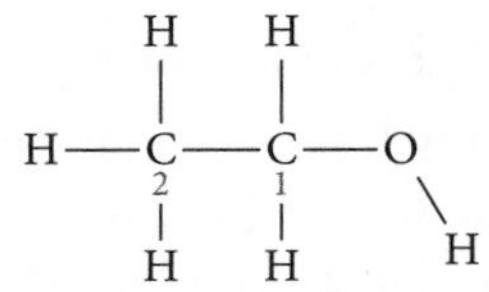

FIGURE 4.12 **The structural formula of ethanol with carbon environments numbered**

This means we should expect two peaks for the carbon ^{13}C NMR spectrum, corresponding with the two different carbon environments. These are shown in Figure 4.13.

- *environment 1*: the $-CH_2OH$ carbon environment is visible as the peak at about 60 ppm
- *environment 2*: the $-CH_3$ carbon environment shows as a peak at about 20 ppm.

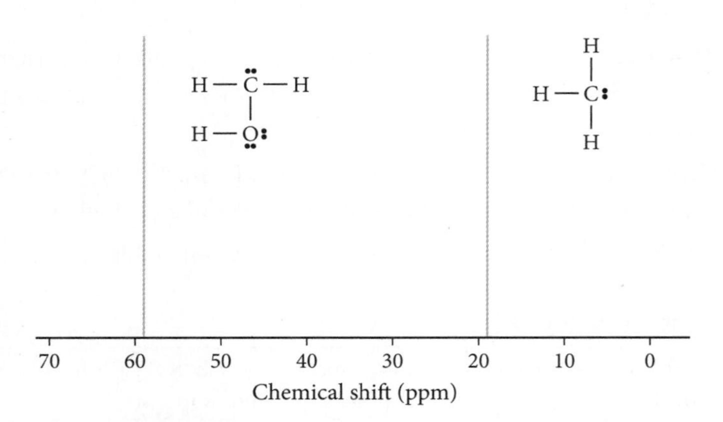

FIGURE 4.13 **The ^{13}C NMR spectrum for ethanol**

Matching the environment to the peak is much easier when we know which compound we are interpreting, but what if we did not know this was ethanol? The data sheet shows several possibilities for the peak at 20 ppm, including alkanes and haloalkanes. The other peak is near 60 ppm. There are again a couple of options including alcohols, ethers and esters.

Combining the ^{13}C NMR spectrum with the ^{1}H NMR spectrum might fill in a few gaps, and there are several other spectra that can help with identification. For example, we could compare ethanol and ethene. The structural formula for ethene is shown in Figure 4.14.

The ^{13}C NMR spectrum shows two carbon atoms, so we might expect two carbon environments. However, the molecule is symmetrical, so there is only one peak.

For the ^{1}H NMR spectrum, there are four hydrogen atoms, and every hydrogen atom is equivalent. They have the same arrangement on their parent carbon atom and the same neighbours, hence they represent a single hydrogen environment. Each hydrogen has two neighbours, so $n + 1 = 3$. This results in a triplet with the ratio 1:2:1.

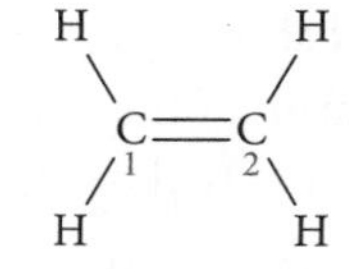

FIGURE 4.14 **The structural formula of ethene with carbons numbered**

Mass spectroscopy

Of all of the analytical techniques we look at, **mass spectroscopy** is probably the most sensitive. It is certainly more sensitive than either IR or NMR spectroscopy. It applies Newton's second law, which states that the acceleration of a particle will be inversely proportional to its mass if the applied force is constant. More simply, smaller particles (of lower mass) will experience a greater acceleration from an external force than more massive particles.

In order to be affected by the force due to a magnetic field, the particles passing through the magnetic field need to be charged. The greater the charge, the greater the force and hence the greater the acceleration, so for this technique we need to consider the mass-to-charge ratio.

During mass spectroscopy, the sample is first vaporised, then ionised, usually as a result of electron loss, making a positive charge (Figure 4.15). These rapidly moving charged particles are exposed to a strong (but variable) magnetic field. The ionised particles are deflected by the magnetic field as shown in Figure 4.16.

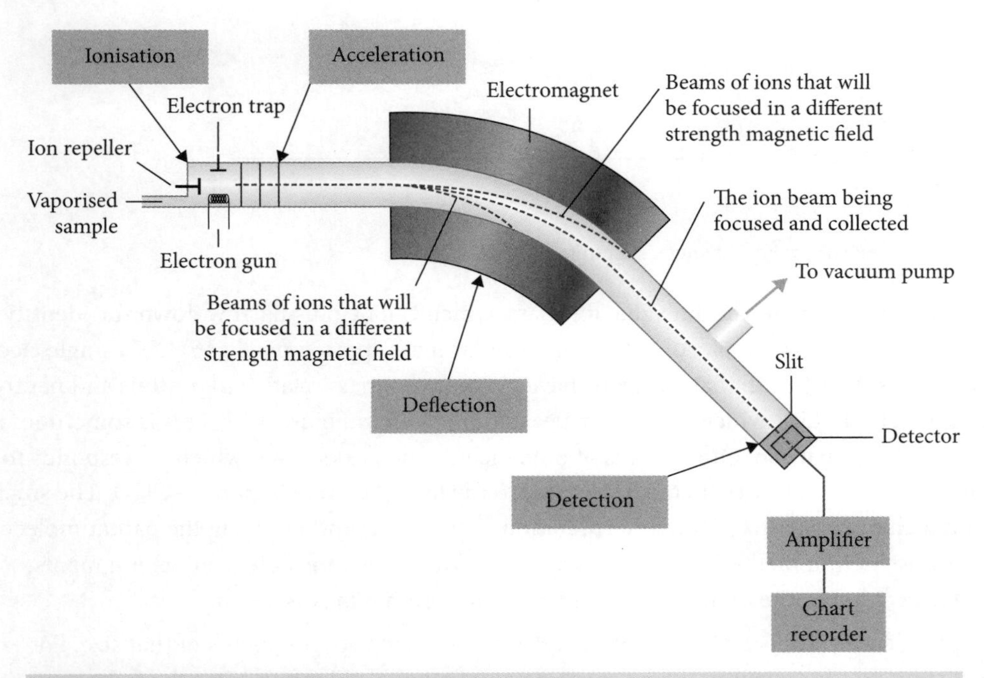

FIGURE 4.15 The components of a mass spectrometer

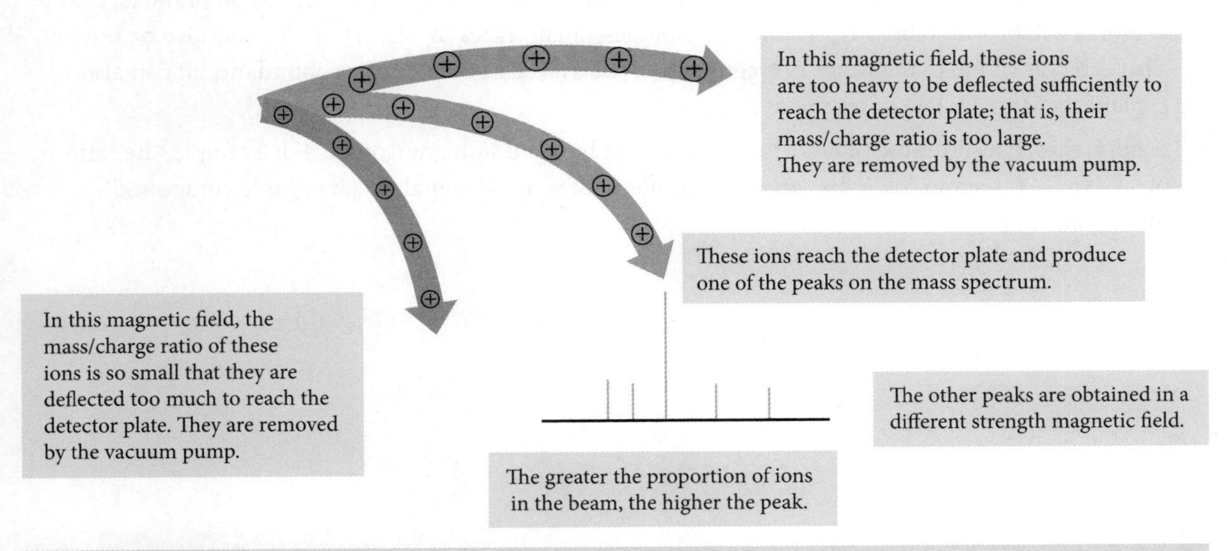

FIGURE 4.16 The effect of mass on ion deflection in a mass spectrometer

9780170465281

The amount of deflection will depend on charge and mass, so particles of different masses must be detected and their charge determined (through the regaining of electrons). This allows the calculation of a mass-to-charge ratio (m/z). Larger molecules may also be fragmented, and they will also be detected.

Working out what sorts of fragments are possible provides valuable information about the structure of molecules. Ethanol is used as an example in Figure 4.17.

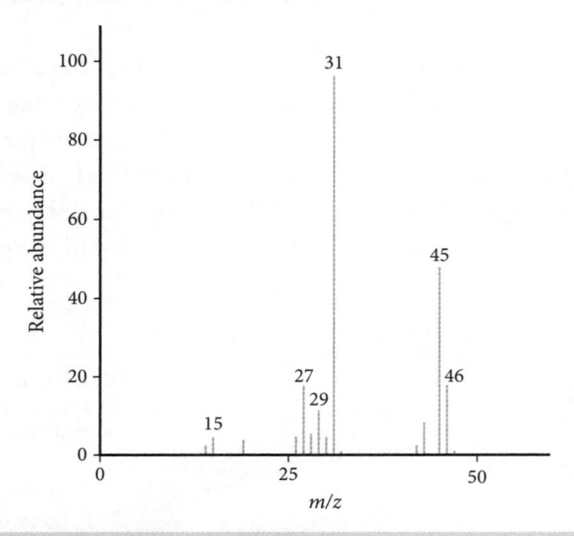

FIGURE 4.17 The mass spectrum of ethanol

Mass spectra provide a maximum value for mass, which can help us narrow down the identity of an organic molecule. If an intact molecule of ethanol (molar mass approximately 46) has a single electron removed at the first step (recall that electrons have virtually no mass relative to protons and neutrons), its mass-to-charge ratio is 46:1, which shows as a peak at $m/z = 46$ in Figure 4.17. This is sometimes referred to as the parent molecular ion. Other peaks of note include the peaks at 45 (which corresponds to the loss of one hydrogen atom) and 31 (which corresponds to the loss of a methyl group, $-CH_3$). The small peak at 15 represents a charged methyl group. Interpretation of the peaks and knowing the parent molecule mass is a useful reconstruction method when trying to identify a whole molecule from its fragments.

We need to remember the following points when interpreting mass spectra.

- The height of each peak is a measure of the relative abundance of fragments of that size. For example, the most abundant fragment in the mass spectrum of ethanol (Figure 4.17) is the fragment with a mass-to-charge ratio of 31.
- Isotopes of elements within a molecule can affect the size of the fragments. Almost all of the C, H and O atoms will be the stable ^{12}C, ^{1}H and ^{16}O isotopes; small traces of ^{13}C, ^{2}H or ^{18}O may also be present. This can contribute to the range of fragment masses as well as their relative abundance. It can also explain the large variety of fragment sizes.
- Halogens such as bromine or chlorine often do not have a dominant isotope. For example, the ratio of ^{35}Cl to ^{37}Cl is about 3:1. This can add a number of fragment signals to an organic compound's mass spectrum.

We can show the effects of fragmentation by looking at propane. The structural formula for propane is shown in Figure 4.18.

When propane is ionised through the loss of an electron, it looks as shown in Figure 4.19.

This ionisation is necessary so propane will be deflected by the magnetic field. Also, several other fragments are commonly produced from propane. A methyl group ($-CH_3$) with $m/z = 15$ may be fragmented off one end, leaving a $-C_2H_5-$ fragment with $m/z = 29$. Hydrogen atoms may also be removed from different sites, as indicated in Figure 4.20.

This range of fragments and their relative abundances produce a mass spectrum similar to the one shown in Figure 4.21.

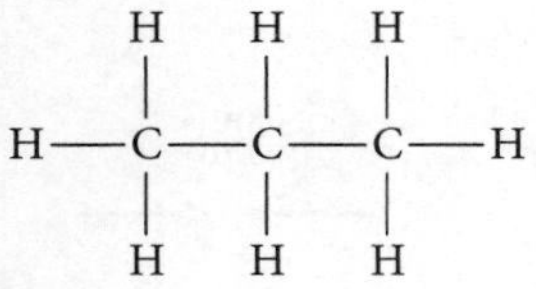

FIGURE 4.18 The structural formula of propane

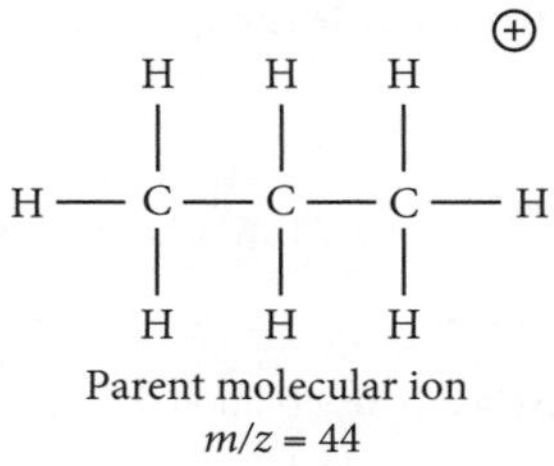

FIGURE 4.19 The structural formula of a propane ion (due to the loss of an electron)

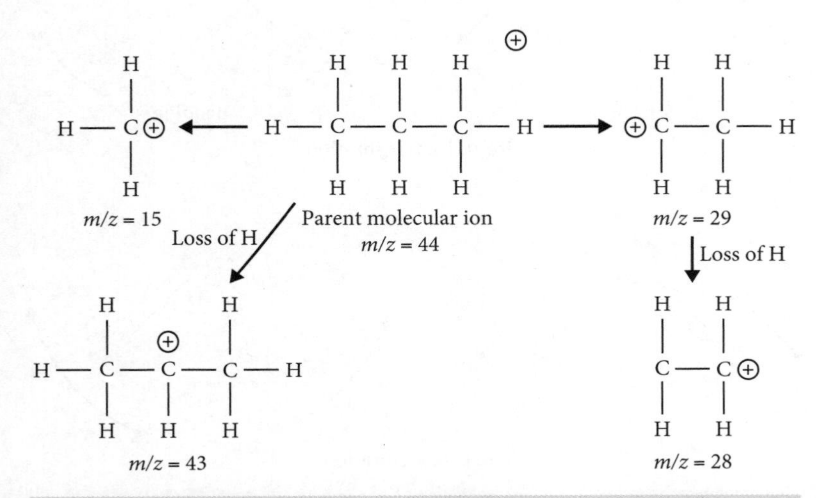

FIGURE 4.20 The structural formulae of different fragments of propane ions

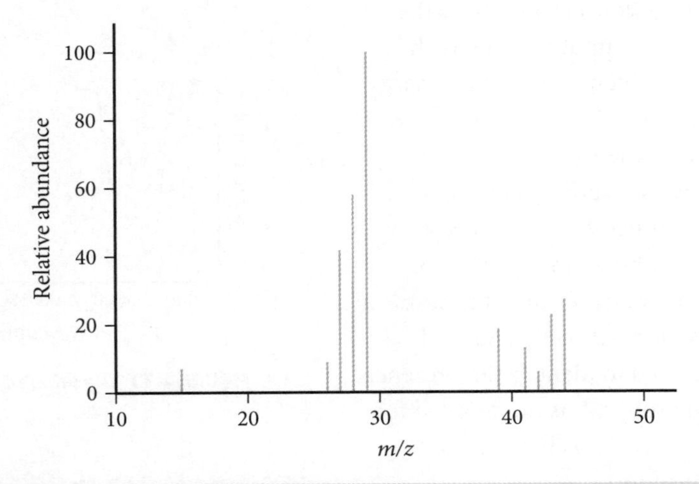

FIGURE 4.21 The mass spectrum of propane

Infrared spectroscopy

In IR spectroscopy, a detector measures the absorption of infrared energy by different bonds within a molecule, so IR spectroscopy provides us with information about bond types. Covalent bonds in infrared radiation vibrate in ways that depend on the strength of the bonds and the mass of the atoms. This bond vibration can be symmetric stretching, asymmetric stretching, scissoring, rocking, wagging or twisting, as shown in Figure 4.22.

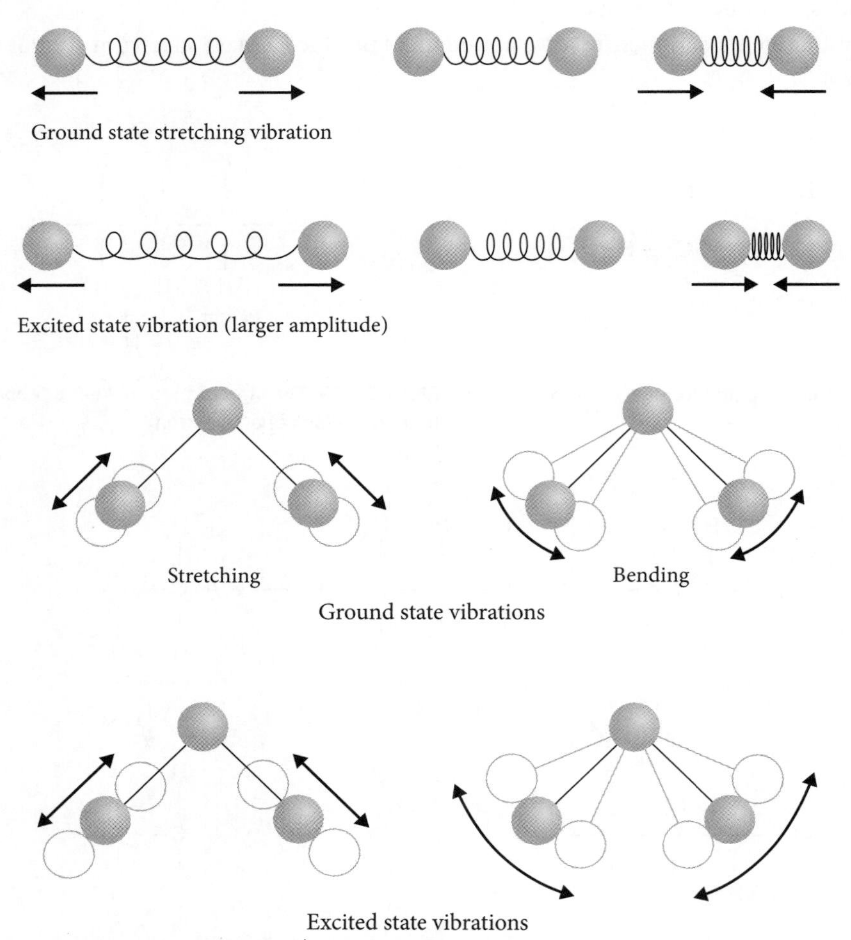

FIGURE 4.22 Infrared vibration types for diatomic and triatomic molecules

The key to interpreting IR spectroscopy is the 'upside-down' display – the output represents the percentage transmission (which relates to IR energy that is not absorbed). The y-axis is 0–100%, while the x-axis represents the wavenumber, which is the inverse of wavelength ($\frac{1}{\lambda}$) or the number of waves per centimetre. The IR spectrum for methanol (CH_3OH) is shown in Figure 4.23.

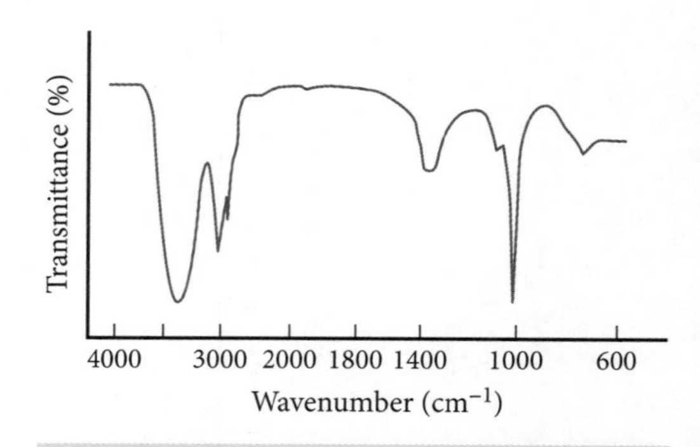

FIGURE 4.23 The infrared spectrum for methanol

The characteristics of bonds relate to the atoms in the bond and any other atoms directly attached, so IR spectroscopy can be used to identify the presence of particular functional groups. It is not as good for identifying the length of a carbon chain or the position of functional groups.

For our purposes, the most important aspect of IR spectroscopy is how we use it to analyse the structure of organic compounds.

If we compare the spectrum for methanol in Figure 4.23 to the IR data in the HSC Chemistry data sheet (Figure 4.24), we can identify some absorption bands or troughs. The most important is the O–H bond. This appears at 3400–3500 cm^{-1} as a broad tongue corresponding to the alcohol (hydroxyl) functional group. There is a small trough at 3000 cm^{-1} (identified as C–H on the data sheet) and a deep trough at about 1000 cm^{-1} (the C–O bond).

This sort of analysis shows both the benefits and the limitations of IR spectra. IR spectroscopy is good for identifying different functional groups, but it doesn't show where in a substance a functional group is located. In the spectrum for methanol, the absorption band at 750–1100 cm^{-1} may even confuse us about the number of carbon atoms in the molecule if we didn't know that the substance was methanol and hence has just one carbon atom.

Here is one more example, this time with an unidentified substance. All the information we have is the IR spectrum shown in Figure 4.25.

Without any other information, we must just look for functional groups.

- The broad band at about 3000 cm^{-1} is quite distinctive. The data in Figure 4.24 suggests an O–H or C–H bond. Because it is broad, it is more likely to be an O–H bond.
- A broad, smooth 'tongue' at about 3000 cm^{-1} indicates an alcohol group; however, this one looks more like a beard. This suggests the O–H bond of an organic acid.
- To confirm an organic acid, we also need to identify a C=O bond. Figure 4.24 shows that the C=O bond occurs at 1680–1750 cm^{-1}. A clear, well-defined trough corresponds to this value, suggestive of a C=O bond.
- Another significant trough occurs at about 1300 cm^{-1}. From Figure 4.24, this suggests a C–O bond.
- Other troughs are consistent with C–C and C–H bonds, so we can be pretty confident that this is a carboxylic acid. In fact, this is the IR spectrum for ethanoic acid.

Infrared absorption data

Bond	Wavenumber/cm^{-1}
N—H (amines)	3300–3500
O—H (alcohols)	3230–3550 (broad)
C—H	2850–3300
O—H (acids)	2500–3000 (very broad)
C≡N	2220–2260
C=O	1680–1750
C=C	1620–1680
C—O	1000–1300
C—C	750–1100

FIGURE 4.24 Infrared absorption data from the HSC Chemistry data sheet

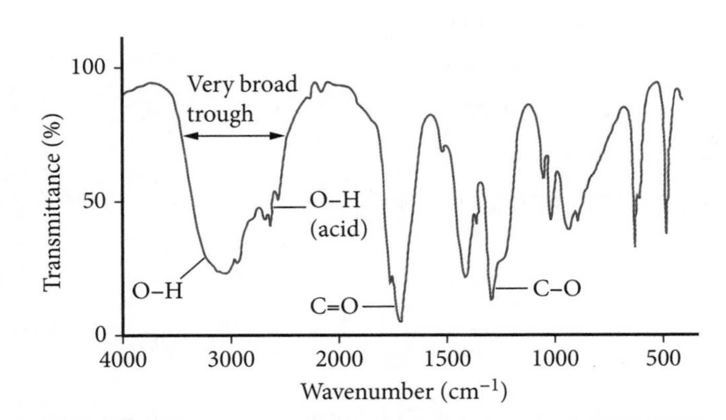

FIGURE 4.25 Infrared spectrum of an unknown organic compound

Ultimately, if we want to identify an unknown organic molecule, we need to put together a range of spectral outputs, including NMR, mass and IR spectra, to help us build a more comprehensive picture of what type of organic molecule may be present in a sample. Table 4.10 compares the advantages and disadvantages of these analytical techniques for organic compounds.

TABLE 4.10 Analytical techniques for organic compounds

Instrumental technique	What it does	Advantages	Disadvantages
UV–visible spectroscopy	Can detect if a species is present Can be quantitative	Useful for coloured organic and inorganic species Simple to operate Very cheap Small sample size Quick sample preparation	Not very sensitive
NMR spectroscopy	Determines C–H backbone of organic molecules	Highly sensitive and precise Small sample size Sample can be in solution	Very expensive to buy and operate
Mass spectroscopy	Determines molar mass, isotopic abundances	Very sensitive Small sample size	Very expensive to buy and operate
Infrared spectroscopy	Determines functional groups present	Huge range of analytes and samples Small sample size	Moderately expensive

4.3 Chemical synthesis and design

Chemical synthesis is the process of making a new chemical substance. In an industrial process, the focus is on scale and economy – large quantities need to be produced while keeping costs manageable. This means managing the various factors that affect equilibrium systems, as well as the cost of materials, disposal of wastes and staff safety.

For any industrial process, you should be able to discuss the:

- availability of reagents
- reaction conditions
- **yield** and **purity**
- industrial uses
- environmental, social and economic issues.

4.3.1 Availability of reagents

CASE STUDY 1

INDUSTRIAL PRODUCTION OF SODIUM CARBONATE

Sodium carbonate is an industrially important compound; for example, in the manufacture of glass and the production of other chemicals. It is an alkaline substance like sodium hydroxide but it is cheaper to produce. The Solvay process is used to produce sodium carbonate. The reagents are sodium chloride and calcium carbonate, and the products are sodium carbonate and calcium chloride (a waste product). Ammonia is a raw material required during this process, and it is recovered for reuse.

The Solvay process involves four main steps:

1. brine (highly concentrated salt solution) purification
2. sodium hydrogen carbonate formation
3. sodium carbonate formation
4. ammonia recovery.

The overall process can be represented as:

$$2NaCl(aq) + CaCO_3(s) \rightarrow CaCl_2(aq) + Na_2CO_3(s)$$

An important factor in industrial processes is scale. Producing commercial quantities of a product requires large quantities of reagents; in this case, sodium chloride is sourced from salt water and calcium carbonate is sourced from limestone.

Brine purification refers to increasing the concentration of sodium chloride in salt water by evaporation. Ions such as magnesium and calcium are removed by precipitation. Ideally an industrial Solvay plant should be close to a limestone quarry (a solid mineral deposit) and/or a brine supply (e.g. near the ocean). If either raw material is not close, costs of transportation significantly increase. It is more cost effective to transport the solid limestone than the liquid brine.

4.3.2 Reaction conditions

CASE STUDY 2

INDUSTRIAL PRODUCTION OF AMMONIA

Ammonia is an important industrial chemical. It is used as a fertiliser, in many alkaline cleaning products and as a refrigerant gas. Ammonia is also used to manufacture fertilisers (such as ammonium sulfate, ammonium nitrate, ammonium hydrogen phosphate and urea), nitric acid (used to manufacture fertiliser, dyes, fibres, plastics and explosives) and cyanides (used to manufacture synthetic polymers and to extract gold from ore bodies).

The synthesis reaction to produce ammonia, known as the Haber process, is:

$$N_2(g) + 3H_2(g) \rightleftharpoons 2NH_3(g) \qquad \Delta H = -92\,kJ$$

The stages of the Haber process are shown in Figure 4.26.

The reaction for the synthesis of ammonia from nitrogen gas and hydrogen gas does not go to completion. It is a reversible reaction that reaches an equilibrium (rate of reactants forming product equals rate of products re-forming reactants). Varying the reaction conditions can significantly affect the behaviour of the reactants and products of this reaction.

Effect of temperature

Higher temperatures increase the kinetic energy of the reactant molecules. A greater proportion of the molecules reach energies greater than the activation energy, so they react. Thus, higher temperatures increase the rate of the reaction.

9780170465281

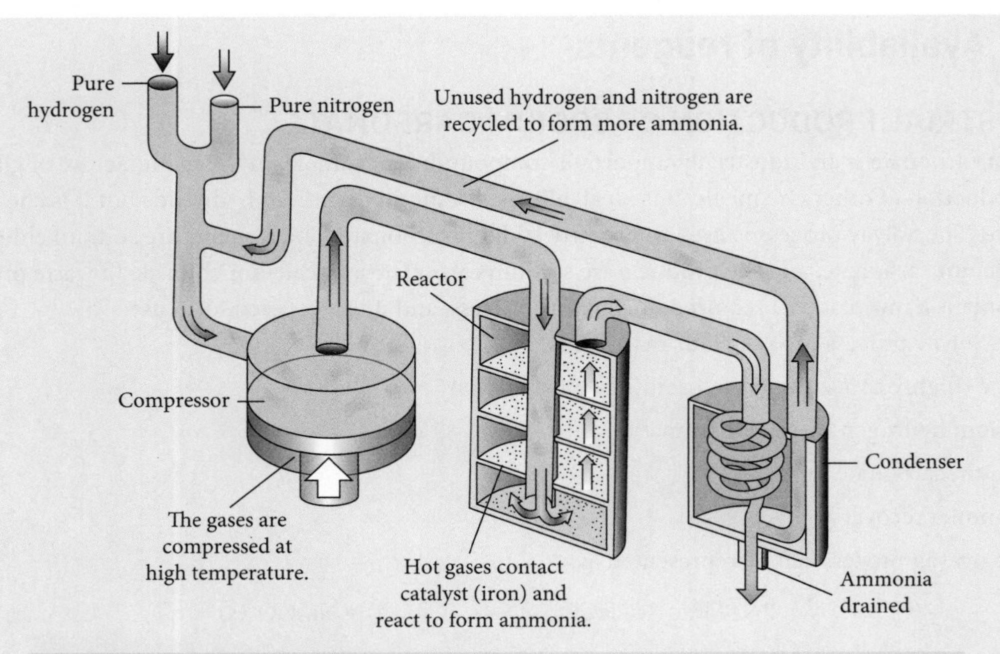

FIGURE 4.26 The industrial production of ammonia

The synthesis of ammonia is an exothermic reaction (i.e. releases heat). This means the reverse reaction is endothermic. Increasing temperature is equivalent to adding energy to the system. This is a change in the equilibrium conditions. Le Chatelier's principle predicts that the system will respond to counteract the change. In this case, the equilibrium will shift to absorb the extra energy, which means it favours the endothermic reaction. Hence, higher temperatures in the reaction reduce the yield of ammonia in the Haber process.

The Haber process is a commercial process. There are costs associated with the manufacture of ammonia and hence it is important to have a high reaction rate and a high yield. Both reaction rate and yield are temperature dependent. As temperature increases, the rate of reaction increases, but yield decreases. At lower temperatures, rate of reaction decreases and yield increases.

A moderate temperature is required to balance the competing demands of reaction rate and yield. This provides the maximum yield of ammonia in the shortest time. Temperatures for this reaction are typically 400–500°C.

Effect of pressure

Gas pressures are also important to a system in equilibrium when one or more species are in the gaseous state. In the Haber process, 4 moles of reactant gases produce 2 moles of ammonia gas (the product). As a result, the forward reaction results in a decrease in the pressure of the system.

An increase in pressure causes the equilibrium to shift in the forward direction, to counteract the change by reducing the pressure. Higher pressures result in an increased yield of ammonia in the Haber process. Typical pressures used in the Haber process are 15–35 MPa.

Effect of a catalyst

Using a catalyst does not change the position of equilibrium, but it can provide a pathway of lower activation energy for reactants. This means the reaction temperature does not need to be as high for the reaction to proceed. One catalyst used in the Haber process is magnetite (Fe_3O_4). The magnetite may be fused with smaller amounts of promoter substances consisting of other metal oxides. The catalyst is ground finely, ensuring it has a large surface area (around $50\,m^2\,g^{-1}$), and the magnetite is reduced to iron. The large surface area allows gaseous molecules to rapidly absorb and react.

The use of the catalyst allows the reaction rate to remain high, despite the reaction occurring at a moderate temperature. All these factors make it critical to monitor the reaction vessel during the Haber process to ensure there is a safe balance of reaction conditions to ensure maximum yield at minimal cost.

4.3.3 Yield and purity

CASE STUDY 3

CHAPTER 4

INDUSTRIAL PRODUCTION OF SODIUM HYDROXIDE

Sodium hydroxide is produced industrially by electrolysis. In electrolytic cells:

1. the cathode (reduction electrode) is negatively charged
2. the anode (oxidation electrode) is positively charged
3. anions carry charge towards the anode
4. cations carry charge towards the cathode.

Electrons flow through the external circuit from the anode to the cathode. We can use the table of standard electrode potentials on the HSC Chemistry data sheet to work out what voltage must be applied to get a reaction to occur, in the same way we work out the electropotential of two half-cells. This technique is used for recharging rechargeable batteries.

Electrolysis of concentrated sodium chloride solutions is used industrially to produce sodium hydroxide. These reactions depend on the concentration of the solution and the particular voltage applied. The NaCl comes from sea water or salt mines.

Chloride ions are oxidised at the anode, forming chlorine gas:

$$2Cl^- \rightarrow Cl_2(g) + 2e^-$$

Note

Chloride ions, rather than water molecules, are oxidised at the anode – mainly due to the high concentration of chloride ions in the solution and the type of electrode used.

Sodium ions are too stable to be reduced at the cathode; instead, water is reduced to hydroxide ions and hydrogen gas:

$$2H_2O(l) + 2e^- \rightarrow H_2(g) + 2OH^-(aq)$$

The sodium ions are spectators and remain in solution, so:

Net ionic equation: $2H_2O(l) + 2Cl^- \rightarrow Cl_2(g) + H_2(g) + 2OH^-(aq)$

Full formulae equation: $2NaCl(aq) + 2H_2O(l) \rightarrow Cl_2(g) + H_2(g) + 2NaOH(aq)$

Electrolysis of molten sodium chloride is much simpler. Sodium chloride, as with chlorides of other group 1 and 2 elements, can be melted and electrolysed to produce the metal and chlorine gas:

$$2Na^+ + 2Cl^- \rightarrow 2Na(l) + Cl_2(g)$$

Three different methods are used in electrolysis of sodium hydroxide from sodium chloride solution: the mercury process, the diaphragm process and the membrane process. As well as some significant environmental and social issues associated with both the mercury process (use of a dangerous heavy metal) and the diaphragm process (use of asbestos), there are issues with yield and purity. Table 4.11 compares these processes.

TABLE 4.11 A comparison of industrial processes used in the production of sodium hydroxide

Process	Yield	Purity
Mercury process	Dependent on flow of mercury and high voltage (energy) to ensure all Na^+ ions are reduced and form an amalgam with Hg(l)	Very pure NaOH No contact between Cl^- ions and NaOH(aq)
Diaphragm process	High electric currents required for high yield leads to high costs	Lower purity due to contamination of NaOH by Cl^- ions
Membrane process	Lower electric current requirements can lead to better yields for a cost similar to diaphragm process	High purity (dependent on purity of brine, particularly other cations because the membrane is selectively permeable)

9780170465281

4.3.4 Industrial uses

Industrial chemical processes are large-scale chemical reactions involving large amounts of reactants and products. Synthesis of a single product may reduce the need to manage unwanted by-products by converting them to something else, selling them or disposing of them as wastes. Industries are businesses driven by the fundamentals of economics: supply and demand. Chemicals must be available and close to or easily accessed by those who need them. Many chemical design processes are developing from the need to replace existing materials with new materials.

Orica manufactures ammonia by the Haber process. Much of the ammonia produced is used in the manufacture of ammonium nitrate, which is sold to mining companies to be used as an explosive. However, carbon dioxide is a by-product. Rather than releasing the carbon dioxide to the atmosphere, Orica recovers some of the carbon dioxide and converts it into dry ice for refrigeration. In this way, the company gains additional income from a by-product while reducing its carbon dioxide emissions.

Pharmaceuticals

Some of our most important **pharmaceuticals** have been sourced from **natural products**. Many pharmaceuticals are synthesised to treat or relieve the symptoms of diseases. Many pharmaceuticals are complex organic compounds whose synthesis may involve many steps. Seeking greater efficiencies in chemical pathways is an ongoing challenge for research and analytical chemists.

One common example is the synthesis of aspirin (acetyl salicylic acid). Aspirin is a complex organic molecule with both a carboxylic acid and an ester functional group. It was originally collected from natural sources such as willow tree bark, but is now synthetically mass produced by a complex esterification reaction. For pharmaceutical products, purity is a critical factor.

Cosmetics

The cosmetics industry is enormous in terms of both chemical production and the commercial market for cosmetic products. Greater awareness of the testing of cosmetics on animals, problems with the small polymer beads in some scrubs and levels of polymer waste in general have been driving change in the cosmetics industry. More consideration is being put into making products that have fewer negative environmental impacts.

An important group of organic compounds in the cosmetic industry are esters. Esters are colourless and volatile so they do not stain the skin, but they vaporise easily and distribute their scent. Some of the more common esters have a fruity scent, which makes them ideal ingredients in many perfumes. Esters also have some properties that make them a good choice as skin conditioners, helping skin to remain soft, and reducing dryness. They also tend to be low-irritant, which reduces the chance of itching or dermatitis.

Cleaning products

CASE STUDY 4

INDUSTRIAL SAPONIFICATION

Cleaning products have significant industrial and domestic applications. Their synthesis is different to the process used in the school laboratory.

During saponification, an ester is substituted by a fat or oil. Fats and oils consist of one alcohol molecule (e.g. glycerol) combined with three acid molecules. The acids are long-chain acids, which may be saturated (produce solid fats) or unsaturated (produce liquid oils). The fat or oil is heated in 30% NaOH solution until it dissolves. Any solid residues are skimmed off during cooling.

The solution is mixed with concentrated brine and then left to stand. Curds of soap form on top of liquid glycerol in brine. The curds are removed and washed, and perfumes or colours are added before they are moulded and packaged. The glycerol can be recovered by neutralisation of the excess sodium hydroxide and distillation. It is then suitable for use in the production of confectionery, cosmetics, pharmaceuticals and explosives. Sodium salts are used to produce bar soaps, whereas potassium salts are used in shampoos and shaving soaps.

There are several main differences between industrial and small-scale laboratory saponification.

- Industrial processes occur at much higher temperatures.
- Industrial processes often continue over days or weeks to reach completion.
- Fats and oils are less volatile than esters and hence do not need to be refluxed.
- Excess sodium hydroxide is neutralised and water is distilled industrially, improving the purity of the final product.
- The industrial 'salting out' process is difficult to replicate in a school laboratory.

Synthetic detergents may be one of several types.

- **Anionic detergents** are the original and still most widely used group of detergents. They are used in laundry detergents and dishwashing liquids. The structure of the molecules is very similar to that of natural soaps. Each has a long, non-polar tail and an anionic head (a sulfonate, $C–O–SO_2–O–$, similar to SO_4^{2-}). They work just like soaps but are slightly more effective. They produce a lot of foam, but this has no effect on the cleaning ability of the surfactant.
- **Cationic detergents** contain a region that is an alkyl derivative of ammonium. The long chains are non-polar and the nitrogen region is water soluble. These are the preferred cleaner for plastics, as well as being used as hair conditioners and fabric softeners because they are absorbed into the fibres, reducing friction and static. They are a component of many disinfectants and antiseptics.
- **Non-ionic detergents** have similar tails to the other types of surfactant, but the type of head is different. It contains a sequence of oxy groups with a terminal alcohol. This creates polarity within the head end and hence solubility in water. These detergents are molecules, not ions. They produce less foam and are used in cosmetics, paints, adhesives and pesticides.

Early detergents were made from benzene derivatives. These had poor biodegradability and produced excessive amounts of foam when washed into waterways. To overcome problems with hard water, detergents may contain additional substances such as phosphate derivatives. These were designed to reduce water hardness and increase pH to optimum levels for the surfactant. Unfortunately, the discharge of phosphates into waterways contributes significantly to eutrophication – algal blooms and their associated problems.

Cationic synthetic detergents can act as antiseptics and biocides. When detergents are rinsed down the sink, this can affect waterways. Sewage treatment plants use bacteria to decompose sewage, and these bacteria may be killed by the biocidic action of these detergents.

Many modern detergents and soaps are biodegradable, reducing their impact on waterways.

9780170465281

Fuels

In Module 7 we looked at biofuels. Cellulose is being closely examined as a potential source of biofuels and biopolymers. Large amounts of land clearing are needed to grow the ideal crops required, and many of these already exist. In the 1970s, in response to large amounts of the country's sugar cane crops going to waste, the Brazilian government guided car manufacturers towards 100% ethanol-fuelled cars. While the rest of the world has not followed Brazil down the path of non-petroleum-based automobile fuels, there have been modifications to the engines of cars in many other nations to run higher ethanol–petrol mixes than are currently available in Australia.

An initiative of the recently elected Australian Government promises more action on reducing greenhouse gas emissions under the Driving the Nation Fund. The government has committed to building a National Electric Vehicle (EV) Charging Network for Australia's major highways, providing subsidies to encourage greater uptake of EVs, and supporting the states to develop Hydrogen Highways for heavy transport.

4.3.5 Social, environmental and economic issues

CASE STUDY 5

INDUSTRIAL PRODUCTION OF SULFURIC ACID

Sulfuric acid is the strongest acid in the school laboratory. It is used to catalyse a number of reactions, such as conversion of ethanol to ethene and the process of esterification.

The industrial production of sulfuric acid involves several important steps. Figure 4.27 outlines the reagents and production steps.

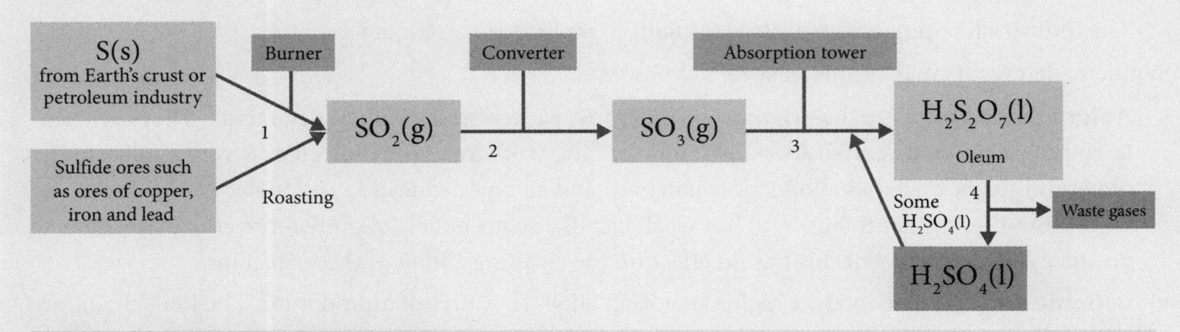

FIGURE 4.27 The industrial production of sulfuric acid

Social issues – sulfuric acid

Social issues may be regarded as considerations that impact on the quality of human life.

Three-quarters of the sulfuric acid produced in Australia is used in the manufacture of superphosphate fertilisers. It is also used in the production of rayon and other synthetic fibres, ethanol (hydrating ethene), paint, plastic and paper pigments, detergents, explosives, drugs, dyes, lead–acid batteries, steel production, oil extraction and ore refining.

One of the main social concerns with the production of sulfuric acid is the health implications upon exposure to sulfuric acid or any of the substances produced during the process. Sulfuric acid is highly corrosive, so it can cause severe burns and breathing difficulties. Misuse of sulfuric acid in industry could cause harm to others.

Environmental issues – the Frasch process

The first step in sulfuric acid production is the Frasch process. The equation for the reaction is:

$$S(l) + O_2(g) \rightarrow SO_2(g) \qquad \Delta H = -297\,kJ\,mol^{-1}$$

This reaction is strongly exothermic, so the products need to be cooled (from 1000°C to 400°C).

Water at high temperature and pressure (160°C) is forced into an underground sulfur deposit through a pipe. This melts the sulfur (melting point 113°C), forming an emulsion. Air is blown down another pipe, forcing the sulfur/water emulsion up a third pipe. The liquid sulfur is sprayed into oxygen-enriched dry air at normal atmospheric pressure. When the mixture reaches the surface, it cools and the sulfur solidifies. It can easily be separated from the water because it is insoluble.

Sulfur is a relatively stable element that in itself does not cause an environmental problem. However, sulfur is readily oxidised to sulfur dioxide or reduced to hydrogen sulfide. These gases pollute the atmosphere, reacting with water and oxygen in the atmosphere to form acid rain. Hence, it is important that neither oxidation nor reduction occurs during the Frasch process. Water used in the Frasch process may be contaminated and hence cannot be released into the environment. Instead it is recycled and reused.

Economic issues – the contact process

The second step is the contact process, and the reaction that occurs is:

$$2SO_2(g) + O_2(g) \rightleftharpoons 2SO_3(g) \qquad \Delta H = -198\,kJ$$

It is an equilibrium reaction, and hence changing the reaction conditions will affect the yield. The exothermic nature of the forward reaction means there is a compromise: increasing the reaction rate with heat favours the endothermic formation of the reactants. To overcome this, the reactants are in contact with a vanadium oxide catalyst at 550°C. The selection of a catalyst helps increase the rate of the reaction, but the cost of the catalyst must be weighed against the benefit it provides in increased yields. Chemists often experiment with catalysts in industrial processes to gain the maximum advantage possible for synthesis reactions.

After about 70% of the SO_2 reacts, the remaining mixture is cooled to 400°C and passed over a second catalyst bed. The reaction at this temperature is slower, but the yield is higher. To further reduce the concentration of sulfur dioxide, and to make further product, a third catalyst bed may be used. The proportion of sulfur dioxide in waste gases must be less than 0.3% before it can be released into the environment.

In the third step of sulfuric acid production, the sulfur trioxide can be dissolved in water droplets to form sulfuric acid:

$$SO_3(g) + H_2O(l) \rightarrow H_2SO_4(l) \qquad \Delta H = -133\,kJ\,mol^{-1}$$

The mist that forms may be difficult to separate from the sulfur trioxide gas. Also, the reaction between sulfuric acid and water is highly exothermic and can be dangerous.

Workplace health and safety is an important social issue in chemical industries. A safer alternative to the reaction is to pass the sulfur trioxide directly into concentrated sulfuric acid to form oleum ($H_2S_2O_7$). Oleum can then be reacted with water to form sulfuric acid.

Glossary

absorption The taking in of energy, usually in the form of electromagnetic radiation, which moves an electron or atomic nucleus, from a low-energy state to a higher energy state

absorption spectrum The different wavelengths of electromagnetic radiation absorbed by an atom as electrons move from the ground state to an excited state

anionic detergent A synthetic detergent with a negatively charged hydrophilic head and hydrophobic tail

atomic absorption spectroscopy (AAS) An analytical technique for determining the concentration of named cations in a solution based on their absorbance of specific wavelengths of electromagnetic radiation

cationic detergent A synthetic detergent with a positively charged = hydrophilic head and hydrophobic tail

chemical synthesis The process of making a new chemical substance

colourimetry An analytical technique used to quantitatively measure the amount of light absorbed by coloured solutions containing certain cations. Usually involves setting up a calibration curve

complex A chemical combination of a central metal atom or ion (usually one of the transition metals) and a Lewis base (often a cyanide ion, hydroxide ion or ammonia molecule) that has been bonded to it. It may remain as an ion in an aqueous solution

emission spectrum The different wavelengths of electromagnetic radiation released by an atom as electrons move from an excited state back to the ground state

environmental monitoring The processes that test the quality of the environment – the air, the water and the land

fuel A chemical substance that can undergo a combustion reaction to release energy

gravimetric analysis A precise analytical technique that is used to determine the proportion, by mass, of a particular chemical substance in a compound or mixture

heavy metal One of a group of metals with high atomic weights and a relative density ≥5.0; arsenic, lead, mercury, chromium and cadmium are important heavy metal pollutants with high toxicity

A+ DIGITAL FLASHCARDS

Revise this topic's key terms and concepts by scanning the QR code or typing the URL into your browser.

https://get.ga/aplus-hsc-chemistry-u34

infrared (IR) spectroscopy An analytical technique for determining the nature of the chemical bonds in an organic molecule based on their absorbance of specific wavelengths of infrared radiation

inorganic substance A chemical substance that is not organic; it is not based on a carbon structure

ligand An ion or molecule that can act as a Lewis base and form a coordinate bond with a metal atom or ion to produce a complex

mass spectroscopy An analytical technique for determining the composition of a chemical substance based on fragments with varying masses and their differential interaction with a magnetic field

monochromator A device that filters out all wavelengths of light except one desired wavelength

natural product A substance produced by a living organism

non-ionic detergent A synthetic detergent with a polar hydrophilic head and hydrophobic tail

nuclear magnetic resonance (NMR) spectroscopy An analytical technique for determining the nature and number of unique chemical environments involving carbon (^{13}C) or hydrogen (^{1}H) in an organic molecule based on the orientation of their nuclei (spin) in an externally applied magnetic field

pharmaceutical A chemical substance that is used in the improvement or maintenance of health

pollutant A substance that has a negative effect on the natural environment, in soil, air or water

purity A measure of the proportion of the desired substance in a chemical mixture; can be expressed as a percentage

qualitative analysis Chemical analysis to detect the presence or absence of a particular chemical substance in a compound or mixture based on general observations, e.g. flame colour, presence of a precipitate, generation of a gas

quantitative analysis Chemical analysis to measure the amount of a particular chemical substance in a compound or mixture based on specific measurements and/or calculations of its mass or concentration

reagent A chemical substance used in a chemical reaction, particularly a synthesis reaction

spectroscopy The study of the interactions between electromagnetic radiation (usually focused on a small range of wavelengths) and matter

spin–spin coupling The magnetic interactions between adjacent nuclei in an organic molecule under the influence of an externally applied magnetic field

titration A technique involving the use of a burette to add a precise volume of a solution to another solution to reach a desired end point

UV–visible spectroscopy An analytical technique for determining the concentration of named ions in a solution based on their preferential absorbance of specific wavelengths of visible and/or ultraviolet radiation

yield A measure of the quantity of the desired product in a chemical process

Exam practice

Multiple-choice questions

Solutions start on page 200.

Question 1

What is the best way to distinguish between calcium and barium ions in a solution?

A Add an acid to each solution and look for bubbles.

B Add some sulfate ions to precipitate white barium sulfate.

C Add some carbonate ions to precipitate white calcium carbonate.

D Compare the flame colours produced by the two solutions.

Question 2

A student was testing a range of unknown solutions with 0.1 M hydrochloric acid. The student was trying to identify the anions present in each solution. Only one of the following anions resulted in bubbles in the presence of the acid. That anion was

A carbonate.

B chloride.

C phosphate.

D acetate.

Question 3

$[FeSCN]^{2+}$ is best classified as

A a complex.

B a molecule.

C an anion.

D a precipitate.

Question 4

An appropriate test to distinguish between a secondary alcohol and a tertiary alcohol in a science laboratory is the

A bromine water test.

B addition of a carbonate.

C addition of an oxidising agent, e.g. potassium permanganate.

D colour of the two liquids.

Question 5

A student carried out a precipitation titration reaction to determine the chloride concentration of a water sample. She used a 10 mL water sample, added a solution of chromate ions as an indicator, and used a 0.050 M silver nitrate solution in the burette. After several repetitions, she calculated an average titre of 25.5 mL. The original concentration (mol L^{-1}) of chloride ions was closest to

A 0.019.

B 0.039.

C 0.064.

D 0.13.

Question 6

Iron(III) thiocyanate provides a sufficiently coloured solution for determining the equilibrium constant in an iron(III) thiocyanate equilibrium. A student set up his experiment and wanted to use the Beer–Lambert law to calculate the concentration of iron(III) thiocyanate ions in his solution. His cuvet had a length of 15 mm and he used his calibration curve to calculate a molar absorptivity constant (ε) of 4500 L mol^{-1} cm^{-1}. When he placed his sample in the colourimeter, he obtained an absorbance reading of 0.038. Using these values to calculate the concentration in ppm, his answer would be closest to

A 0.56.

B 0.64.

C 5.6.

D 6.4.

Question 7

The technique of atomic absorption spectroscopy uses which part of the electromagnetic spectrum?

A Radio waves **B** Microwaves **C** Infrared rays **D** Visible light

Question 8 ©NESA 2020 SI Q16 

Compounds X, Y and Z are in equilibrium. The diagram shows the effects of temperature and pressure on the equilibrium yield of compound Z.

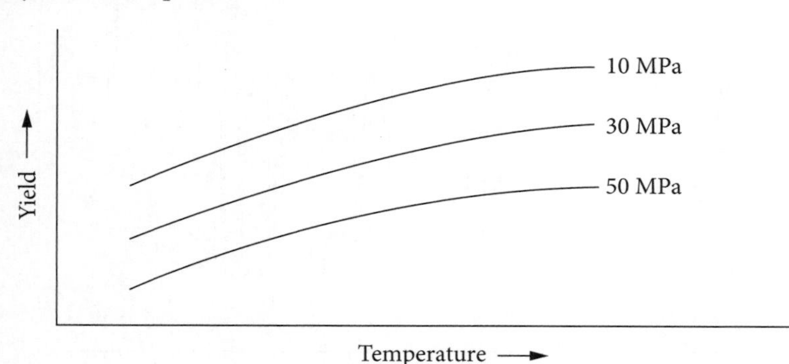

Which equation would be consistent with this data?

A $X(g) + 3Y(g) \rightleftharpoons 2Z(g)$ $\Delta H > 0$

B $X(g) + 3Y(g) \rightleftharpoons 2Z(g)$ $\Delta H < 0$

C $2X(g) \rightleftharpoons 2Y(g) + Z(g)$ $\Delta H > 0$

D $2X(g) \rightleftharpoons 2Y(g) + Z(g)$ $\Delta H < 0$

Question 9

The compound butanal was analysed by ^{13}C NMR spectroscopy. How many peaks would you expect for butanal in the spectrum?

A One **B** Two **C** Three **D** Four

Question 10

The Haber process is used to synthesise ammonia from its constituent elements. The chemical equation is:

$$N_2(g) + 3H_2(g) \rightleftharpoons 2NH_3(g) \qquad \Delta H = -92\,kJ$$

Which of the following reaction conditions would most favour an increased yield of ammonia?

A Low pressure and high temperature

B Low pressure and low temperature

C High pressure and high temperature

D High pressure and low temperature

Short-answer questions

Solutions start on page 202.

Question 11 (5 marks)

Discuss why it is important for chemists to monitor the environment.

Question 12 (7 marks)

A student collected a sample of sea water and evaporated the water in order to collect the salt. Her sample had a mass of 6.60 g. She wanted to determine the proportion of chloride ions in the salt, so she added sufficient silver nitrate to precipitate the chloride ions from the solution. She dried and weighed the silver chloride precipitate and obtained a mass of 9.60 g.

a Determine the proportion of chloride ions in the original seawater sample. 3 marks

b Suggest a possible source of error in the student's experimental procedure. 1 mark

c Suggest a possible improvement to her procedure which may address the error you identified in part **b**. Justify your change. 3 marks

Question 13 (7 marks) ©NESA 2020 SII Q30

A chemist discovered a bottle simply labelled '$C_5H_{10}O_2$'.

To confirm the molecular structure of the contents of the bottle, a sample was submitted for analysis by infrared spectroscopy and ^{1}H and ^{13}C NMR spectroscopy. The resulting spectra are shown.

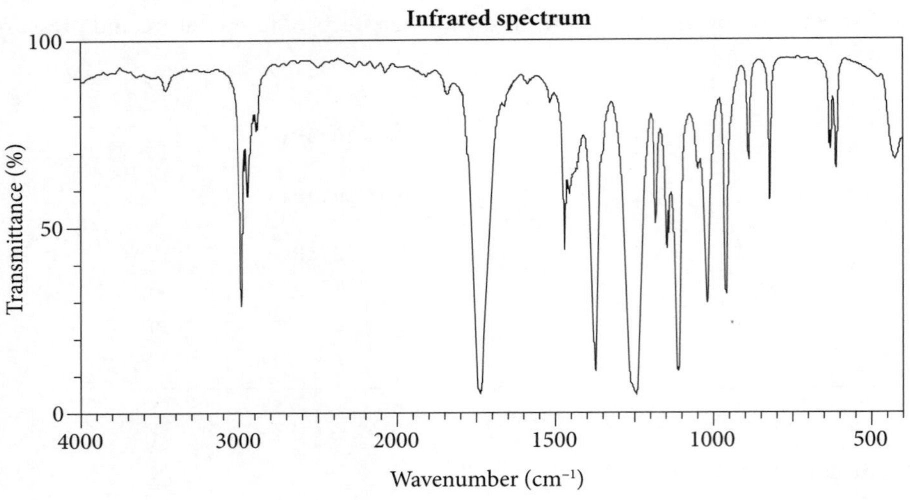

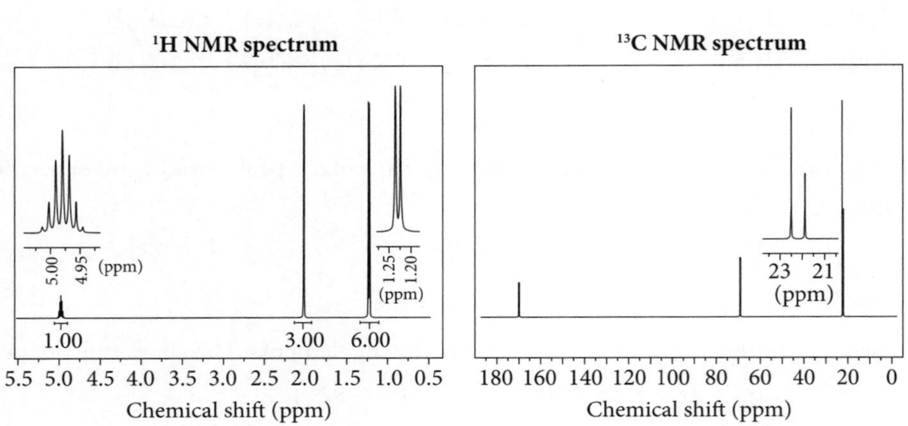

Data from ^{1}H NMR spectrum

Chemical shift	*Relative peak area*	*Splitting pattern*
1.2	6	Doublet (2)
2.0	3	Singlet (1)
5.0	1	Septet (7)

^{1}H NMR chemical shift data

Type of proton		δ/ppm
$Si(CH_3)_4$	(TMS)	0
R—CH_3		0.7–1.3
R—CH_2—R		1.2–1.5
R—CHR_2		1.5–2.0
H_3C—CO—	(aldehydes, ketones or esters)	2.0–2.5
—CH—CO—	(aldehydes, ketones or esters)	2.1–2.6
H_3C—O—	(alcohols or esters)	3.2–4.0
—CH—O—	(alcohols or esters)	3.3–5.1
R_2—CH_2—O—	(alcohols or esters)	3.5–5.0
R—OH		1–6
R_2C=CHR	(alkene)	4.5–7.0
R—CHO	(aldehyde)	9.4–10.0
R—COOH		9.0–13.0

Draw a structural formula for the unknown compound that is consistent with all of the information provided. Justify your answer with reference to the information provided.

Question 14 (5 marks)

Explain how the principles of mass spectroscopy can be used to identify an organic compound such as ethanamine.

Question 15 (7 marks)

Select a chemical synthesis process with which you are familiar and evaluate this process on the basis of reaction conditions, industrial use and environmental issues.

Question 16 (7 marks) ©NESA 2021 SII Q29

A chemist obtained spectral data of pentane-1,5-diamine ($C_5H_{14}N_2$).

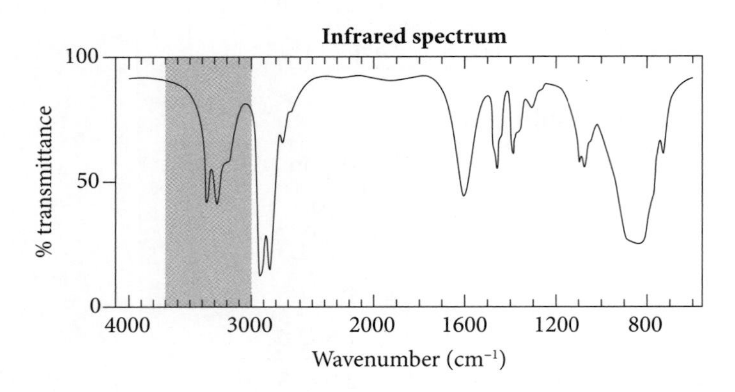

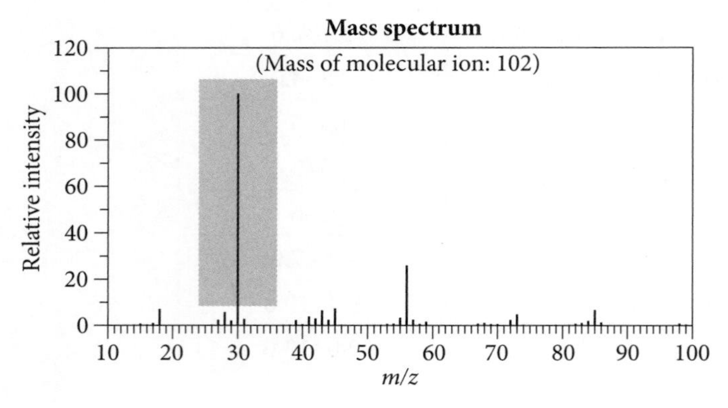

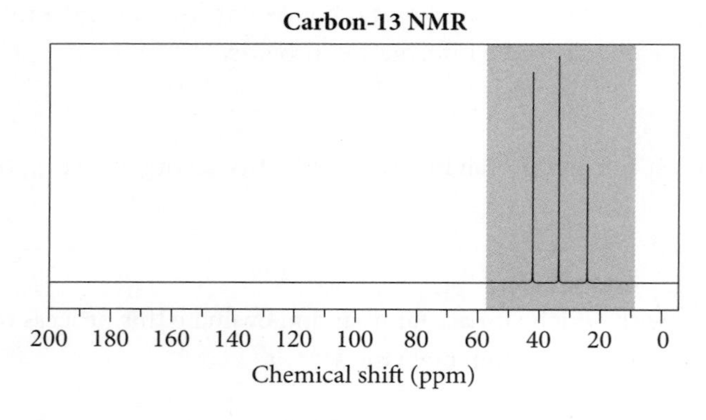

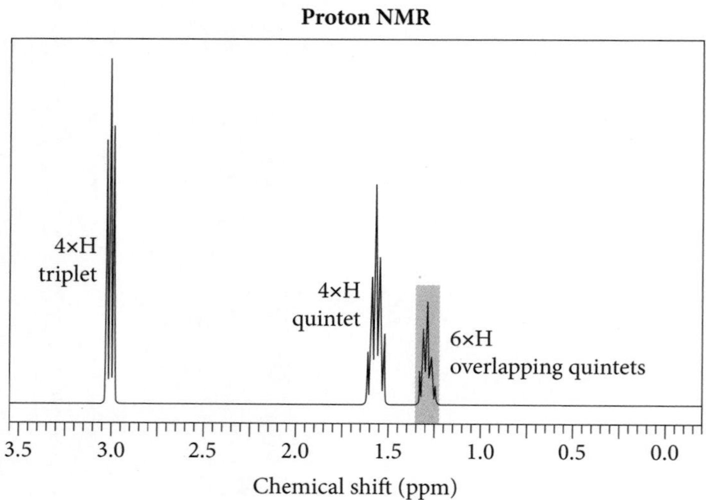

Relate the highlighted features of the spectra to the structure of pentane-1,5-diamine.

SOLUTIONS

CHAPTER 1 MODULE 5

Equilibrium and acid reactions

Multiple-choice solutions

1 B

When magnesium reacts with acid, it generates hydrogen. In a beaker, the hydrogen escapes from the system; hence, this is an open system.

A saturated solution of sodium chloride with a visible precipitate in a test tube is open, but the equilibrium system exists between the solid and the ions in solution; hence, there is no exchange of matter between the system and the surroundings, and **A** is not correct. A combustion reaction taking place inside a bomb calorimeter is ideally an isolated system because neither matter nor energy can escape, so **C** is not correct. An equilibrium established between iron(III) ions, thiocyanate ions and ferric thiocyanate ions in a sealed flask is a closed system, so **D** is not correct.

2 D

Increasing the pressure favours the side of the reaction with fewer moles of gas. In this case, this is the product side, so it increases the yield. The reaction is exothermic as written, so an increase in temperature favours the reverse endothermic reaction, decreasing the yield, so **A** is not correct. A catalyst increases the rate of the reaction, but does not change the yield, so **B** is not correct. Removing nitrogen dioxide causes the system to shift to the left to replace the lost NO_2, again decreasing the yield, so **C** is not correct.

3 C

Adding hydroxide ions (OH^-) causes a neutralisation reaction with the H^+ ions.

According to Le Chatelier's principle, reducing $[H^+]$ will cause the equilibrium to shift to the left to increase $[H^+]$ and this will also increase $[CrO_4^{2-}]$, which will cause the solution to become more yellow.

4 C

Apply the NAGSAG rules. Group I metals are all soluble, so **A** is not correct. All ammonium salts are soluble, so **B** is not correct. All nitrates and ethanoates (acetates) are soluble, so **D** is not correct. **C** is correct because, although all sulfates are soluble the exceptions include CaStroBear, one of which is barium. So barium sulfate is insoluble and **C** will form a precipitate.

5 C

A negative ΔH indicates an exothermic reaction and the states indicate the sign of ΔS. A solid is reacting to form two ions in solution, so the disorder of the system is increasing, and ΔS is positive. Since $\Delta G = \Delta H - T\Delta S$, if ΔH is negative and ΔS is positive, ΔG must be negative. A negative ΔG corresponds to a spontaneous reaction, so **C** is correct.

6 C

The best explanation is one that identifies the shift towards the endothermic reaction. However, the most appropriate explanation is one that explains it in terms of changes to the average kinetic energy of all particles and the relationship between average kinetic energy of particles and activation energy. Increasing the temperature raises the average kinetic energy of all particles but significantly affects those of the endothermic reaction more, increasing their reaction rate more than those of the exothermic reaction. **C** is the best option because it identifies both the correct reaction and draws on a relationship between average kinetic energy and energy required for the reaction.

SOLUTIONS

7 A

If you are asked to identify the *most* correct statement, you can assume that there may be more than one correct statement. This is the case here. Seeds were ground and pounded to increase the surface area for the reaction (**C**). Washing was also used to remove toxins (**D**), as was leaching. However, the critical process, and hence best answer, is that leaching has been traditionally used to effectively remove toxins through dissolution by ensuring that an equilibrium between the solid and the ions was not established (**A**). Because an equilibrium system was not established, the open system did not allow an exchange between the toxins in solution and those in the seeds. Instead, the continual flow of water shifts the equilibrium so that the dissolved toxins are replaced and more toxin is removed.

8 D

From the data sheet, $K_{sp}(Ag_3PO_4) = 8.89 \times 10^{-17}$.

The concentration of silver ions in a saturated solution of silver phosphate can be found using a chemical equation and the K_{sp} formula.

$$Ag_3PO_4(s) \rightleftharpoons 3Ag^+(aq) + PO_4^{3-}(aq)$$

Mole ratio 1 3 1

$$\begin{aligned} K_{sp} &= [Ag^+]^3[PO_4^{3-}] \\ &= (3x)^3 \times x \\ 8.89 \times 10^{-17} &= 27x^4 \\ x^4 &= \frac{8.89 \times 10^{-17}}{27} \\ &= 3.29 \times 10^{-18} \\ x &= \sqrt[4]{3.29 \times 10^{-18}} \\ &= 4.26 \times 10^{-5}\ \text{mol L}^{-1} \end{aligned}$$

But the concentration of silver ions is 3x, so:

$$\begin{aligned} [Ag^+] &= 3 \times 4.26 \times 10^{-5} \\ &= 1.28 \times 10^{-4}\ \text{mol L}^{-1} \end{aligned}$$

9 A

This question involves calculating Q_{sp} and comparing it with the published value of K_{sp}.

First, write the equation:

$$Ca_3(PO_4)_2(s) \rightleftharpoons 3Ca^{2+}(aq) + 2PO_4^{3-}(aq)$$

Next, calculate the concentration of each ion, based on the mass of the dissolved solid:

$$n = \frac{m}{MM}\ (MM \text{ from periodic table: } 3 \times 40.08 + 2 \times 30.97 + 8 \times 16.00 = 310.18\ \text{g mol}^{-1})$$

$$n = \frac{1}{310.18} = 3.22 \times 10^{-3}\ \text{mol, but this is in 1 L.}$$

$$c = \frac{n}{V} = \frac{3.22 \times 10^{-3}}{1} = 3.22 \times 10^{-3}\ \text{mol L}^{-1}$$

Now, calculate the individual concentrations of the ions.

$$\begin{aligned} [Ca^{2+}] &= 3 \times 3.22 \times 10^{-3} = 9.67 \times 10^{-3}\ \text{mol L}^{-1} \\ [PO_4^{3-}] &= 2 \times 3.22 \times 10^{-3} = 6.45 \times 10^{-3}\ \text{mol L}^{-1} \\ Q_{sp} &= [Ca^{2+}]^3[PO_4^{3-}]^2 \\ &= (9.67 \times 10^{-3})^3 \times (6.45 \times 10^{-3})^2 \\ &= 3.76 \times 10^{-11} \end{aligned}$$

From the data sheet, K_{sp} for calcium phosphate is 2.07×10^{-29}.

Since $Q_{sp} > K_{sp}$, a precipitate would form.

10 A

First, work out whether a precipitate would have formed from the first process.

$$CaSO_4(s) \rightleftharpoons Ca^{2+}(aq) + SO_4^{2-}(aq)$$

$n = \frac{m}{MM}$ (MM from periodic table: $1 \times 40.08 + 1 \times 32.07 + 4 \times 16 = 136.15\,g\,mol^{-1}$)

$n = \frac{0.5}{136.158} = 3.67 \times 10^{-3}$ mol, but again, this is in 1 L.

$c = \frac{n}{V}$

$= \frac{3.67 \times 10^{-3}}{1} = 3.67 \times 10^{-3}\ mol\,L^{-1}$

Now, calculate the individual concentrations of the ions; this is easy because the ratio is 1 : 1 : 1.

$$[Ca^{2+}] = 3.67 \times 10^{-3}\,mol\,L^{-1}$$
$$[SO_4^{2-}] = 3.67 \times 10^{-3}\,mol\,L^{-1}$$
$$Q_{sp} = [Ca^{2+}][SO_4^{2-}]$$
$$= (3.67 \times 10^{-3}) \times (3.67 \times 10^{-3})$$
$$= 1.35 \times 10^{-5}$$

From the data sheet, K_{sp} for calcium sulfate is 4.93×10^{-5}.

Since $Q_{sp} < K_{sp}$, no precipitate would form.

Adding potassium sulfate would have no effect on the concentration of the calcium ions, but there might be additional sulfate ions. Working through the same steps but just for the sulfate ion:

$$K_2SO_4(s) \rightleftharpoons 2K^+(aq) + SO_4^{2-}(aq)$$

$n = \frac{m}{MM}$ (MM from periodic table: $2 \times 39.10 + 1 \times 32.07 + 4 \times 16 = 174.27\,g\,mol^{-1}$)

$n = \frac{0.5}{174.27} = 2.87 \times 10^{-3}$ mol, but again, this is in 1 L.

$c = \frac{n}{V}$

$= \frac{2.87 \times 10^{-3}}{1} = 2.87 \times 10^{-3}\ mol\,L^{-1}$

Now, calculate the individual concentration of the sulfate ion; this is easy because the ratio is 1 : 1.

$$[SO_4^{2-}] = 2.87 \times 10^{-3}\,mol\,L^{-1}$$

Recalculate Q_{sp} using the new values:

$$[Ca^{2+}] = 3.67 \times 10^{-3}\,mol\,L^{-1}$$
$$[SO_4^{2-}] = 3.67 \times 10^{-3} + 2.87 \times 10^{-3} = 6.54 \times 10^{-3}\,mol\,L^{-1}$$
$$Q_{sp} = [Ca^{2+}][SO_4^{2-}]$$
$$= (3.67 \times 10^{-3}) \times (6.54 \times 10^{-3})$$
$$= 2.40 \times 10^{-5}$$

From the data sheet, K_{sp} for calcium sulfate is 4.93×10^{-5}.

Since Q_{sp} is still $< K_{sp}$, no precipitate forms.

Assume that the chemist did not observe the extra salt being added to the solution. Also assume that the chemist's judgement is only based on their observations of the reaction vessel. There is no visible change to the solution (i.e. no colour change and no precipitate) so the chemist would not know there had been a change. To see a precipitate, a larger quantity of potassium sulfate would have had to be added.

Short-answer solutions

11 The carbon dioxide equilibrium in a soft drink is affected whenever the bottle is opened. Some of the gas escapes and the system becomes an open system. As carbon dioxide is lost from the system, carbon dioxide in solution (aq) is converted to gas (g), according to Le Chatelier's principle. The equilibrium shifts to the right as the system adjusts to restore the lost carbon dioxide. As long as the top of the bottle is off, the system remains open and equilibrium can never be re-established. However, when the bottle top is replaced, the system becomes a closed system again and it adjusts to produce a new equilibrium.

Mark breakdown

- 1 mark for correctly identifying the change to the system (i.e. drop in pressure, loss of carbon dioxide)
- 1 mark for quoting and applying Le Chatelier's principle to counter the change (shift towards the gaseous side; right)
- 1 mark for re-establishing a closed system when the bottle is sealed again

12 The addition of potassium chloride increases the concentration of chloride ions in the solution. Le Chatelier's principle suggests that when a change is experienced by an equilibrium system, the system responds to counter the change. So, the increased chloride ion concentration causes the equilibrium to shift to the right. This reduces the chloride ion concentration while at the same time decreasing the concentration of the hydrated cobalt ion and increasing the cobalt chloride ion concentration. This is a shift towards the blue ion away from the pink ion and hence the solution becomes more blue.

Mark breakdown

- 1 mark for correctly identifying the change to the system (i.e. increase in chloride ions)
- 1 mark for quoting and applying Le Chatelier's principle to counter the change (shift towards the right)
- 1 mark for identifying the increase in the cobalt chloride ion
- 1 mark for identifying the change in colour of the solution (becomes more blue)

13

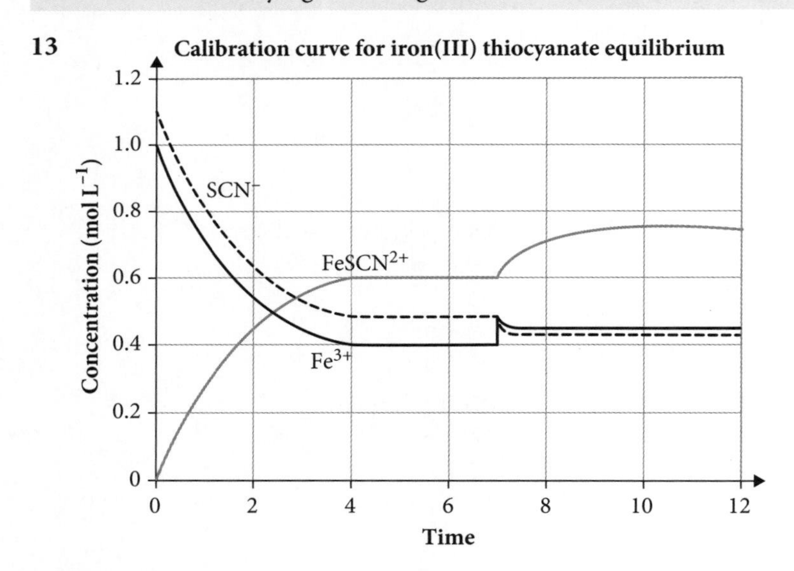

Mark breakdown

- 1 mark for correctly identifying the change to the system (i.e. increase in ferric ions)
- 1 mark for correctly showing how the lines of each species change (ferric ions and thiocyanate ions decrease and ferric thiocyanate ions increase)
- 1 mark for showing equilibrium re-establishing through parallel horizontal lines

14 When an ionic solid is added to water, the ions in the solid begin to separate. This is a result of the strong forces applied by water molecules to ions on the edges of the solid. When these forces are greater than the forces of ionic electrostatic attraction holding the ions of the solid together, the ions leave the solid and diffuse through the water. This is shown for sodium chloride in the following diagram.

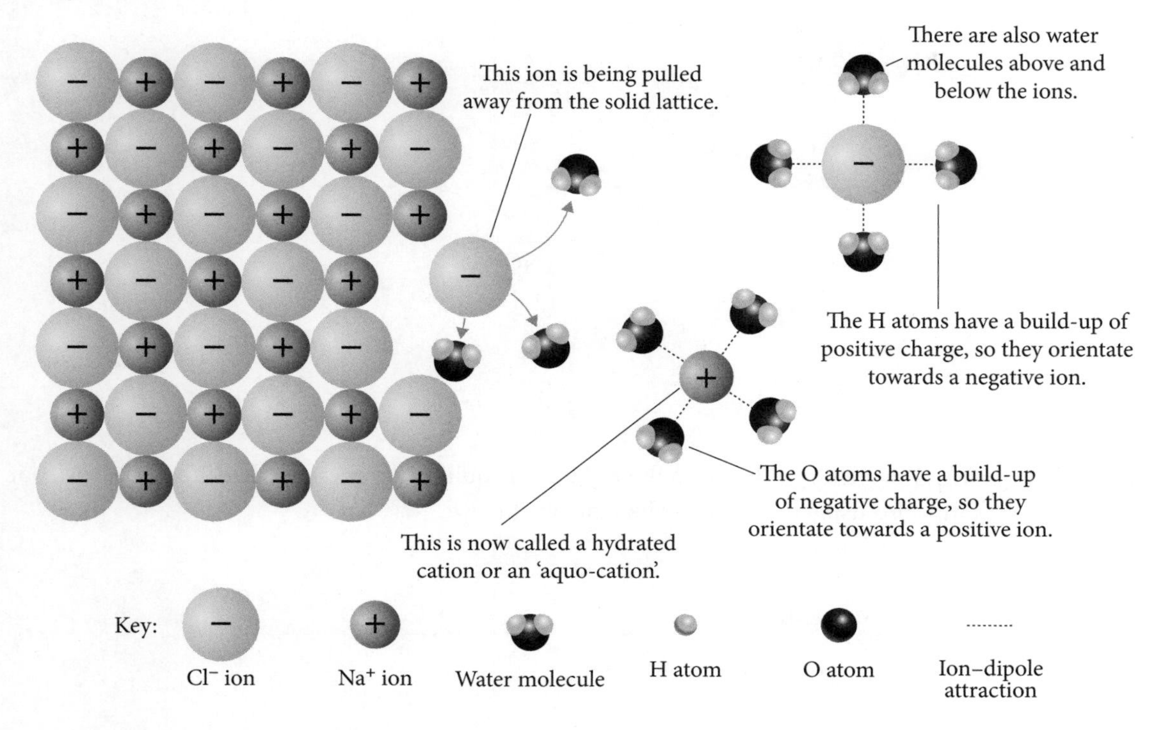

The equation for the sodium chloride is:

$$NaCl(s) \rightleftharpoons Na^+(aq) + Cl^-(aq)$$

The equilibrium expression is:

$$K_{eq} = \frac{[Na^+][Cl^-]}{[NaCl]}$$

However, the reactant is a solid, so we do not calculate its concentration. Twice the amount of solid takes up twice the volume, so this value will be a constant. This also means we can multiply it through both sides of the equation to derive the following relationship for the reaction:

$$K_{eq} = [Na^+] \times [Cl^-]$$

This is equivalent to the solubility product K_{sp}.

Mark breakdown

- 1 mark for correctly identifying the equation for the dissolution of an ionic solid (general or specific)
- 2 marks for explaining what happens to ionic solids when they dissolve in water (a labelled diagram is desired)
- 1 mark for displaying the equilibrium constant expression (full or derived)
- 1 mark for identifying why the solid should not appear in the final equilibrium expression

15 a The equation for this process is: $2NO_2(g) \rightleftharpoons N_2O_4(g)$

$$K_{eq} = 215.5$$
$$= \frac{[N_2O_4]}{[NO_2]^2}$$

You need to create an equation to find out whether the system is at equilibrium. The following table shows how to set this out.

Equation	$2NO_2(g)$	$\rightleftharpoons$	$N_2O_4(g)$
Mole ratio	2		1
n_i (mol)	3		2
V_i (L)	3		3
$c_i = \frac{n_i}{V_i}$ (mol L^{-1})	$= \frac{3}{3}$ $= 1$		$= \frac{2}{3}$

$$Q_{eq} = \frac{[N_2O_4]}{[NO_2]^2} = \frac{\frac{2}{3}}{1^2} = \frac{2}{3}$$

Since $Q_{eq} < K_{eq}$, the system will shift to the right until equilibrium is established. You can work out these concentrations using an ICE table like the one below.

Equation	$2NO_2(g)$	$\rightleftharpoons$	$N_2O_4(g)$
Mole ratio	2		1
c_i (mol L^{-1})	$= 1$		$= \frac{2}{3}$
c_c	$-2x$		$+x$
c_e	$1 - 2x$		$\frac{2}{3} + x$

So $K_{eq} = \frac{[N_2O_4]}{[NO_2]^2} = 215.5$

$$215.5 \times (1 - 2x)^2 = \frac{2}{3} + x$$

$$215.5 \times (1 - 4x + 4x^2) = \frac{2}{3} + x$$

$$215.5 - \frac{2}{3} - (4 \times 215.5)x - x + (4 \times 215.5)x^2 = 0$$

$$862x^2 - 863x + 214.83 = 0$$

Use the quadratic formula to solve this problem (only one real solution).

$$x = \frac{-b \pm \sqrt{b^2 - 4ac}}{2a}$$

$$= \frac{-(-863) \pm \sqrt{(-863)^2 - 4 \times 862 \times 214.83}}{2 \times 862}$$

$x = 0.537$ (non-sense: $1 - 2x$ must be +ve) and 0.464

Hence, at equilibrium:

$$[N_2O_4] = \frac{2}{3} + 0.464 = 1.130 = 1.1 \text{ mol L}^{-1}$$

$$[NO_2] = 1 - 2 \times 0.464 = 0.072 \text{ mol L}^{-1}$$

Mark breakdown

- 1 mark for correctly calculating initial concentrations
- 1 mark for determining whether the system is originally at equilibrium
- 1 mark for creating an ICE table to determine equilibrium concentrations
- 1 mark for correct calculation of equilibrium concentrations

b We can adjust our ICE table from part **a** to perform this calculation. This time we know the change will be going to the left to counter the addition of 2.0 mol of N_2O_4. Use the following table to set up the equation.

Equation	$2NO_2(g)$	$\rightleftharpoons$	$N_2O_4(g)$
Mole ratio	2		1
c_i **(mol L^{-1})**	= 0.072		$= 1.13 + \frac{2}{3}$
c_c	$+2x$		$-x$
c_e	$0.072 + 2x$		$1.80 - x$

$$K_{eq} = \frac{[N_2O_4]}{[NO_2]^2} = 215.5$$

So $K_{eq} = \dfrac{1.80 - x}{(0.072 + 2x)^2} = 215.5$

$$215.5 \times (0.00052 + 0.29x + 4x^2) = 1.80 - x$$
$$1.13 - 1.80 + (0.29 \times 215.5)x + x + (4 \times 215.5)x^2 = 0$$
$$862x^2 + 63.43x - 0.670 = 0$$

Use the quadratic formula to solve this problem (only one real solution).

$$x = \frac{-b \pm \sqrt{b^2 - 4ac}}{2a}$$
$$= \frac{-63.43 + \sqrt{(63.43)^2 - 4 \times 862x - 0.670}}{2 \times 862}$$
$$= 0.0936$$

Hence, at the new equilibrium:

$$[N_2O_4] = 1.80 - 0.00936 = 1.79\ \text{mol L}^{-1}$$
$$[NO_2] = 0.072 + 2 \times 0.00936 = 0.091\ \text{mol L}^{-1}$$

Mark breakdown

- 1 mark for correctly calculating initial concentrations
- 1 mark for determining how the equilibrium will shift
- 1 mark for creating an ICE table to determine equilibrium concentrations
- 1 mark for correct calculation of new equilibrium concentrations

16 The treatment involves the addition of calcium hydroxide ($Ca(OH)_2$) to the stormwater. Calcium hydroxide is reasonably soluble in water ($K_{sp} = 5.02 \times 10^{-6}$) so it will dissociate into the calcium cation (Ca^{2+}) and the hydroxide anion (OH^-) as follows:

$$Ca(OH)_2(aq) \rightleftharpoons Ca^{2+}(aq) + 2OH^-(aq)$$

Both copper(II) and lead(II) cations have much lower solubility than calcium cations with hydroxide anions. From the data table provided, we can see the K_{sp} for copper(II) hydroxide is 2.2×10^{-20}, while the K_{sp} for lead(II) hydroxide is 1.43×10^{-15}. The result of the addition of calcium hydroxide is that almost all of the lead or copper ions that are in the solution will precipitate out of the solution when the hydroxide ions are added. The precipitation of lead(II) hydroxide is shown below:

$$Pb^{2+}(aq) + 2OH^-(aq) \rightleftharpoons Pb(OH)_2(s)$$

This will significantly reduce their concentration in the stormwater to the required levels.

Mark breakdown

- 1 mark for correctly linking the treatment (addition of calcium hydroxide) to the relatively low solubility of the undesired cations
- 1 mark for a correct and relevant chemical equation

CHAPTER 2 MODULE 6

Acid/base reactions

Multiple-choice solutions

1 D

$$[H^+] = 2.3 \times 10^{-6}\ \text{mol L}^{-1}$$
$$\begin{aligned}pH &= -\log_{10}[H^+]\\ &= -\log_{10}(2.3 \times 10^{-6})\\ &= 5.6\end{aligned}$$

2 C

Rinsing the burette with the solution you are going to add to it is good practice. So is rinsing the reaction flask with distilled water prior to adding the aliquot because this will not change the number of moles of reactant in the flask. Rinsing the graduated pipette with the solution you are planning to add to it is also good procedure, so the only source of error is rinsing the reaction flask with the solution you are planning to add because this will add additional moles to the flask that need to be neutralised, affecting your calculations.

3 D

Sodium hydroxide is not a suitable primary standard. Its mass is quite low although it is soluble in water, forming an alkaline solution. The main concern with sodium hydroxide is it absorbs chemicals (water and carbon dioxide) from the atmosphere which change its concentration and hence require standardisation every time it is used.

4 C

This question is tricky because it is a double negative; not inconsistent is equivalent to consistent. This means we need an acid that contains oxygen, and sulfuric acid fulfils this requirement.

5 C

The K_a value represents products over reactants, so small K_a means small product concentration, which means weak acid. However, when we take the negative log of a very small number, it becomes a bigger number, so this means a weak acid with a small K_a has a large pK_a.

6 B

When the dihydrogen phosphate ion is acting as an acid, it is donating a proton to water and hence would become its conjugate base, the hydrogen phosphate ion. Water would accept the donated proton and form the hydronium ion, so the correct reaction is:

B $H_2PO_4^-(aq) + H_2O(l) \rightleftharpoons HPO_4^{2-}(aq) + H_3O^+(aq)$

7 D

A strong acid–weak base titration would have an equivalence point below 7. The conjugate acid from the weak base will cause the pH at neutralisation to decrease. This means we need an indicator that changes in the acid region, and the best option available is methyl orange. Phenolphthalein changes at about pH 8, which is too high; universal indicator has too many colours so it is unsuitable in a titration; and bromothymol blue changes colour through the neutral range 6–7.

8 D

A 0.02 M solution of sulfuric acid diluted by a factor of 10 would now be a 0.002 M solution. The ionisation of sulfuric acid is:

$$H_2SO_4(aq) + 2H_2O(l) \rightarrow 2H_3O^+(aq) + SO_4^{2-}(aq)$$

This assumes complete ionisation and a ratio of 1 : 2 for the hydrogen ions. So a sulfuric acid concentration of 0.002 M would liberate 2 × 0.002 = 0.004 M hydrogen ions.

$$\begin{aligned} pH &= -\log_{10}[H^+] \\ &= -\log_{10}(0.004) \\ &= 2.4 \end{aligned}$$

9 C

This question is most easily solved with a table, such as the one below. A monoprotic acid requires an equal number of moles of sodium hydroxide for neutralisation.

Equation	Acid(aq) +	NaOH(aq)	→ salt(aq) + $H_2O(l)$
Mole ratio	1	1	
n (= cV) ↓	$= 3.98 \times 10^{-3}$ ←	$= 0.21 \times 0.01895$ $= 3.98 \times 10^{-3}$ ↑	
V (L)	0.01	0.01895	
c (mol L^{-1})	$= \frac{3.98 \times 10^{-3}}{0.01}$ = 0.0398	0.21	

This value is closest to 0.40 M.

10 A

The major improvement on the Arrhenius definition that was addressed by the Brønsted–Lowry definition of acids relates to the Arrhenius definitions that were based on H^+ ions for acids and OH^- ions for bases in aqueous solutions. So the correct response is acid–base reactions that did not occur in aqueous solutions.

Short-answer solutions

11 a $Al(OH)_3(aq) + 3HCl(aq) \rightarrow AlCl_3(aq) + 3H_2O(l)$

Mark breakdown

- 1 mark for correctly balanced equation.

b The key values are 400 mg of $Al(OH)_3$ and 400 mg of $Mg(OH)_2$ per tablet.

The second equation is:

$$Mg(OH)_2(aq) + 2HCl(aq) \rightarrow MgCl_2(aq) + 2H_2O(l)$$

So every mole of $Al(OH)_3$ can neutralise 3 mol of HCl.

While every mole of $Mg(OH)_2$ can neutralise 2 mol of HCl.

Since $n = \frac{m}{MM}$

For $Al(OH)_3$:

$$\begin{aligned} n &= \frac{0.4}{78.004} \ (MM = 26.98 + 3 \times 16 + 3 \times 1.008 = 78.004) \\ &= 5.13 \times 10^{-3}\,\text{mol} \end{aligned}$$

For $Mg(OH)_2$:

$$n = \frac{0.4}{58.326} \; (MM = 24.31 + 2 \times 16 + 2 \times 1.008 = 58.326)$$
$$= 6.86 \times 10^{-3} \text{ mol}$$

So the number of moles of acid that could be neutralised are:

$$3 \times 5.13 \times 10^{-3} + 2 \times 6.86 \times 10^{-3}$$
$$= 0.0291 \text{ mol of acid}$$

Mark breakdown

- 1 mark for correctly balanced equations
- 1 mark for calculating the number of moles of each base
- 1 mark for using the correct multipliers
- 1 mark for correct answer

c If $n = 0.0291$ moles and $V = 2$ L:

$$c = \frac{n}{V}$$
$$= \frac{0.0291}{2} = 0.0146 \text{ mol L}^{-1}$$

Mark breakdown

- 1 mark for using the correct equation with substitution
- 1 mark for correct answer

12 Universal indicator provides a reasonable level of accuracy for determining the pH of a solution. It changes colour over a range of pH values from strong acids through weak acids to neutral substances, and through weak bases to strong bases. However, universal indicator is only effective in colourless solutions. Universal indicator cannot be more accurate than the range of colours provided as some of the colours persist over a range of pH values. Universal indicator does not have the capacity of a digital meter or probe to display the pH of a solution to 1 or 2 decimal places. For this reason, it does not have the accuracy for precision measurements of solutions. However, for general measures of acidity or basicity, universal indicator is an adequate indicator.

Mark breakdown

- 1 mark for a judgement
- 1 mark for a supporting argument consistent with the judgement
- 1 mark for any alternative evidence

13 Amphiprotism is the ability of a substance to act as acid or base under different conditions. One substance that can do this is the hydrogen carbonate ion, HCO_3^-.

Hydrogen carbonate acting as an acid:

$$HCO_3^-(aq) + OH^-(aq) \rightleftharpoons H_2O(l) + CO_3^{2-}(aq)$$

Hydrogen carbonate acting as a base:

$$HCO_3^-(aq) + H_2O(l) \rightleftharpoons H_2CO_3(aq) + OH^-(aq)$$

Mark breakdown

- 1 mark for a definition of amphiprotism
- 1 mark for each equation that supports the argument for amphiprotism

14 It is useful to use a table, such as the following, to set out your work.

Equation	$H_2SO_4(aq)$ +	$NaOH(aq)$	→ $Salt(aq) + H_2O(l)$
Mole ratio	1	2	
n **(=** ***cV*****)**	$= 0.2 \times 0.025$ $= 5 \times 10^{-3}$	$= 0.25 \times 0.025$ $= 6.25 \times 10^{-3}$	
V **(L)**	0.025	0.025	
c **(mol L^{-1})**	= 0.20	0.25	

From the table: 5×10^{-3} mol of acid would need $2 \times 5 \times 10^{-3} = 0.01$ mol of NaOH to react. There is insufficient NaOH to neutralise all the acid, so the acid is in excess.

6.25×10^{-3} mol of NaOH would neutralise $0.5 \times 6.25 \times 10^{-3} = 3.125 \times 10^{-3}$ mol of H_2SO_4.

This leaves $(5 - 3.125 = 1.875) \times 10^{-3}$ mol of H_2SO_4 in excess.

But each mole of H_2SO_4 liberates 2 mol of H^+ ions.

So there would be $2 \times 1.875 \times 10^{-3} = 3.75 \times 10^{-3}$ mol of H^+.

Final volume would be 25 + 25 = 50 mL of solution or 0.05 L

$$\text{Final } [H^+] = \frac{3.75 \times 10^{-3}}{0.05} = 0.075 \text{ mol L}^{-1}$$

$$\begin{aligned} pH &= -\log10[H^+] \\ &= -\log10(0.075) \\ &= 1.12 \end{aligned}$$

Note: An important rule to remember is that the number of significant figures in the concentration should be equal to the number of decimal places in the calculated pH.

Mark breakdown

- 1 mark for a balanced equation
- 1 mark for calculating the number of moles of each substance
- 1 mark for calculating the excess acid and/or hydrogen ion concentration
- 1 mark for correct calculation of the final pH

15 a Many possible responses. Correct choice of indicator for each experiment because not accurately indicating the end point of the titration would affect the calculations. Correct rinsing of the glassware is also important to minimise errors.

Mark breakdown

- 1 mark for a reasonable source of error
- 1 mark for linking the error to the titration

b Crushing the tablets increases the surface area, making it easier for the tablets to dissolve.

Mark breakdown

- 1 mark for identifying a faster reaction rate for a larger surface area

c The best choice of indicator would be phenolphthalein. This is a reaction between a strong acid and a strong base, so the change through the equivalence point would involve a large pH range through a pH of 7. As phenolphthalein changes between 8.3 and 10, this is an appropriate choice.

Mark breakdown

- 1 mark for a correct identification of phenolphthalein (or bromothymol blue)
- 1 mark for a supporting argument consistent with the choice of indicator

d With a back titration, we start at the end, with the hydrochloric acid/sodium hydroxide titration, as shown in the table.

Equation	HCl(aq) +	NaOH(aq)	→ NaCl(aq) + $H_2O(l)$
Mole ratio	1	1	
n (= cV)	= 3.808×10^{-3} ←	= 0.14×0.0272 = 3.808×10^{-3} ↑	Mole ratio is 1:1
V (L)		0.0272	Value for HCl from average titre
c (mol L^{-1})	= 0.3	0.14	

Now we have the number of moles of HCl that were left over after all of the bicarbonate from the antacid tablets had reacted with the original acid solution.

The original HCl solution was a 0.3 M solution and we added 42 mL. Therefore, the original number of moles of acid was:

$$n = cV$$
$$n = 0.3 \times 0.042 = 0.0126\,\text{mol}$$

We can now calculate the number of moles of acid that reacted with the bicarbonate ion in the tablet.

$$n = 0.0126 - 3.808 \times 10^{-3}$$
$$n = 8.79 \times 10^{-3}\,\text{mol}$$

So 8.79×10^{-3} mol of acid were used to neutralise the antacid tablet.

Mark breakdown

- 1 mark for a correct equation
- 1 mark for a correct calculation of the moles of sodium hydroxide (or correct working shown)
- 1 mark for a correct calculation of the moles of HCl for neutralising the NaOH (or correct working shown)
- 1 mark for correct calculation of the moles of HCl for neutralising the antacid (or correct working shown)

e The ratio of acid to base in the antacid neutralisation was 1:1. Hence, if 8.79×10^{-3} mol of acid were used for neutralisation, there must have been 8.79×10^{-3} mol of bicarbonate ions in the original antiacid tablet. However, this was in the form of sodium bicarbonate. We can get to the mass of the $NaHCO_3$ via the molar mass.

$$m = n \times MM$$
$$= 8.79 \times 10^{-3} \times (22.99 + 1.008 + 12.01 + 3 \times 16) = 84.008$$
$$= 0.74\,\text{g}$$

So there was a mass of 0.74 g of sodium bicarbonate in the original tablet.

To find the percentage of sodium hydrogen carbonate in the original tablet, we divide this value by the mass of the original tablet and multiply by 100.

$$\frac{0.74}{0.8} \times 100 = 92.3\%$$

Mark breakdown

- 1 mark for a correct calculation of the moles of sodium hydrogen carbonate (or correct working shown)
- 1 mark for a correct calculation of the mass of sodium hydrogen carbonate (or correct working shown)
- 1 mark for a correct calculation of the percentage of sodium hydrogen carbonate in the tablet (or correct working shown)

16 a The reaction between zinc and hydrochloric acid is:

$$Zn(s) + 2HCl(aq) \rightarrow ZnCl_2(aq) + H_2(g)$$

We also have the reaction between copper and nitric acid:

$$Cu(s) + 4HNO_3(aq) \rightarrow Cu(NO_3)_2(aq) + 2H_2O(l) + 2NO_2(g)$$

Similarities:

- In each reaction, a salt is formed between the metal and the anion of the acid, i.e. zinc chloride ($ZnCl_2$) and copper nitrate ($Cu(NO_3)_2$).
- In each reaction, a gas is liberated.

Differences:

- In the first reaction the gas is hydrogen, whereas in the second reaction the gas is nitrogen dioxide.
- Water is a product only of the copper and nitric acid reaction.

Mark breakdown

- 1 mark for a correct, balanced equation for the reaction between zinc and hydrochloric acid
- 1 mark for a correct similarity between the two reactions
- 1 mark for a correct difference between the two reactions

b First, write the full equation:

$$Cu(s) + 4HNO_3(aq) \rightarrow Cu(NO_3)_2(aq) + 2H_2O(l) + 2NO_2(g)$$

Then, expand each substance to form the ionic equation:

$$Cu(s) + 4H^+(aq) + \cancel{4NO_3^-(aq)} \rightarrow Cu^{2+}(aq) + \cancel{2NO_3^-(aq)} + 2H_2O(l) + 2NO_2(g)$$

Then, eliminate the common ions to form the net ionic equation:

$$Cu(s) + 4H^+(aq) + 2NO_3^-(aq) \rightarrow Cu^{2+}(aq) + 2H_2O(l) + 2NO_2(g)$$

Mark breakdown

- 1 mark for a correct, balanced, net ionic equation
- 1 mark for a correct, balanced ionic equation including all correct states

Note: If only the net ionic equation is shown and it is correct, it could score 2 marks.

CHAPTER 3 MODULE 7

Organic chemistry

Multiple-choice solutions

1 C

The longest chain has 3 carbon atoms, so 'prop-'. There are no double or triple bonds, so '-an-'.

There is a hydroxyl group and it is on the second carbon, so '2-ol'. There is also a methyl group but it can only be located on the second carbon so no locant is required.

Hence, the name is methylpropan-2-ol (**C**).

2 C

But-1-ene has the structural formula shown on the right.

```
      H   H     H
      |   |    /
  H—C—C—C
      |   |    \\
      H   H     C—H
               /
              H
```

A chain isomer has the same molecular formula (C_4H_8) and the same functional group, but a different chain length. The easiest way to change the length of the parent chain is to remove a terminal carbon and create a methyl group. This gives **C**.

But-2-ene is a positional isomer (the double bond is between two different carbon atoms), cyclobutane is a functional group isomer (an alkane not an alkene) and butan-1-ol is not an isomer of but-1-ene.

3 D

Dispersion forces hold long hydrocarbon chains together and must be overcome for a substance to change from the liquid phase to the gas phase. However, since they are the weakest intermolecular forces, pentane (**A**) would not have the highest boiling point. The presence of an oxygen results in a dipole, and dipole–dipole interactions are stronger than dispersion forces, requiring more energy to be overcome. **B–D** all include oxygen. However, if the oxygen is also bonded to a hydrogen, then hydrogen bonding can occur between molecules. This is the strongest type of intermolecular force and pentanoic acid (**D**) has both the C–O dipole and the O–H dipole, resulting in two potential sites for hydrogen bonding. Hence, pentanoic acid would have the highest boiling point.

4 B

The most stable arrangement of electrons around a central carbon atom is a tetrahedron. All four valence electrons form bonds with other atoms and a tetrahedral arrangement is the most stable way for them to exist.

However, when a double bond forms, four electrons occupy the region of the double bond and this leaves only four other bonded electrons around the carbon atom, divided between two bonds. In this case, the most stable arrangement is when each of these three bonds form the vertices of a triangle in the same plane. Hence, they form a trigonal planar shape (**B**).

5 C

First, write the balanced equation for fermentation:

$$C_6H_{12}O_6(aq) \rightarrow 2C_2H_5OH(aq) + 2CO_2(g)$$

This shows a ratio of 1 mole of glucose producing 2 moles of carbon dioxide.

To use this ratio, convert the 18 g of glucose into moles. This requires the molar mass of glucose:

$$MM = 6 \times 12.01 + 12 \times 1.008 + 6 \times 16 = 180.156\,\text{g mol}^{-1}.$$

Convert mass to moles:

$$n = \frac{m}{MM} = \frac{18}{180.156} = 0.10\,\text{mol}$$

From the mole ratios, if all of the glucose fermented, it would produce $2 \times 0.1 = 0.20$ mol of carbon dioxide.

From the data sheet, at 100 kPa and 298 K, 1 mol of gas occupies 24.79 L.

Calculate the final volume by multiplying the number of moles by the molar volume at SLC.

$V = n \times MV$

$= 0.2 \times 24.79$ (keep the full answer in your calculator when you calculate each step)

$= 4.95$ L

This is **C**.

6 D

Propene has the structural formula shown on the right.

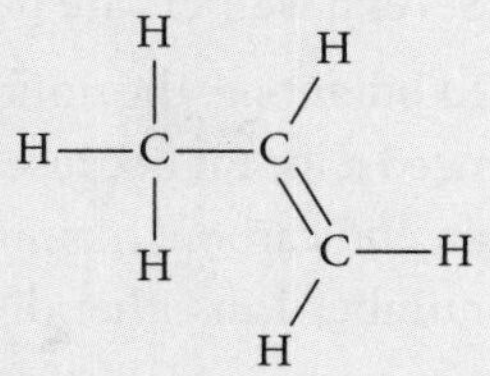

On reaction with hydrogen chloride, the double bond breaks and there is an addition reaction. A hydrogen atom attaches to one of the carbons of the double bond and a chlorine atom attaches to the other carbon. We could expect there to be an equal chance of the chlorine atom attaching to either carbon; however, one carbon has two hydrogens already attached and the other carbon only has one. Markovnikov's rule says that the hydrogen is more likely to attach to the end carbon and the chlorine to the middle carbon. This means the product would consist of more 2-chloropropane than 1-chloropropane (**D**).

7 C

An orange solution of dichromate ions changes colour to green on reduction to the chromium ion. Dichromate ions are reduced because they oxidise another substance. To determine which of the reactants would not change the colour of the dichromate, we need to know what happens with each reactant.

- Ethanol (**A**) is a primary alcohol. It is oxidised to ethanal or ethanoic acid and will change the colour of the dichromate ion.
- Butan-2-ol (**B**) is a secondary alcohol. It is oxidised to butanone and will change the colour of the dichromate ion.
- Pentanal (**D**) is an aldehyde. It is oxidised to pentanoic acid and will change the colour of the dichromate ion.
- 2-Methylbutan-2-ol (**C**) is a tertiary alcohol. Tertiary alcohols cannot be oxidised and hence will not change the colour of the dichromate ion.

8 A

There are two clues in this question. The starting materials are ethanol (a 2-carbon compound) and pentanoic acid (a 5-carbon compound), so you can eliminate any answer that does have not a 5-carbon chain on one side of the ester group and 2 carbons on the other. Hence, **C** (1 and 5), and **D** (2 and 3) can be eliminated

To distinguish between **A** and **B**, you need to remember that the C=O was part of the original acid so the C=O carbon must have been originally part of the pentanoic acid. In **B**, the C=O is on a carbon that is part of a chain of 2, so this is incorrect.

In **A**, the C=O bond is part of a 5-carbon chain. On the other side of the COOC are the 2 carbons from the original ethanol alcohol, so **A** is the correct answer.

9 B

There is no trick to this question, but you should recall that carbon dioxide is a product of complete combustion. Solid carbon (soot) and carbon monoxide are produced when oxygen concentrations are low.

Look for a balanced equation containing carbon dioxide as a product (**B**).

10 A

When matching a structure to a name, or vice versa, remember that there is a hierarchy for functional groups.

- The carboxylic acid group takes precedence over the amine group. So the acid group (COOH) is carbon number 1.
- The name 'hexanoic' implies 6 carbons in the chain.
- The amine group (NH_2) is located on carbon 6 at the opposite end from the carboxyl group.

The only option with 6 carbons in the chain, a COOH at one end and an NH_2 group on the other is **A**.

9780170465281

Short-answer solutions

11 Several isomers are possible, including a ketone (butanone) and an aldehyde (butanal).

In butanone, the double-bonded oxygen is on a middle carbon, making this a ketone, but it does not need a locant because it is carbon 2 in either direction. In butanal, the double-bonded oxygen is on an end carbon, making this an aldehyde. It also does not need a locant because it would be carbon number 1 in either direction.

Butanone   Butanal

Mark breakdown

- 1 mark for each correctly drawn isomer
- 1 mark for each name that correctly matches the structure drawn

12 There are many possible responses, one of which is shown below.

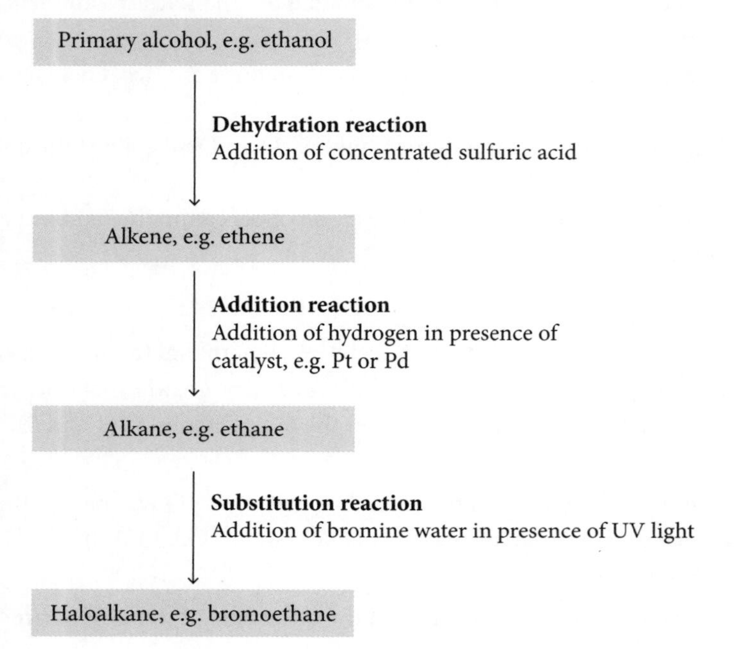

Mark breakdown

- 1 mark for a diagrammatic pathway rather than a series of reactions (does not have to be linear as shown)
- 3 marks for the correct reaction conditions
- 2 marks for logical sequence of reactions, including each intermediate

13 Soap molecules have two significant regions: an anionic (hydrophilic) head and a non-polar (hydrophobic) tail. An example is shown below.

$$CH_3-CH_2-CH_2-CH_2-CH_2-CH_2-CH_2-CH_2-CH_2-CH_2-CH_2-CH_2-CH_2-CH_2-CH_2-CH_2-CH_2-C(=O)-O^-Na^+$$

Soaps and detergents can clean because they can bond with both polar and non-polar substances. Their heads attract polar substances (e.g. water) and their tails attract non-polar substances (e.g. oil or grease). To separate a spot of grease from the surface of a plate or fabric, individual soap molecules bond to the grease spot via their tails. Their heads stay in contact with the water. More soap molecules bond to the grease and begin to lift it away from the surface. As the grease starts to lift, more soap molecules surround the non-polar molecule by attaching with their tails and having heads pointing out. This forms a micelle. This process repeats to produce many micelles, which keep the non-polar molecules separate from one another. This is shown below.

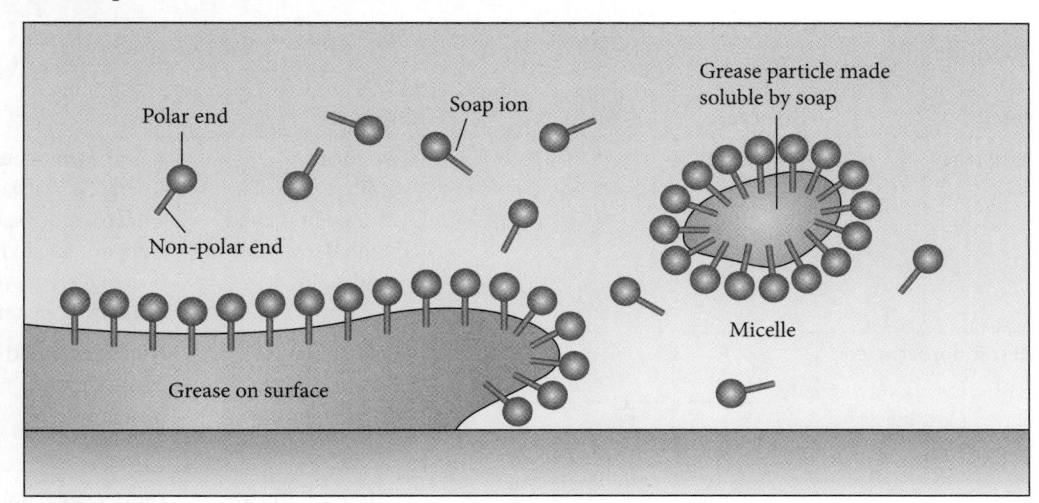

Soaps are very good cleaning agents because of their ability to bond to polar or ionic and non-polar substances to produce emulsions.

Mark breakdown

- 1 mark for correct description of the structure of soaps
- 1 mark for identifying the hydrophilic and hydrophobic ends
- 1 mark for including part or all of the description of the formation of a micelle
- 1 mark for a labelled diagram of the structure of a soap and/or micelle
- 1 mark for referring to the effectiveness of soap as cleaning agents

14 Refluxing is the use of a vertical condenser over a reaction vessel. The purpose of refluxing is to allow heated reactants that have vaporised to condense and drip back into the reacting vessel.

The esterification reaction is an equilibrium reaction and takes place very slowly at room temperature. Heating the mixture increases the reaction rate, but low boiling point substances in the reacting mixture can vaporise and escape through the mouth of the reaction vessel. The reaction vessel cannot be sealed because the build-up of pressure would be too great.

Refluxing allows the reaction to proceed at a higher temperature. It keeps volatile organic compounds, especially alcohols, away from a heat source where they could combust, and it recycles unreacted reactant molecules back into the reaction vessel where they continue to drive the forward reaction. The process of refluxing results in a higher yield of the organic product and is a safer method for esterification at higher temperatures.

$$\text{Ethanol + propanoic acid} \rightleftharpoons \text{ester (ethyl propanoate) + water}$$

This can be shown as structural formulae, as in the following diagram.

ethanol + propanoic acid ⇌ ethyl propanoate + water

$CH_3CH_2OH + CH_3CH_2COOH \rightleftharpoons CH_3CH_2COOCH_2CH_3 + H_2O$

Mark breakdown

- 1 mark for discussing the technique of refluxing
- 1 mark for stating a reason for using refluxing
- 1 mark for giving a benefit of refluxing for the esterification reaction
- 1 mark for a correct example of reactants and products for an esterification reaction

15 For example:

Polymer	Structure	Property	Use
Polyethylene	—(CH_2—CH_2)— chain: —C—C—C—C—C—C—C—C—C—C— with H above and below each C	Two forms: • soft and flexible, highly branched • high density, strong and rigid	• Soft form – cling wrap, carry bags • Harder, denser form – kitchen utensils, toys and building materials
Polytetrafluoroethylene (PTFE)	—C—C—C—C—C— with F above and below each C	Low friction, non-stick, high chemical and electrical resistance	Non-stick cooking pans (Teflon), insulator
Polyester	—O—C(=O)—C(CH_3)(H)—O—C(=O)—C(CH_3)(H)—O—C(=O)—C(CH_3)(H)—O—C(=O)—	High strength and elasticity, electrical resistance, stable in aqueous solutions	Fabrics, PET bottles for beverage storage

Mark breakdown

- 1 or 2 marks for correct structural formula (or description) for each polymer (maximum of 1 mark if one of the structures is incorrect)
- 1 or 2 marks for correct, relevant properties of each polymer (maximum of 1 mark if one of the structures is incorrect)
- 1 or 2 marks for correct uses specifically linked to at least one listed property for each polymer (maximum of 1 mark if one of the structures is incorrect)

16 Equation for fermentation is:

$$C_6H_{12}O_6(aq) \rightarrow 2C_2H_5OH(aq) + 2CO_2(g)$$

Mole ratio 1 : 2 : 2

You can construct a table to solve these types of problems.

	Glucose	Ethanol	Carbon dioxide
Mole ratio	1	2	2
***n* (number of moles)**		0.041 ←	0.041
***m* (mass)**		↓	↑
***MM* (molar mass)**		46.068	
***V* (volume)**			1.006
***MV* (molar volume)**			24.79

Use the total gas produced to represent the volume of CO_2: $V = 1006\,mL = 1.006\,L$.

The experiment was conducted under SLC. From the data sheet, at SLC the molar volume of a gas is 24.79 L.

$$n = \frac{V}{MV}$$

$$= \frac{1.006}{24.79}$$

$$= 0.041\,mol \text{ (add this value to the table)}$$

(Remember to keep the full number in your calculator for the next calculation.)

Mole ratio of CO_2 to ethanol is 2 : 2 or 1 : 1.

So 0.041 mol of carbon dioxide means 0.041 mol of ethanol.

To calculate the mass of ethanol, you need the molar mass.

C_2H_6O: $MM = 2 \times 12.01 + 6 \times 1.008 + 1 \times 16 = 46.068\,g\,mol^{-1}$

Add this value to the table and we can perform the final calculation:

$$m = n \times MM$$

$$= 0.041 \times 46.068$$

$$= 1.869\,g$$

The minimum number of significant figures in the question is 4, so give your answer to 4 significant figures.

Mark breakdown

- 1 mark for correct step in calculating moles of carbon dioxide
- 1 mark for correct step in calculating mass of ethanol
- 1 mark for correct mole ratio of carbon dioxide to ethanol
- 1 mark for correct balanced equation for fermentation

CHAPTER 4 MODULE 8

Applying chemical ideas

Multiple-choice solutions

1 D

Barium and calcium are both in group 2 in the periodic table (alkaline earth metals) so behave quite similarly in solutions. Neither causes bubbles in the presence of an acid. The addition of sulfate ions precipitates both barium sulfate and calcium sulfate, and both precipitates would be white. Likewise, the addition of carbonate ions precipitates white barium carbonate as well as white calcium carbonate. However, calcium produces a red flame while barium produces a yellow-green flame, so this is the best test to distinguish between the two cations.

2 A

When testing for anions, the addition of acid only produces one positive result; all of the other results are negative. Acids combine with carbonates to produces bubbles of carbon dioxide. So bubbling could only occur with the carbonate anion (**A**).

3 A

$[FeSCN]^{2+}$ is a complex formed in an equilibrium reaction of the ferric cation (Fe^{3+}) and the thiocyanate anion (SCN^-). It is a positive ion so could be regarded as a cation but not an anion, and it is not a precipitate.

4 C

The test for alcohols is an oxidation reaction with a coloured oxidising agent, such as potassium permanganate or potassium dichromate. When the oxidising agent is reduced, it changes colour. Primary alcohols can be oxidised to aldehydes or carboxylic acids whereas secondary alcohols can be oxidised to ketones. Tertiary alcohols cannot be oxidised so they show no reaction with an oxidising agent and will not change the colour of the solution.

5 D

First write an equation for the precipitation reaction:

$$Ag^+(aq) + Cl^-(aq) \rightarrow AgCl(s)$$

From the balanced equation, you can see that the mole ratio is 1 : 1.

Use the values for silver nitrate to calculate the number of moles of silver ions.

The concentration is 0.05 mol L^{-1} and the average volume of the titre is 25.5 mL

$$n = CV$$

$$n = 0.05 \times 0.0255\text{ L}$$

$$= 0.001\,275\text{ mol}$$

The mole ratio is 1 : 1, so the number of moles of chloride ions is also 0.001 275 mol.

This is the number of moles in the original 10 mL sample, so calculate the concentration:

$$C = \frac{n}{V}$$

$$= \frac{0.001275}{0.01}$$

$$= 0.128\text{ mol L}^{-1}$$

The closest value from those provided is **D**.

6 B

This is a straight substitution into an equation, but we do need to select the correct equation. The data sheet provides the following formula for the Beer–Lambert law

$$A = \varepsilon lc$$

$$c = \frac{A}{\varepsilon l}$$

Extract the relevant variables. $A = 0.038$, $\varepsilon = 4500$ and $l = 15$ mm. Convert the value for l into cm, so $l = 1.5$ cm.

$$c = \frac{0.038}{4500 \times 1.5}$$

$$= 5.63 \times 10^{-6} \text{ mol L}^{-1}$$

Convert mol L^{-1} to ppm by converting the number of moles to a mass in mg.

$$MM = 55.85 + 32.07 + 12.01 + 14.01 = 113.94 \text{ g mol}^{-1}$$

$$m = n \times MM$$

$$= 5.63 \times 10^{-6} \times 113.94$$

$$= 6.4 \times 10^{-4} \text{ g or } 0.64 \text{ mg}$$

Hence, the concentration of iron(III) thiocyanate in ppm is 0.64 (**C**).

7 D

The technique of atomic absorption spectroscopy uses the ultraviolet and visible light parts of the electromagnetic spectrum. This is because AAS is based on the movement of electrons from their ground state to an excited state.

8 C

When you are given a graph such as this, keep one variable constant and see what happens to the yield as the other one varies.

First, look at pressure; this trend is quite easy, but you have to consider a particular temperature, so pick a temperature, e.g. halfway along the x-axis. As you go up, the yield increases as you pass through the 50 MPa line, the 30 MPa line and finally the 10 MPa line. So at the same temperature, as pressure decreases, yield increases. In an equilibrium system involving gases, this means the side with the most gas moles is favoured. So, **A** and **B** are incorrect.

Now, look at temperature. Choose any of the pressure lines, e.g. 30 MPa. As temperature increases (to the right) the 30 MPa line goes up, so the yield increases with increasing temperature. Increasing temperature favours the endothermic reaction, so the forward reaction must be endothermic. This means it has a positive ΔH value, or $\Delta H > 0$. Since we only have **C** and **D** left to choose from, the answer is **C**.

9 D

Butanal has the following structural formula.

```
       H    H    H
       |    |    |        O
                         //
  H —  C —  C —  C —  C
       |    |    |        \
       H    H    H         H
```

Determine the number of carbon environments. There are four carbon environments:

- a terminal carbon attached to a CH_2 group
- a carbon attached to a CH_3 and a CH_2 group
- a carbon attached to a CH_2 and a CHO group
- a carbon in the CHO group, which is attached to a CH_2 group.

This is 4 separate carbon environments so we would expect four separate peaks on the ^{13}C NMR spectrum (**D**).

SOLUTIONS

10 D

The Haber process is based on a delicate balance of temperature and pressure. In the equation, there are 4 mol of gaseous reactants and only 2 mol of gaseous products. If we want to increase the yield of the product, we need to shift the equilibrium to favour the product side. Increasing the pressure favours the side with the smaller number of gaseous moles, hence we would want a high pressure to favour the yield of product.

Likewise, as written the reaction is exothermic because the enthalpy change is negative. If we want to favour the forward reaction, we need to lower the temperature.

So the correct combination to favour an increased yield is high pressure and low temperature (**D**).

Short-answer solutions

11 This statement could generate many responses; however, a greater depth of understanding is demonstrated by focusing on monitoring of the air, water and land (or at least two of these). You should identify specific chemicals of interest , e.g. ozone in the lower atmosphere, particulates from incomplete combustion, NO_x or SO_x levels that lead to acid rain. You should also discuss potential consequences associated with the chemicals of interest and why it is important to monitor them. For example, discuss the health implications related to ozone in the lower atmosphere; the link between brain damage and heavy metals in soils or waterways; or the effect of plastic waste on aquatic life. When a question asks why something is important, you can assume that it is important. Discuss the specifics of the process, provide examples of what is monitored and why, and aim for a response that includes a range of good examples rather than focusing on one (e.g. air pollution).

Mark breakdown

- 2 marks for identifying at least two different examples of environmental monitoring
- 1 mark for each logical reason linked to a specific, named example of environmental monitoring (maximum 2 marks)
- 1 mark for justifying the importance of the named examples

12 a One method is:

$$Ag^+(aq) + Cl^-(aq) \rightleftharpoons AgCl(s)$$

The proportion of chloride ions in the silver chloride precipitate is:

$$\frac{35.45}{35.45 + 107.9} = 0.247$$

Hence, the mass of chloride ions in the precipitate is $0.247 \times 9.6 = 2.37$ g.

Assume the mass of chloride ions in the sample is also 2.37 g.

So, the proportion of chloride ions in the sample is:

$$\frac{2.37}{6.60} = 0.360 \text{ or } 36.0\% \text{ of the sea water sample.}$$

Mark breakdown

- 1 mark for correct calculation of the mass of chloride ions in the silver chloride precipitate
- 1 mark for correctly linking the mass of chloride in the precipitate to the mass of chloride ions in the sample
- 1 mark for correct calculation of the proportion of chloride ions in the sample

b There are several possible responses, e.g. one assumption the student has made is that the precipitate formed by the addition of silver nitrate is all silver chloride. This may not be the case.

Mark breakdown

- 1 mark for a reasonable source of error

c This should link to part **b**, but 2 marks could still be obtained if it does not. So you could discuss repetition with consistent results as a way of increasing reliability, which is correct but doesn't specifically address the error identified in part **b**, so it would only gain a maximum of 2 marks. If you discussed some specific tests that could be performed on the sample first, to eliminate the presence of any other anions that might form a similar precipitate with silver cations, then this could be worth 3 marks.

Mark breakdown

- 1 mark for a reasonable procedural improvement
- 1 mark for linking the improvement to the identified error
- 1 mark for justifying the change on the basis of increased validity or reliability

13 These spectra are of propan-2-yl ethanoate (propan-2-yl acetate). It has the following structural formula.

$$H_3C—C(=O)—O—CH(CH_3)—CH_3$$

This was a particularly challenging question from the 2020 HSC Chemistry exam because although the structure was an ester, it was produced from a secondary alcohol rather than a primary alcohol, which most students would have suspected. Therefore, it was critical to look for specific clues among the various spectra included. Some examples are:

- The IR spectrum has a trough at 1750 cm^{-1}, which suggests a carbonyl group.
- The IR spectrum lacks both the broad tongue (indicative of an O–H) and the hairy beard (indicative of an acid group) in the 2500–3300 cm^{-1} region, so it is not an acid.
- The ^{13}C NMR spectrum has a peak at about 170 ppm, suggesting a carbonyl bond, which is consistent with an ester (we have eliminated an acid).
- The ^{13}C NMR spectrum shows four separate carbon environments, including a peak around 70 ppm consistent with a C–O bond in an ester group.
- The ^{1}H NMR spectrum includes a septet, indicating a hydrogen with six neighbouring H atoms, which is the two methyl (CH_3) groups either side of the central carbon with the oxygen.

You could also refer to the ^{1}H NMR doublet (one neighbouring H atom) and a ^{1}H NMR singlet (no neighbouring H atoms (the methyl group from the original ethanoic acid)).

Mark breakdown

- 1 mark for correct identification of propan-2-yl ethanoate
- 1 mark for correct structural formula
- 1 mark for each relevant reference to one of the three spectra shown that are consistent with the named organic compound
- 1 mark for using data correctly from all three spectra
- 1 mark for logical construction of answer

9780170465281

SOLUTIONS

14 There are several ways to answer this question. The best way is to include an annotated diagram or labelled sketch of the process of mass spectrometry and then explain what you have drawn and link this explanation to the identification of organic compounds, specifically ethanamine. Ethanamine has the structural formula on the right.

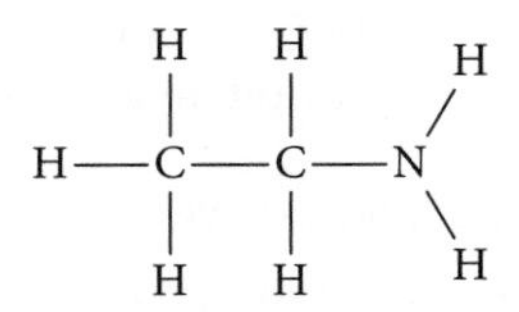

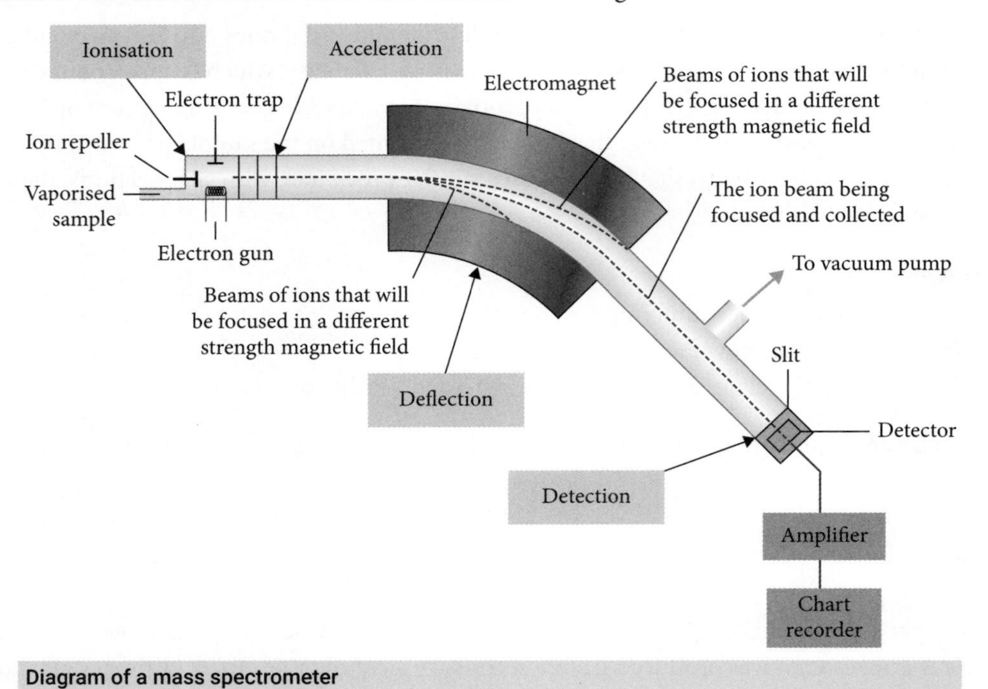

Diagram of a mass spectrometer

The diagram indicates several important steps.

1. The sample is vaporised (this allows for independent motion of the molecules).
2. The sample is ionised (this means they will be influenced by the external magnetic field).
3. The ionised particles are accelerated.
4. The high-speed ions are exposed to a strong magnetic field.
5. The ionised particles are deflected by the magnetic field according to size: smaller particles have greater deflection (this also depends on their charge).

During this process, the molecule can fragment into smaller pieces and the detector identifies all of these fragments on the basis of their mass to charge ratio. The maximum value corresponds to the molar mass of the compound; for ethanamine ($CH_3CH_2NH_2$) this is 45. The fragments can also be used to identify common groups, e.g. 15 could indicate a methyl group (CH_3), 16 an amine group (NH_2) and 29 an ethyl group (C_2H_5).

Mark breakdown

- 1 mark for each of the key steps in the process of mass spectrometry (maximum 4 marks)
- 1 mark for linking the process specifically to the identification of ethanamine

15 There are many possible responses. For extended response questions it is important to address all components of the question and ensure you use evidence to support your points. Chemical synthesis processes you could discuss include the Haber process for the production of ammonia, esterification, polymerisation, fermentation and saponification. Some key points to remember:

1. Reaction conditions for industrial synthesis often involve high temperatures and/or high pressures. They may also involve catalysts. However, do not mention these unless you know the specifics. A specifically named catalyst is one way to demonstrate your specific knowledge. It is also easier to discuss reaction conditions for an equilibrium synthesis, like the Haber process or esterification.
2. Industrial use is basically code for large scale for profit. It means not just scaling up a process like saponification but adding specific features to increase the marketability (e.g. adding perfumes and shape to soaps), increase profit (e.g. can you sell or recycle the by-products, can you locate close to a source of raw materials to increase the availability of reagents, can you improve the quality or purity of your product?) and reduce environmental waste (e.g. by testing, recycling). Be as specific as you can when discussing what your product is used for.
3. Environmental issues are primarily about the release of waste products into the environment but could include thermal waste, high temperature water entering waterways or furnaces heating the air. The product itself may have environmental implications, such as polymers or lead–acid batteries.

Finally, when you are asked to evaluate, your opinion must be obvious. It is better to state it at the start and reinforce it again at the end, rather than assume it is obvious in your writing.

Mark breakdown

- 1 mark for an identified chemical synthesis process
- 1 mark for an explanation, including chemical equation for the process
- 1 mark for a relevant discussion of the reaction conditions associated with your chosen process
- 1 mark for a relevant discussion of the industrial uses associated with your chosen process
- 1 mark for a relevant discussion of the environmental issues associated with your chosen process
- 2 marks for a clear evaluative statement associated with your chosen chemical synthesis process along with specific evidence from your response to support your position

16 Pentane-1,5-diamine has the following structural formula.

```
H      H   H   H   H   H      H
 \     |   |   |   |   |     /
  N——C——C——C——C——C——N
 /     |   |   |   |   |     \
H      H   H   H   H   H      H
```

Drawing the full structural formula helps determine the functional groups for IR spectroscopy, common fragments for mass spectroscopy, and different environments for NMR spectroscopy. Then you can analyse each spectrum.

1. Infrared spectrum

   The highlighted trough occurs at a wavenumber of 3300–3400 cm^{-1}. From the data sheet, this could be an OH or NH group. The lack of a broad tongue eliminates the alcohol functional group, leaving the amine functional group. Pentane-1,5-diamine has two of these groups.

2. Mass spectrum

   The highlighted peak occurs at a ratio of 30 m/z. The NH_2^+ end fragments have a mass of 16. The $CH_2NH_2^+$ fragment, which could occur from either end, has a mass of 30. This corresponds to the highlighted peak, confirming the presence of this fragment of the molecule.

3 ^{13}C NMR spectrum

The highlighted region encompasses the entire spectra, so you need to consider the whole molecule. Even though there are 5 carbon atoms in the compound, the compound has symmetry, which reduces the number of carbon environments. These are numbered in the following diagram.

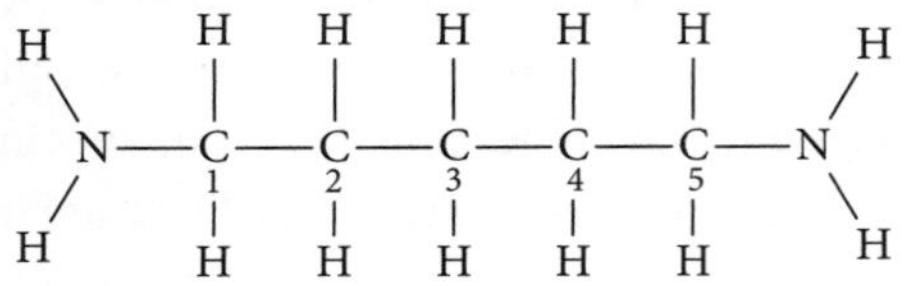

```
H        H    H    H    H    H        H
 \       |    |    |    |    |       /
  N —— C —— C —— C —— C —— C —— N
 /       1    2    3    4    5       \
H        |    |    |    |    |        H
         H    H    H    H    H
```

Carbon 1 is in a similar environment to carbon 5. Carbon 2 is in a similar environment to carbon 4, so only carbon 3 is unique. The large peak at about 42 ppm corresponds to carbon 1 and is a result of the presence of the amine group (values on the data sheet for amine groups (25–60 ppm)). The other two peaks, at about 25 ppm and 33 ppm, are within the range of values on the data sheet for ethyl groups (5–40 ppm).

4 ^{1}H NMR spectrum

The highlighted signal results from similar chemical shifts of protons in two different environments: an overlap of a 2H signal and a 4H signal, giving 6 H. Hence the highlighted region is the signals from carbons 2 and 4.

Mark breakdown

- 1 mark for correct interpretation of highlighted region of the infrared spectrum
- 1 mark for correct interpretation of highlighted region of the mass spectrum
- 1 mark for correct interpretation of highlighted region of ^{13}C NMR spectrum
- 1 mark for correct interpretation of highlighted region of ^{1}H NMR spectrum
- 1 mark for specific use of data from the question
- 2 marks for providing clear links between the interpretation of a particular spectra and the structure of pentane-1,5-diamine

HIGHER SCHOOL CERTIFICATE EXAMINATION

Formulae sheet

$n = \frac{m}{MM}$	$c = \frac{n}{V}$	$PV = nRT$
$q = mc\Delta T$	$\Delta G^\circ = \Delta H^\circ - T\Delta S^\circ$	$\text{pH} = -\log_{10}[\text{H}^+]$
$pK_a = -\log_{10}[K_a{}^+]$	$A = \varepsilon lc = \log_{10}\frac{I_o}{I}$	

Avogadro constant, N_A		$6.022 \times 10^{23}\ \text{mol}^{-1}$
Volume of 1 mole ideal gas: at 100 kPa and		
	at 0°C (273.15 K)	22.71 L
	at 25°C (298.15 K)	24.79 L
Gas constant		$8.314\ \text{J mol}^{-1}\ \text{K}^{-1}$
Ionisation constant for water at 25°C (298.15 K), K_w		1.0×10^{-14}
Specific heat capacity of water		$4.18 \times 10^3\ \text{J kg}^{-1}\ \text{K}^{-1}$

Data sheet

Solubility constants at 25°C

Compound	K_{sp}
Barium carbonate	2.58×10^{-9}
Barium hydroxide	2.55×10^{-4}
Barium phosphate	1.3×10^{-29}
Barium sulfate	1.08×10^{-10}
Calcium carbonate	3.36×10^{-9}
Calcium hydroxide	5.02×10^{-6}
Calcium phosphate	2.07×10^{-29}
Calcium sulfate	4.93×10^{-5}
Copper(II) carbonate	1.4×10^{-10}
Copper(II) hydroxide	2.2×10^{-20}
Copper(II) phosphate	1.40×10^{-37}
Iron(II) carbonate	3.13×10^{-11}
Iron(II) hydroxide	4.87×10^{-17}
Iron(III) hydroxide	2.79×10^{-39}
Iron(III) phosphate	9.91×10^{-16}

Compound	K_{sp}
Lead(II) bromide	6.60×10^{-6}
Lead(II) chloride	1.70×10^{-5}
Lead(II) iodide	9.8×10^{-9}
Lead(II) carbonate	7.40×10^{-14}
Lead(II) hydroxide	1.43×10^{-15}
Lead(II) phosphate	8.0×10^{-43}
Lead(II) sulfate	2.53×10^{-8}
Magnesium carbonate	6.82×10^{-6}
Magnesium hydroxide	5.61×10^{-12}
Magnesium phosphate	1.04×10^{-24}
Silver bromide	5.35×10^{-13}
Silver chloride	1.77×10^{-10}
Silver carbonate	8.46×10^{-12}
Silver hydroxide	2.0×10^{-8}
Silver iodide	8.52×10^{-17}
Silver phosphate	8.89×10^{-17}
Silver sulfate	1.20×10^{-5}

9780170465281

Infrared absorption data

Bond	Wavenumber/cm^{-1}
N—H (amines)	3300–3500
O—H (alcohols)	3230–3550 (broad)
C—H	2850–3300
O—H (acids)	2500–3000 (very broad)
C≡N	2220–2260
C=O	1680–1750
C=C	1620–1680
C—O	1000–1300
C—C	750–1100

^{13}C NMR chemical shift data

Type of carbon		δ/ppm
—C—C— (sp³ carbons)		5–40
R—C—Cl or Br		10–70
R—C(=O)—C—		20–50
R—C—N<		25–60
—C—O—	alcohols, ethers or esters	50–90
>C=C<		90–150
R—C≡N		110–125
benzene ring		110–160
R—C(=O)—	esters or acids	160–185
R—C(=O)—	aldehydes or ketones	190–220

UV absorption

(This is not a definitive list and is approximate.)

Chromophore	λ_{max} (nm)
C—H	122
C—C	135
C=C	162

Chromophore	λ_{max} (nm)
C≡C	173 178 196 222
C—Cl	173
C—Br	208

9780170465281

Some standard potentials

$K^+ + e^-$	$\rightleftharpoons$	$K(s)$	–2.94 V
$Ba^{2+} + 2e^-$	$\rightleftharpoons$	$Ba(s)$	–2.91 V
$Ca^{2+} + 2e^-$	$\rightleftharpoons$	$Ca(s)$	–2.87 V
$Na^+ + e^-$	$\rightleftharpoons$	$Na(s)$	–2.71 V
$Mg^{2+} + 2e^-$	$\rightleftharpoons$	$Mg(s)$	–2.36 V
$Al^{3+} + 3e^-$	$\rightleftharpoons$	$Al(s)$	–1.68 V
$Mn^{2+} + 2e^-$	$\rightleftharpoons$	$Mn(s)$	–1.18 V
$H_2O + e^-$	$\rightleftharpoons$	$\frac{1}{2}H_2(g) + OH^-$	–0.83 V
$Zn^{2+} + 2e^-$	$\rightleftharpoons$	$Zn(s)$	–0.76 V
$Fe^{2+} + 2e^-$	$\rightleftharpoons$	$Fe(s)$	–0.44 V
$Ni^{2+} + 2e^-$	$\rightleftharpoons$	$Ni(s)$	–0.24 V
$Sn^{2+} + 2e^-$	$\rightleftharpoons$	$Sn(s)$	–0.14 V
$Pb^{2+} + 2e^-$	$\rightleftharpoons$	$Pb(s)$	–0.13 V
$H^+ + e^-$	$\rightleftharpoons$	$\frac{1}{2}H_2(g)$	0.00 V
$SO_4^{2-} + 4H^+ + 2e^-$	$\rightleftharpoons$	$SO_2(aq) + 2H_2O$	0.16 V
$Cu^{2+} + 2e^-$	$\rightleftharpoons$	$Cu(s)$	0.34 V
$\frac{1}{2}O_2(g) + H_2O + 2e^-$	$\rightleftharpoons$	$2OH^-$	0.40 V
$Cu^+ + e^-$	$\rightleftharpoons$	$Cu(s)$	0.52 V
$\frac{1}{2}I_2(s) + e^-$	$\rightleftharpoons$	I^-	0.54 V
$\frac{1}{2}I_2(aq) + e^-$	$\rightleftharpoons$	I^-	0.62 V
$Fe^{3+} + e^-$	$\rightleftharpoons$	Fe^{2+}	0.77 V
$Ag^+ + e^-$	$\rightleftharpoons$	$Ag(s)$	0.80 V
$\frac{1}{2}Br_2(l) + e^-$	$\rightleftharpoons$	Br^-	1.08 V
$\frac{1}{2}Br_2(aq) + e^-$	$\rightleftharpoons$	Br^-	1.10 V
$\frac{1}{2}O_2(g) + 2H^+ + 2e^-$	$\rightleftharpoons$	H_2O	1.23 V
$\frac{1}{2}Cl_2(g) + e^-$	$\rightleftharpoons$	Cl^-	1.36 V
$\frac{1}{2}Cr_2O_7^{2-} + 7H^+ + 3e^-$	$\rightleftharpoons$	$Cr^{3+} + \frac{7}{2}H_2O$	1.36 V
$\frac{1}{2}Cl_2(aq) + e^-$	$\rightleftharpoons$	Cl^-	1.40 V
$MnO_4^- + 8H^+ + 5e^-$	$\rightleftharpoons$	$Mn^{2+} + 4H_2O$	1.51 V
$\frac{1}{2}F_2(g) + e^-$	$\rightleftharpoons$	F^-	2.89 V

Aylward and Findlay, *SI Chemical Data* (5th Edition) is the principal source of data for the standard potentials.
Some data may have been modified for examination purposes.

9780170465281

Periodic table of the elements

KEY

Atomic Number	79
Symbol	Au
Standard Atomic Weight	197.0
Name	Gold

1 H 1.008 Hydrogen																	2 He 4.003 Helium
3 Li 6.941 Lithium	4 Be 9.012 Beryllium											5 B 10.81 Boron	6 C 12.01 Carbon	7 N 14.01 Nitrogen	8 O 16.00 Oxygen	9 F 19.00 Fluorine	10 Ne 20.18 Neon
11 Na 22.99 Sodium	12 Mg 24.31 Magnesium											13 Al 26.98 Aluminium	14 Si 28.09 Silicon	15 P 30.97 Phosphorus	16 S 32.07 Sulfur	17 Cl 35.45 Chlorine	18 Ar 39.95 Argon
19 K 39.10 Potassium	20 Ca 40.08 Calcium	21 Sc 44.96 Scandium	22 Ti 47.87 Titanium	23 V 50.94 Vanadium	24 Cr 52.00 Chromium	25 Mn 54.94 Manganese	26 Fe 55.85 Iron	27 Co 58.93 Cobalt	28 Ni 58.69 Nickel	29 Cu 63.55 Copper	30 Zn 65.38 Zinc	31 Ga 69.72 Gallium	32 Ge 72.64 Germanium	33 As 74.92 Arsenic	34 Se 78.96 Selenium	35 Br 79.90 Bromine	36 Kr 83.80 Krypton
37 Rb 85.47 Rubidium	38 Sr 87.61 Strontium	39 Y 88.91 Yttrium	40 Zr 91.22 Zirconium	41 Nb 92.91 Niobium	42 Mo 95.96 Molybdenum	43 Tc Technetium	44 Ru 101.1 Ruthenium	45 Rh 102.9 Rhodium	46 Pd 106.4 Palladium	47 Ag 107.9 Silver	48 Cd 112.4 Cadmium	49 In 114.8 Indium	50 Sn 118.7 Tin	51 Sb 121.8 Antimony	52 Te 127.6 Tellurium	53 I 126.9 Iodine	54 Xe 131.3 Xenon
55 Cs 132.9 Caesium	56 Ba 137.3 Barium	57–71 Lanthanoids	72 Hf 178.5 Hafnium	73 Ta 180.9 Tantalum	74 W 183.9 Tungsten	75 Re 186.2 Rhenium	76 Os 190.2 Osmium	77 Ir 192.2 Iridium	78 Pt 195.1 Platinum	79 Au 197.0 Gold	80 Hg 200.6 Mercury	81 Tl 204.4 Thallium	82 Pb 207.2 Lead	83 Bi 209.0 Bismuth	84 Po Polonium	85 At Astatine	86 Rn Radon
87 Fr Francium	88 Ra Radium	89–103 Actinoids	104 Rf Rutherfordium	105 Db Dubnium	106 Sg Seaborgium	107 Bh Bohrium	108 Hs Hassium	109 Mt Meitnerium	110 Ds Darmstadtium	111 Rg Roentgenium	112 Cn Copernicium	113 Nh Nihonium	114 Fl Flerovium	115 Mc Moscovium	116 Lv Livermorium	117 Ts Tennessine	118 Og Oganesson

Lanthanoids

57 La 138.9 Lanthanum	58 Ce 140.1 Cerium	59 Pr 140.9 Praseodymium	60 Nd 144.2 Neodymium	61 Pm Promethium	62 Sm 150.4 Samarium	63 Eu 152.0 Europium	64 Gd 157.3 Gadolium	65 Tb 158.9 Terbium	66 Dy 162.5 Dysprosium	67 Ho 164.9 Holmium	68 Er 167.3 Erbium	69 Tm 168.9 Thulium	70 Yb 173.1 Ytterbium	71 Lu 175.0 Lutetium

Actinoids

89 Ac Actinium	90 Th 232.0 Thorium	91 Pa 231.0 Protactinium	92 U 238.0 Uranium	93 Np Neptunium	94 Pu Plutonium	95 Am Americium	96 Cm Curium	97 Bk Berkelium	98 Cf Californium	99 Es Einsteinium	100 Fm Fermium	101 Md Mendelevium	102 No Nobelium	103 Lr Lawrencium

Standard atomic weights are abridged to four significant figures.

Elements with no reported values in the tables have no stable nuclides.

Information on elements with atomic numbers 113 and above is sourced from the International Union of Pure and Applied Chemistry Periodic Table of Elements (November 2016 version).

The International Union of Pure and Applied Chemistry Periodic Table of the Elements (February 2010 version) is the principal source of all other data. Some data may have been modified.